PSI Handbook of Business Security

PRAEGER SECURITY INTERNATIONAL ADVISORY BOARD

Board Cochairs

Loch K. Johnson, Regents Professor of Public and International Affairs, School of Public and International Affairs, University of Georgia (U.S.A.)

Paul Wilkinson, Professor of International Relations and Chairman of the Advisory Board, Centre for the Study of Terrorism and Political Violence, University of St. Andrews (U.K.)

Members

Anthony H. Cordesman, Arleigh A. Burke Chair in Strategy, Center for Strategic and International Studies (U.S.A.)

Thérèse Delpech, Director of Strategic Affairs, Atomic Energy Commission, and Senior Research Fellow, CERI (Fondation Nationale des Sciences Politiques), Paris (France)

Sir Michael Howard, former Chichele Professor of the History of War and Regis Professor of Modern History, Oxford University, and Robert A. Lovett Professor of Military and Naval History, Yale University (U.K.)

Lieutenant General Claudia J. Kennedy, USA (Ret.), former Deputy Chief of Staff for Intelligence, Department of the Army (U.S.A.)

Paul M. Kennedy, J. Richardson Dilworth Professor of History and Director, International Security Studies, Yale University (U.S.A.)

Robert J. O'Neill, former Chichele Professor of the History of War, All Souls College, Oxford University (Australia)

Shibley Telhami, Anwar Sadat Chair for Peace and Development, Department of Government and Politics, University of Maryland (U.S.A.)

Fareed Zakaria, Editor, Newsweek International (U.S.A.)

PSI Handbook of Business Security

VOLUME TWO

SECURING PEOPLE
AND PROCESSES

EDITED BY W. TIMOTHY COOMBS

Praeger Security International
Westport, Connecticut • London

Library of Congress Cataloging-in-Publication Data

PSI handbook of business security / edited by W. Timothy Coombs.
 2 v. ; cm.
 Includes bibliographical references and index.
 Contents: v. 1. Securing the enterprise—v. 2. Securing people and processes.
 ISBN 978-0-275-99394-8 (set : alk. paper) — ISBN 978-0-275-99395-5 (vol. 1 : alk. paper) — ISBN 978-0-275-99396-2 (vol. 2 : alk. paper)
 1. Industries—Security measures. I. Coombs, W. Timothy. II. Title: Handbook of business security.
HV8290.P77 2008
658.4'7—dc22   2007040469

British Library Cataloguing in Publication Data is available.

Copyright © 2008 by W. Timothy Coombs

All rights reserved. No portion of this book may be reproduced, by any process or technique, without the express written consent of the publisher.

Library of Congress Catalog Card Number: 2007040469

ISBN-13: 978-0-275-99394-8 (set)
     978-0-275-99395-5 (vol. 1)
     978-0-275-99396-2 (vol. 2)

First published in 2008

Praeger Security International, 88 Post Road West, Westport, CT 06881
An imprint of Greenwood Publishing Group, Inc.
www.praeger.com

Printed in the United States of America

The paper used in this book complies with the
Permanent Paper Standard issued by the National
Information Standards Organization (Z39.48-1984).

10 9 8 7 6 5 4 3 2 1

CONTENTS

Preface ix

VOLUME 1: SECURING THE ENTERPRISE

INFORMATION (CYBER) PROTECTION

Information Security	1
Security Models	6
Security Documentation	13
Developing, Publishing, and Maintaining Information Security Policies and Standards	20
Internet and E-mail Use Policies	32
Portable Device Security	38
Insider Threat	40
Suspicious Cyber Activities	42
Social Engineering: Exploiting the Weakest Link	43
Data Backup	51
US-CERT (United States Computer Emergency Readiness Team)	54

TERRORISM AS A BUSINESS SECURITY AND SAFETY CONCERN

Terrorism	59
Ecoterrorism	62
Agroterrorism	65
Strategic Partnership Program Agroterrorism (SPPA)	73
Food Security	75
Terrorism and Chemical Facilities	77
Customs-Trade Partnership Against Terrorism (C-TPAT)	81

GENERAL SAFETY CONCERNS

Benefits of Emergency Management	85
Emergency Preparedness and Response Component	90

Community Emergency Response Team	98
Emergency Response Training and Testing: Filling the Gap	99
Disaster Recovery Management	104
Business Continuity	120
Risk Management	132
Risk Communication	136
Vulnerability Assessment Team	141
Types of Risk	150
Crisis Management	152
Types of Crises	164
Crisis Communications: External and Internal	180
Crisis Sensing Mechanism	187
Crisis Management Plan	194
Crisis Management Team	200
Crisis Spokesperson	203

UNCOMMON BUSINESS SECURITY CONCERNS

Corruption as a Business Security Concern	209
Competitive Intelligence	217
Ethics as a Business Security Concern	223
Ethical Conduct Audit	227
Employee Theft and Fraud	242
Reputation Management	244

VOLUME 2: SECURING PEOPLE AND PROCESSES

PHYSICAL PROTECTION

Physical Security	251
Security Guards/Officers	255
Workplace Violence Prevention and Policies	257
Workplace Aggression	272
Contributors to Workplace Aggression	278
Employee Background Screening and Drug Testing	280
Video Surveillance	288
Countersurveillance	290
Radio Frequency Identity	292
Biometrics	294
Shelter-in-Place	296
Evacuation in Large and Multiple Tenant Buildings	298
Integrating Physical and Information Security	300

CONTENTS

SECURITY ON A GLOBAL SCALE

Pandemics	303
Pandemic Communication	312
Outsourcing and Security	321
Supply Chain Security	323
Supply Chain Continuity	327
Travel Overseas	335
Corporate or Industrial Espionage	338

ENHANCING THE HUMAN SIDE OF SECURITY AND SAFETY

Improving Team Effectiveness	343
Avoiding the Silo Effect to Improve Business Security	346
Winning Acceptance for Security or Other New Programs	353
Managing Organizational Culture and Change Acceptance	356
Culture of Integrity	359
Exercise and Training Basics	361

GUIDANCE APPENDIX — 377

RESOURCE APPENDIX — 405

DOCUMENT APPENDIX — 425

GLOSSARY — 697

INDEX — 713

ABOUT THE EDITOR AND CONTRIBUTORS — 729

PREFACE

In general, security means freedom from risk, fear, doubt, or anxiety. For organizations, business security comprises the measures taken to protect people, data, physical assets, and financial and other assets. These measures help to create a safe workplace and to reduce risks. Business security further seeks to combat the various security risks faced by organizations. When these risks become a reality, people and organizations suffer. Examples include workplace violence, terror attacks, computer hacking/data loss, disruption of the supply chain, and top management engaged in illegal acts. Moreover, these "problems" draw intense coverage from the news media, which are drawn to singular, negative events. Negative publicity resulting from incidences that might have been avoided further harms an organization by damaging its reputation, which is a valuable, if intangible, resource. Business security clearly involves high stakes.

When most people think about business security, a security guard at a front desk or gate comes to mind. Others may think of the password they have to enter when they use their computers at work. Business security is much more extensive and involved than these two common reference points. The diversity of business security is a reflection of the multitude of risks that organizations face. The security guard and the computer password represent two broad areas of business security—physical security and information security—but many more areas of concern exist. There is obviously a concern for people at a location (employees, vendors, and visitors), the grounds, the building itself, and the materials in the building. Yet business security extends outside of the facility to people and businesses located near the facility, cyberspace, vendors and others in the supply chain, and so forth. These two volumes try to capture the range and complexity of business security.

The objectives of the *PSI Handbook of Business Security* are twofold. The primary objective is to create a reference tool for people involved in business security. The entries and resources identify key security concerns and provide guidance on how to handle them. This work thus serves as a resource for anyone looking to improve security in an organization.

The secondary objective is to raise people's awareness of their role in security. If employees fail to commit to business security, the organization is at risk. What are the odds that new hires, such as fresh college graduates, know much about security or appreciate its value? Does your orientation program properly educate them on physical and information security? Do these new people read and commit themselves to abiding by the security policies in the employee handbook? Do your long-time employees understand and appreciate their role in business security? Anyone who reads the *PSI Handbook of Business Security* should realize that every employee plays an active role in keeping an organization secure.

Business security is complex and multifaceted. This set is comprised of two volumes that explore this diverse area. Volume 1 focuses on information security, terrorism, and topics related to business security that often are not included in discussions of business security. Volume 2 focuses on physical security, the growing global concerns of business security, and the role people play in making business security a success.

I am grateful to the many experts who took time from their busy lives to contribute to this project. They embraced it and realized the importance of sharing what they know with others to help improve business security. Hopefully, readers will benefit from the information compiled in these two volumes and use it to make their workplaces safer.

Volume 1: *Securing the Enterprise*

The workplace is very different than it was just ten years ago. Computers and the Internet have become common business tools. The tragic events of 9/11 still resonate in organizations, as terrorism remains a concern. Corporate misdeeds from Enron, Tyco, and others have resulted in new regulations, creating renewed interest in business ethics. This changing workplace environment requires revisions in how we approach business security. The entries in Volume 1: *Securing the Enterprise* reflect this evolving workplace.

Volume 1 begins with the Information (Cyber) Protection section. While computers and the Internet increase business productivity, they also create enormous security risks. Protecting data and access to a computer system must be a priority for all organizations and a concern that all employees share. The entries reflect the range of issues and concerns that emerge as information security grows in importance. This section starts with a general discussion of information security and then moves to core concepts in information security:

- Security models
- Security documentation
- Security policies and standards

These core concepts are followed by a collection of additional information security concerns including the security threats and ways to protect Internet and e-mail use, portable device use, insider threat, and social engineering.

Terrorism is a reality that organizations face all around the globe and is broader than most people might think. The Terrorism as a Business Security and Safety Concern section begins with an analysis of terrorism including the global

PREFACE   xi

locations at greatest risk. The entry Ecoterrorism reflects the broad nature of terrorism. The breadth and scope of terrorism have implications for business security. Organizations anywhere, for example, might have to deal with environmental groups targeting a facility. Another feature of this section is its consideration of special terrorist risks for agriculture (Agroterrorism, Strategic Partnership Program Agroterrorism, and Food Security) and the chemical industry (Terrorism and Chemical Facilities). Both industries are prime terror targets and need to address terror concerns as part of business security.

The General Safety Concerns section relates to the connections between security and safety. This section examines the role of business security in protecting people and things both inside and outside of the organization. Efforts to protect employees and others in and around an organization must address the concerns generated by significant negative events such as disasters, crises, emergencies, and business disruptions. These negative events are often interconnected and can be threats inside and outside of the organization. The entries cover these important areas:

- Emergency preparedness and response component
- Disaster recovery management
- Business continuity
- Crisis management

Preparations for these negative events are a key to safety, and security can contribute significantly to these efforts. These entries concentrate on crisis management and emergency management because of the strong focus these two processes have on protecting people.

Finally the Uncommon Business Security Concerns section reflects the growing concern over ethics in organizations (Ethics as a Business Security Concern and Ethical Conduct Audit) and the issues of corruption (Corruption as a Business Security Concern) that arise as organizations compete in a global business environment. Ethical and corruption issues do relate to business security. Competitive intelligence is included because organizations can step over the ethical line when seeking competitive intelligence and should know how to make it more difficult for others to collect intelligence on them. The final entry, Reputation Management, is included because many other entries note the negative effect that security lapses can have on an organization's reputation. This entry explains why damage to a reputation is such a strong business concern.

Volume 2: *Securing People and Processes*

As the workplace changes, the core need for physical protection remains but its nature changes as well. The Physical Protection section contains a variety of entries related to protecting people, equipment, and buildings. The main topics include the following:

- Security guards/officers
- Workplace violence
- Employee background screening and drug testing
- Video surveillance
- Radio frequency identity
- Biometrics
- Shelter-in-place
- Evacuation

The last entry in this section explains how physical and information security are beginning to merge and what the benefits of that synergy are.

The Security on a Global Scale section recognizes the unique security demands of global organizations. The Terrorism entries touched on the subject, but this section's entries look at the problems and possible security solutions when an organization is connected to various locations around the world. The topics include the following:

- Pandemics
- Outsourcing
- Supply chain security
- Travel overseas
- Corporate or industrial espionage

A theme that echoes throughout this introduction and collection is the importance of all employees in security. The entries in the Enhancing the Human Side of Security and Safety section concentrate on how to work with employees to embrace programs that will make a more secure workplace. The focus is on how to integrate new programs into an organization by avoiding the silo thinking that prevents integration, winning acceptance, and integrating the new programs into the organizational culture. The entries also provide tips on how to create teams and the value of exercises and training.

Volume 2 also includes the Guidance Appendix, Resource Appendix, Documents Appendix, and Glossary. The Guidance Appendix provides "how-to" advice on a number of the topics in the two volumes; whereas the Resource Appendix lists books, magazines, and web sites that can provide additional information on a variety of the entries. The Document Appendix offers a collection of government documents that are useful reference tools for those interested in business security. Last, the Glossary provides an extensive list of terms related to business security. It is a useful tool for clarifying the meaning of terms or concepts and learning the diverse vocabulary of business security.

PREFACE

Business security is a complex topic that involves a wide range of activities and preventative measures. Hopefully the *PSI Handbook of Business Security* does justice to the complicated nature of the topic. Many departments in an organization have a role to play in business security. These two volumes illustrate how the various elements of business security in an organization must work in concert and include all employees in that effort. This project will be a success if people use the ideas in this book to improve security in their organizations.

ACKNOWLEDGMENTS

I would like to thank the many attendees and presenters at CPM West that I have talked to and heard over the years. Your ideas about and insights into business security helped to guide the development of this project. I would also like to thank Jeff Olson at Praeger Publishers, Greenwood Publishing Group, for his dedicated attention to and editing of this project.

PHYSICAL PROTECTION

The most visible part of business security is physical security. People can see security guards, fences, gates, and video cameras. This section examines a range of concerns related to physical security. It includes procedures and methods for improving physical security for people, facilities, and products.

PHYSICAL SECURITY
W. Timothy Coombs

Any organization that has a building or an office has a need for physical security. Many employees think of physical security when they hear the term *security* because it is usually the most visible aspect of business security. We see security guards, card readers, gates, and closed circuit television cameras (CCTV). (The term *CCTV* is used generically for all camera systems, including the newer digital systems.) Physical security includes efforts to monitor and control the facility's exterior and interior perimeters. Its scope includes mail service security, lock and key controls, and perimeter and interior alarms. Physical security is a layering process of overlapping activities designed to keep bad people out and to keep your people, information, and physical assets safe.

AREAS AND SECURITY MEASURES

Physical security begins at the site perimeter. There need to be physical barriers to prevent access to your facility, even if it is an office building rather than a production or transportation facility. A fence with detection systems, electronic gates with card readers, other barriers, and security officer checkpoints can separate your facility from the outside world. The guards and electronic gates help to monitor who is actually inside your perimeter, such as employees, delivery personnel, vendors, or other visitors. Facilities that are likely terrorist targets may want to consider examining all vehicles entering the parking area, especially underground parking. Security officers should be looking for any indications the vehicle may contain explosives. Also consider rearranging physical barriers periodically to change the traffic flow into a facility. Predictability is a friend to a possible intruder.

Make sure the perimeter is easily visible with plenty of lighting and removal of anything that could block sight lines such as vegetation or signs. Use technology to enhance coverage of the perimeter. Install motion detectors, electronic beams, CCTV cameras, and microwave perimeter intrusion systems. Be certain all of your security officers are trained in the use of the systems and the desired security protocols.

The main entrance to the building should have high-security locks and doors, CCTV cameras, and security officers. The entrance can also be equipped with two-way voice communication and biometric card readers. The two-way communication allows outsiders contact with people inside without a person gaining access, and the biometric cards help ensure only the proper people get inside. Refer to the Biometrics entry for more details on that system. Secondary personnel entrances should be secured as well with high-security locks, strong door closers, CCTV cameras, card readers with biometrics, and motion-sensitive lighting.

Shipping and receiving is a third means of entry into a building. Security officers and CCTV cameras should be positioned there. It is important that drivers do not leave their designated areas and that all shipments are verified. To check cargo, the shipping and receiving area should have X-ray equipment, radio frequency identification (RFID) equipment, and bar code scanners. Your people need to verify that the proper cargo is being delivered and nothing "extra" is in the shipment.

Inside a facility, there is the need to monitor interior doorways and elevators. Card readers and biometrics should be used to limit access to different areas of the building. Not all employees require access to all areas of a facility. In high-security areas, make sure only authorized personnel have access. Support the access control system with CCTV cameras to monitor who is actually entering and leaving, and reinforce the door jambs and locks to prevent people from forcing their way into a secure area. One of the areas you must treat as high security is the security/life safety control room, which contains your CCTV monitors and

PHYSICAL PROTECTION

recorders, alarm panel, lighting control, fire suppression systems, HVAC controls, and the communication system. Compromise of any of these systems means a facility is at risk. Make sure the doors have reinforced jambs and there are biometric card scanners.

Finally, make certain to store, lock, and inventory keys, access cards, uniforms, badges, and vehicles. Any of these assets can be used to compromise an organization's security by providing a means of access to a facility.

PHYSICAL SECURITY IS EVERYONE'S CONCERN

Your employees should not think that physical security is taken care of by the security staff. Reinforce that everyone has a role to play in security. Television news reports have demonstrated how easy it is for a well-dressed person to enter an office and take a laptop computer. All employees should know the need for badge and area challenges. Employees should confront people without badges or who are not in a designated area. Challenge involves asking the persons who they are and why they are there. Security should be contacted if the wrong answers are given. Employees also need to make sure all windows and doors are properly

---

Suspicious Activity

This information comes from the Department of Homeland Security's "Report Suspicious Behavior and Activity" brochure.

- *Surveillance.* Someone is recording or monitoring activities at your facility. People are seen with cameras, taking notes, or using binoculars.
- *Deploying assets.* There are abandoned vehicles, packages, or luggage near a facility.
- *Suspicious people.* There are people who do not seem to belong in the workplace, in the office, or near the facility.
- *Suspicious questioning.* Someone is trying to get information about your facility over the phone, in person, or through e-mail.
- *Tests of security.* There are actions taken to test the physical security of the facility.
- *Acquiring supplies.* There are attempts to buy the ingredients for explosives.
- *Dry runs.* Someone observes what appear to be preparations for an attack such as mapping routes or timing traffic lights.[1]

closed and locked. Unlocked windows and doors provide easy egress for would-be intruders. Be sure to report any broken windows, doors, or locks as soon as they are spotted. Train employees on what constitutes suspicious behavior so that they can identify and report those behaviors. Key points for suspicious activities include entrances, loading docks, parking areas, and garages. See the Suspicious Activity box for additional information.

EVOLUTION OF PHYSICAL SECURITY

A trend in physical security is its growing use of technology and integration with IT security. This entry has mentioned biometrics, CCTV, and microwave perimeter intrusion systems—all technology-driven features of physical security. Although attentive and well-trained security officers are still the heart of physical security, their efforts are supplemented and enhanced by technology. The entry Integrating Physical and Information Security elaborates on this topic.

CONCLUSION

You must assume that certain people out there would like to access your facility for a variety of nefarious activities ranging from theft to terrorism. Physical security is a critical defense to keep these people out of your facility or from inflicting injury on your personnel. Security is a process of integrating various protective systems and devices to create a web of safety around a facility and its personnel. Your employees must realize that physical security, like information security, is everyone's responsibility and that they have a role to play in keeping a facility or building safe.

See also Biometrics; Integrating Physical and Information Security; Supply Chain Security; Terrorism; Terrorism and Chemical Facilities; and Travel Overseas.

NOTE

1. U.S. Department of Homeland Security, "Protect Your Workplace," online at http://www.us-cert.gov/reading_room/brochure_securityguidance.pdf (accessed 14 March 2007).

SECURITY GUARDS/OFFICERS

W. Timothy Coombs

Organizations often have a legal duty to provide security to employees, customers, and visitors. Security guards have a variety of responsibilities including perimeter security, heating and cooling systems, communication, alarm systems, receiving visitors and providing information, and responding in emergency situations. Beyond keeping people and facilities safe, security guards are essential elements in business continuity, emergency management, and crisis management. The job can be both routine and dangerous. Hiring security guards is an important consideration for management. This entry reviews key issues related to employing security guards.

GENERAL TYPES OF SECURITY GUARD

Because organizations vary in terms of what they need from a security guard, there are different types of security guards. Some organizations simply need a friendly deterrent. Their security guard is a visual deterrent to shoplifters and possible miscreants. People expect the security guard to be friendly and helpful. The guard provides information, directions, or even calls when a patron needs assistance with automobile problems in the parking lot.

A second type of security guard is one to prevent employee theft. The friendly but firm security officer is more of an authority figure, and this seriousness must be reflected in the security officer's looks and attitude. The security guard is approachable but serious, primarily there to prevent loss. The final type of security guard is one that watches over the movement of money. These armed guards should convey a simple message: stay away. The organization must be sure the security officer's characteristics match the its needs.

Certain qualities are required of all security guards. Security guards should look neat, clean, and project a professional image. They must have uniforms that are clean and be neatly groomed. They need to understand that confrontation is to be resolved, not escalated. Security guards should be trained in methods for handling and de-escalating conflicts. If security guards are to carry a weapon, they require the necessary registration and training.[1]

HIRING SECURITY GUARDS

Thorough background checks are required before hiring security guards. Refer to the entry Employee Background Screening and Drug Testing for additional information on this topic. Management must be certain the person is actually qualified for the position, including a clean criminal record. Hiring a security guard with a criminal record is a recipe for a negligent hiring suit. Check all references, previous work history, and credentials. You must be sure the person can do the job, has the proper attitude for the specific position, and has the necessary credentials. Organizations must also have an official policy on the use of force and weapons. An organization risks liability if such policies are lacking and someone is injured due to a use of force or weapons.

Organizations have three basic hiring options: full-time officers, contracted officers, and part-time officers drawn from law enforcement personnel or firefighters who are moonlighting. Full-time officers will have a familiarity with the facility, and management can be certain of training and screening. The disadvantages are the costs associated with training and supervising full-time officers. It falls to the organization to ensure officers are trained to current standards. Using contracted officers moves the training costs and responsibility to the security company. The disadvantage is that the officer may be less committed to the organization, and management cannot be certain of credentials. Management must carefully evaluate the quality of the officers being provided by a vendor.

The third option is law enforcement people and firefighters who are moonlighting. These well-trained professionals may have insights about community threats the organization is not yet aware of. There are cost savings, but the issues of integration and commitment remain a concern. An organization cannot rely solely on part-time security. There need to be full-time security guards who serve as team trainers, supervisors, and leaders. Full-time security officers also make sure there is integration between the organization and the part-time security force. A key point of that integration is informing the part-time officers about the unique security needs of the organization, such as hazardous waste issues.[2]

CONCLUSION

Whatever mix of security guards an organization uses, management must be diligent in vetting all security officers. Security guards should reduce, not increase, an organization's liability. Careful evaluation and training of all security personnel will ensure this is the case.

See also Employee Background Screening and Drug Testing; Integrating Physical and Information Security; and Physical Security.

NOTES

1. David C. Tryon, "Limiting Liability: Guarding Against Legal Problems," online at https://www.schinnerer.com/risk_mgmt/security/pdfs/stnd0901.pdf (accessed 15 March 2007).

2. Robert E. Uhorchak and Stanley N. Parker, "Solving the Security/Safety Staffing Paradox, *Risk Management Magazine*, Sept. 2005, pp. 60–61.

WORKPLACE VIOLENCE PREVENTION AND POLICIES

Geary Sikich

It seems that the subject of workplace violence has been overshadowed in recent years by the events of September 11, 2001, the war in Iraq, Iranian nuclear aspirations, global warming, and a host of other issues. However, the "Survey of Workplace Violence Prevention, 2005," published by the Bureau of Labor Statistics (BLS), U.S. Department of Labor, for the National Institute for Occupational Safety and Health (NIOSH), Centers for Disease Control and Prevention, reports:

> Nearly 5% of the 7.1 million private industry business establishments in the United States had an incident of workplace violence within the 12 months prior to completing a new survey on workplace violence prevention. Although about a third of these establishments reported that the incident had a negative impact on their workforce, the great majority of these establishments did not change their workplace violence prevention procedures after the incident; almost 9% of these establishments had no program or policy addressing workplace violence.[1]

The Survey of Workplace Violence Prevention looked at the prevalence of security features, the risks facing employees, employer policies and training, and related topics associated with maintaining a safe work environment.

Over 128 million workers were employed at the 7.4 million establishments represented by the survey. In an average week in U.S. workplaces, one employee is killed and at least 25 are seriously injured in violent assaults by current or former coworkers.

Many of the incidences of workplace violence can be prevented. According to *USA TODAY*:

> In nearly eight of 10 cases, killers left behind clear warning signs—sometimes showing guns to co-workers, threatening their bosses or talking about attacking. But in the majority of cases, employers ignored, downplayed or misjudged the threat, according to a *USA TODAY* analysis of 224 instances of fatal workplace violence.[2]

It is interesting that less than 20 percent of the companies targeted in such attacks took any action to enhance security or put internal prevention steps in place.

A PRESCRIPTION FOR PREVENTION

Imagine the following scenario unfolding.

> Your phone rings—4:00 p.m. Friday.
> A simple phone call and all your weekend plans are canceled.
> 4:15 p.m.: Another phone call. An employee you thought you knew well has just redefined your weekend.
> The phone rings again. You now have a full-scale crisis on your hands. The employee you thought to be so stable will become a quick study for you and your staff.
> Another phone call. You can't believe the amount of emotion, confusion, and misinformation generated by the incident.

And this is just the beginning.

When I first began creating and conducting workshops on workplace violence in the 1990s, the phenomenon of violence in the workplace was not new; it was and still is potentially one of the most serious problems facing organizations and individuals. And, we continue to turn a blind eye to a critical aspect of the problem—nonreporting of events. In the 1990s, not much literature existed on the subject of workplace violence. Today, while more literature exists, managers and human resource departments have yet to come up with answers. Outside consultants help to develop and implement appropriate prevention policies and programs, but the incidence of violence continues unabated.

The Bureau of Labor Statistics (BLS) entered into an interagency agreement with NIOSH to conduct a survey, mentioned above, of U.S. employers regarding their policies and training on workplace violence prevention. The survey defined workplace violence as violent acts directed toward a person at work or on duty (i.e., physical assaults, threats of assault, harassment, intimidation, or bullying). Workplace violence was classified as four types of situations:

- *Criminal:* When the perpetrator has no legitimate relationship to the business or its employees and is usually committing a crime in conjunction with the violence (e.g., robbery, shoplifting, or trespassing)
- *Customer or Client:* When the perpetrator has a legitimate relationship with the business and becomes violent while being served by the business (e.g., customers, clients, patients, students, inmates, or any other group to which the business provides services)
- *Coworker:* When the perpetrator who attacks or threatens another employee is an employee, past employee, or contractor who works as a temporary employee of the business
- *Domestic violence:* When the perpetrator, who has no legitimate relationship to the business but has a personal relationship with the intended victim, threatens or assaults the intended victim at the workplace (e.g., family member, boyfriend, or girlfriend)[3]

PHYSICAL PROTECTION 259

Nokia's CEO, Jorma Ollila, seems to sum it up quite well in an excerpt reported by the *Moscow Times* in 2005: "[P]eople are more concerned about individual rights than taking responsibility for their actions and trying to have a positive influence on society."[4]

Why do we consistently fail to identify risks or teach employees how to defuse workplace situations indicating that an attack may be imminent? One reason is "culture"; we as a society frequently fail to react when we are scared. Think about how many times you have had a premonition and did not take extra precautions to enhance your safety and security. Organizations consistently fail, even after an event such as a firing or disciplinary hearing that could trigger an attack. Indeed, *USA TODAY*'s findings are supported by other research. "In more than 100 instances that I studied, in every case there was evidence to suggest this person was hurting and had a potential for aggression," said Jeff Landreth, a senior vice president at New York-based Guardsmark, a security services company. "We found the threats were ignored."[5]

Here are some of the key elements for an effective workplace violence prevention program:

- *Establish a threat management policy.* Make sure that it is adhered to and that regular reviews are conducted in order to assure that it works.
- *Conduct regular threat analysis reviews.* Periodic threat analysis can facilitate early recognition and possibly defuse a situation just waiting to erupt.
- *Develop a threat management team.* Include management, staff, human resources, and external entities that can bring expertise to bear on the threat analysis reviews.
- *Communicate effectively with all employees.* Communication is such a key element and so mismanaged. Tremendous resources are available to organizations to help them with the communication process. Listen to what is being said—then hear what is being said.
- *Develop procedures that provide early warning and clear instructions on what to do if an incident occurs.* I have developed and facilitated the use of a simple system called *IPAC*^{TM}, which has been proven effective in a number of settings, from heavy industry to financial services. The acronym stands for *I*dentify, *P*rotect, *A*lert, *C*ommunicate.[6]
- *Develop a system that provides continual assessment of and feedback on actual or potential situations.* Active analysis can be effectively implemented to provide information from many sources that creates a mosaic resulting in a clearer picture of a situation.
- *Develop training programs for employees to reduce the potential for violence in the workplace and elsewhere.* Well-informed and -educated personnel can be your best deterrent to workplace violence.
- *Develop wellness programs including stress management.* Ensure that the programs are implemented, as these can also defuse situations before they become a crisis.

- *Develop an employee assistance program.* Utilize third-party expertise and ensure that there are no negative ramifications for using the program. Again, implement the program to make sure that employees at all levels are aware and can take advantage of it.
- *Ensure fair compensation and promotional practices.* Involve human resources and other applicable elements of your organization in the process to ensure fair treatment of employees. Stress that employees are personally responsible for their actions too.
- *Provide reasonable sick leave and vacation policies.* This can be as creative as your human resource department wishes. You may also benefit from involving legal specialists early on to ensure that appropriate standards of care are addressed.
- *Improve termination policies and procedures.* Termination policies should be clearly defined and uniformly enforced.

PROGRAMS AND POLICIES

Over 70 percent of workplaces in the United States do not have a formal policy addressing workplace violence. In establishments that reported having a workplace violence program or policy, private industry most frequently reported addressing coworker violence (82 percent). Customer or client violence was the next most frequent subject of private industry policies or programs (71 percent), followed by criminal violence (53 percent) and domestic violence (44 percent). While addressing customer/client and coworker workplace violence the most, state governments dealt with domestic violence (66 percent) more than criminal violence (53 percent), whereas equal numbers of local governments addressed domestic violence and criminal violence (47 percent).[7]

Although most occurrences of workplace violence center on a single employee, companies involved in restructuring, layoffs, and mass termination should pay close attention. How an organization handles the restructuring, layoffs, and/or termination can create a significant potential for acts of violence. When employees are being terminated for any reason, management should demonstrate caring and a sincere interest in the future welfare of these individuals.

Employers may consider providing employees being terminated with such assistance as outplacement services, psychological counseling, severance benefits, skills development, and training and educational assistance. However, it is important to emphasize that outplacement services and any other assistance should be provided off-site, so that terminated employees are not encouraged to return to their former place of employment.

Although there are often displays of certain characteristics and behavior prior to a violent act occurring, management generally is not trained to detect such warning signs. Missing these apparently obvious signals from the discontented employee often leads to self-blame after the fact. A training program to help personnel identify warning signs should be developed and implemented. Certain characteristics and behavior constitute a profile that personifies the likely candidates and should trigger a red flag:

- Male between thirty-five and fifty-five years old
- Midlife transition, dissatisfied with life
- Loner without a true support system
- Low self-esteem
- Generally works in jobs with high turnover
- History of being disgruntled during employment
- Tends to project one's own shortcomings onto others
- History of intimidating coworkers and supervisors
- Feels persecuted and views efforts to help with suspicion
- Watches others for violations and may keep records
- Interested in weapons; may be a collector or marksman
- Probably does not have a police record

Take potential warning signs seriously. When a person displays several characteristics or behaviors outlined here, an employer should pay attention. Management, at all levels, that lacks the necessary experience and expertise to handle this type of potentially explosive situation should rely on appropriate outside resources—for example, specialized psychological counseling or extra security measures—on a temporary or even permanent basis.

Organizations must have these resources identified and in place in order to preempt a potentially devastating situation effectively. There is no time to plan or seek these resources as an incident is unfolding.

DESIGNING A PLAN

Taking the time to develop a detailed plan for what to do in a crisis is essential for getting an organization back on its feet in a more expedient and effective manner. The plan should establish important guidelines for accomplishing critical tasks efficiently during a time of crisis. For violence in the workplace, these tasks may include the following:

- Notification and communication procedures
- Humanitarian assistance for injured employees
- Humanitarian assistance for victims' families
- Real-time counseling of distressed employees
- Organizing professional counseling services
- Emergency repair and damage control activities
- Modified work schedules and temporary services

- Communicating with community and clients
- Media management

Specific tasks involved in developing and implementing the plan include building a threat management team, creating appropriate procedures for implementing the plan, preparing contingency scenarios, training personnel, and conducting crisis simulation drills and exercises.

Clearly, having a well-conceived program consisting of the policy, the plan, implementing procedures, and a trained staff to implement it will place your company in the best position to protect its assets (employees and property) from the potentially devastating effects of an incidence of workplace violence.

Here are some sample workplace violence prevention program and policy documents that may prove useful for your program development.

SAMPLE WORKPLACE VIOLENCE PREVENTION PROGRAM

POLICY STATEMENT
(EFFECTIVE DATE FOR PROGRAM)

Our establishment, [*Employer Name*], is concerned and committed to our employees, and their safety and health. We refuse to tolerate violence in the workplace and will make every effort to prevent violent incidents from occurring by implementing a Workplace Violence Prevention (WPVP) Program. We will provide adequate authority and budgetary resources to responsible parties so that our goals and responsibilities can be met.

All managers and supervisors are responsible for implementing and maintaining our WPVP Program. We encourage employee participation in designing and implementing our program. We require prompt and accurate reporting of all violent incidents whether or not physical injury has occurred. We will not discriminate against victims of workplace violence.

A copy of this Policy Statement and our WPVP Program is readily available to all employees from each manager and supervisor.

Our program ensures that all employees, including supervisors and managers, adhere to work practices that are designed to make the workplace more secure, and do not engage in verbal threats or physical actions that create a security hazard for others in the workplace.

PHYSICAL PROTECTION

All employees, including managers and supervisors, are responsible for using safe work practices, for following all directives, policies, and procedures, and for assisting in maintaining a safe and secure work environment.

The management of our establishment is responsible for ensuring that all safety and health policies and procedures involving workplace security are clearly communicated and understood by all employees. Managers and supervisors are expected to enforce the rules fairly and uniformly.

Our program will be reviewed and updated annually.

WORKPLACE VIOLENCE PREVENTION PROGRAM

THREAT ASSESSMENT TEAM

A Threat Assessment Team will be established, and part of its duties will be to assess the vulnerability to workplace violence at our establishment and reach agreement on preventive actions to be taken. They will be responsible for auditing our overall Workplace Violence Program.

The Threat Assessment Team will consist of:

Name:	Title:	Phone:
Name:	Title:	Phone:
Name:	Title:	Phone:
Name:	Title:	Phone:

The team will develop employee training programs in violence prevention and planning for responding to acts of violence, and will communicate this plan internally to all employees. The Threat Assessment Team will begin its work by reviewing previous incidents of violence at our workplace. It will analyze and review existing records identifying patterns that may indicate causes and severity of assault incidents and identify changes necessary to correct these hazards. These records include but are not limited to OSHA 200 logs, past incident reports, medical records, insurance records, workers' compensation records, police reports, accident

(continued)

investigations, training records, grievances, minutes of meetings, and so on. The team will communicate with similar local businesses and trade associates concerning their experiences with workplace violence.

Additionally, the team will inspect the workplace and evaluate the work tasks of all employees to determine the presence of hazards, conditions, operations, and other situations that might place our workers at risk of occupational assault incidents. Employees will be surveyed to identify the potential for violent incidents and to identify or confirm the need for improved security measures. These surveys shall be reviewed, updated, and distributed as needed or at least once within a two-year period. Periodic inspections to identify and evaluate workplace security hazards and threats of workplace violence will be performed by the following representatives of the Assessment Team, in the following areas of our workplace:

Representative: Area:
Representative: Area:
Representative: Area:

Periodic inspections will be performed according to the following schedule: frequency (daily, weekly, monthly, etc.).

SAMPLE SELF-INSPECTION SECURITY CHECKLIST

Facility:
Inspector:
Date of Inspection

1. Security Control Plan: Yes   No

   If yes, does it contain:
   (A) Policy statement   Yes   No
   (B) Review of employee incident exposure   Yes   No
   (C) Methods of control   Yes   No
     If yes, does it include:
       Engineering   Yes   No
       Work practice   Yes   No
       Training   Yes   No
       Reporting procedures   Yes   No

PHYSICAL PROTECTION

Record keeping	Yes	No
Counseling	Yes	No
(D) Evaluation of incidents	Yes	No
(E) Floor plan	Yes	No
(F) Protection of assets	Yes	No
(G) Computer security	Yes	No
(H) Plan accessible to all employees	Yes	No
(I) Plan reviewed and updated annually	Yes	No
(J) Plan reviewed and updated when tasks added or changed	Yes	No
2. **Policy Statement by Employer**	Yes	No
3. **Work Areas Evaluated by Employer**	Yes	No
If yes, how often?		
4. **Engineering Controls**	Yes	No
If yes, does it include:		
(A) Mirrors to see around corners and in blind spots	Yes	No
(B) Landscaping to provide unobstructed view of the workplace	Yes	No
(C) "Fishbowl effect" to allow unobstructed view of the interior	Yes	No
(D) Limiting the posting of sale signs on windows	Yes	No
(E) Adequate lighting in and around the workplace	Yes	No
(F) Parking lot well lighted	Yes	No
(G) Door control(s)	Yes	No
(H) Panic button(s)	Yes	No
(I) Door detector(s)	Yes	No
(J) Closed circuit TV	Yes	No
(K) Stationary metal detector	Yes	No
(L) Sound detection	Yes	No
(M) Intrusion detection system	Yes	No
(N) Intrusion panel	Yes	No
(O) Monitor(s)	Yes	No
(P) Videotape recorder	Yes	No
(Q) Switcher	Yes	No
(R) Handheld metal detector	Yes	No
(S) Handheld video camera	Yes	No
(T) Personnel traps ("Sally traps")	Yes	No
(U) Other	Yes	No
5. **Structural Modifications** (Plexiglas, glass guard, wire glass, partitions, etc.)	Yes	No
If yes, comment:		

(continued)

6. **Security Guards**
   (A) If yes, are there an appropriate number for
       the site?   Yes   No
   (B) Are they knowledgeable of the company WPVP
       policy?   Yes   No
   (C) Indicate if they are:
       Contract guards (1)   Yes   No
       In-house employees (2)   Yes   No
   (D) At entrance(s)   Yes   No
   (E) Building patrol   Yes   No
   (F) Guards provided with communication?   Yes   No
       If yes, indicate what type:
   (G) Guards receive training on workplace violence
       situations?   Yes   No
       Comments:

7. **Work Practice Controls**   Yes   No
   If yes, indicate:
   (A) Desks clear of objects that may become missiles   Yes   No
   (B) Unobstructed office exits   Yes   No
   (C) Vacant (bare) cubicles available   Yes   No
   (D) Reception area available   Yes   No
   (E) Visitor(s)/client(s) sign in/out   Yes   No
   (F) Visitor(s)/client(s) escorted   Yes   No
   (G) Barriers to separate clients from work area   Yes   No
   (H) One entrance used   Yes   No
   (I) Separate interview area(s)   Yes   No
   (J) ID badges used   Yes   No
   (K) Emergency numbers posted by telephones   Yes   No
   (L) Internal phone system   Yes   No
       If yes, indicate:
       Does it use 120 VAC building lines?   Yes   No
       Does it use phone lines?   Yes   No
   (M) Internal procedures for conflict (problem)
       situations   Yes   No
   (N) Procedures for Employee Dismissal   Yes   No
   (O) Limit Spouse and family visits to designated
       areas   Yes   No
   (P) Key control procedures   Yes   No
   (Q) Access control to the workplace   Yes   No
   (R) Objects that may become missiles removed
       from area   Yes   No

PHYSICAL PROTECTION

(S) Parking prohibited in fire zones — Yes No
Other:

7a. Off-Premises Work Practice Controls
(For staff who work away from fixed workplace, such as social services, real estate, utilities, policy/fire/sanitation, taxi/limo, construction, sales/delivery, messengers, and others)

(A) Trained in hazardous situation avoidance — Yes No
(B) Briefed about areas where they work — Yes No
(C) Have reviewed past incidents by type and area — Yes No
(D) Know directions and routes for day's schedule — Yes No
(E) Previewed client/case histories — Yes No
(F) Left an itinerary with contact information — Yes No
(G) Have periodic check-in procedures — Yes No
(H) After hours contact procedures — Yes No
(I) Partnering arrangements if deemed necessary — Yes No
(J) Know how to control/defuse potentially violent situations — Yes No
(K) Supplied with personal alarm/cellular phone/radio — Yes No
(L) Limit visible clues of carrying money/valuables — Yes No
(M) Carry forms to record incidents by area — Yes No
(N) Know procedures if involved in incident (see also Training section) — Yes No

8. Training Conducted — Yes No
If yes, is it:
(A) Prior to initial assignment — Yes No
(B) At least annually thereafter — Yes No
(C) Does it include:
   Components of security control plan — Yes No
   Engineering and workplace controls instituted at workplace — Yes No
   Techniques to use in potentially volatile situations — Yes No
   How to anticipate/read behavior — Yes No
   Procedures to follow after an incident — Yes No
   Periodic refresher for on-site procedures — Yes No
   Recognizing abuse/paraphernalia — Yes No
   Opportunity for Q and A with instructor — Yes No
   On hazards unique to job tasks — Yes No

9. Written Training Records Kept — Yes No

(continued)

10. **Are Incidents Reported?**	Yes	No
If yes, are they:		
(A) Reported in written form	Yes	No
(B) First report of injury form (if employee loses time)	Yes	No
11. **Incidents Evaluated**	Yes	No
(A) EAP counseling offered	Yes	No
(B) Other action (reporting requirements, suggestions, reporting to local authorities, etc.)	Yes	No
(C) Are steps taken to prevent recurrence?	Yes	No
12. **Floor Plans Posted Showing Exits, Entrances, Location of Security Equipment, Etc.**	Yes	No
If yes, does it:		
(A) Include an emergency action plan, evacuation plan, and/or a disaster contingency plan?	Yes	No
13. **Do Employees Feel Safe?**	Yes	No
(A) Have employees been surveyed to find out their concerns?	Yes	No
(B) Has the employer utilized the crime prevention services and/or lectures provided by the local or state police?	Yes	No

Comments:

General Comments/Recommendations:

SAMPLE INCIDENT REPORT FORM

1. VICTIM'S NAME:
   JOB TITLE:
2. VICTIM'S ADDRESS:
3. HOME PHONE NUMBER:
   WORK PHONE NUMBER:
4. EMPLOYER'S NAME AND ADDRESS:
5. DEPARTMENT/SECTION:
6. VICTIM'S SOCIAL SECURITY NUMBER:
7. INCIDENT DATE:
8. INCIDENT TIME:
9. INCIDENT LOCATION:
10. WORK LOCATION (if different):

11. TYPE OF INCIDENT (circle one):
    Assault, Robbery, Harassment, Disorderly Conduct, Sex Offense, Other (Please Specify)

12. WERE YOU INJURED: (circle): Yes No
    If yes, please specify your injuries and the location of any treatment:

13. DID POLICE RESPOND TO INCIDENT: Yes No
14. WHAT POLICE DEPARTMENT:
15. POLICE REPORT FILED: Yes No
    REPORT NUMBER:
16. WAS YOUR SUPERVISOR NOTIFIED: Yes No
17. SUPERVISOR'S NAME:
18. WAS THE LOCAL UNION/EMPLOYEE REPRESENTATIVE NOTIFIED: Yes No
    Whom should be notified?
19. WAS ANY ACTION TAKEN BY EMPLOYER (specify):
20. ASSAILANT/PERPETRATOR (circle one):
    Intruder, Customer, Patient, Resident, Client, Visitor, Student, Coworker, Former Employee, Supervisor, Family/Friend, Other (specify):
21. ASSAILANT/PERPETRATOR—NAME/ADDRESS/AGE (if known):
22. PLEASE BRIEFLY DESCRIBE THE INCIDENT:
23. INCIDENT DISPOSITION (circle all that apply):
    No Action Taken, Arrest, Warning, Suspension, Reprimand, Other (specify):
24. DID THE INCIDENT INVOLVE A WEAPON: Yes No
    (Please Specify)
25. DID YOU LOSE ANY WORK DAYS:
    (Please Specify)
26. WERE YOU SINGLED OUT OR WAS THE VIOLENCE DIRECTED AT MORE THAN ONE INDIVIDUAL:
27. WERE YOU ALONE WHEN THE INCIDENT OCCURRED:
28. DID YOU HAVE ANY REASON TO BELIEVE THAT AN INCIDENT MIGHT OCCUR: Yes No
    Why:
29. HAS THIS TYPE OR SIMILAR INCIDENT(S) HAPPENED TO YOU OR YOUR COWORKERS: Yes No
    (Please Specify)

(continued)

30. HAVE YOU HAD ANY COUNSELING OR SUPPORT SINCE THE
    INCIDENT: Yes No
    (Please Specify)
31. WHAT DO YOU FEEL CAN BE DONE IN THE
    FUTURE TO AVOID SUCH AN INCIDENT:
32. WAS THIS ASSAILANT INVOLVED IN PREVIOUS
    INCIDENTS:
33. ARE THERE ANY MEASURES IN PLACE TO
    PREVENT SIMILAR INCIDENTS: Yes No
    (Please Specify):
34. HAS CORRECTIVE ACTION BEEN TAKEN:
    (Please Specify):
35. COMMENTS:

SAMPLE EMPLOYEE SECURITY SURVEY

This survey will help detect security problems in your building or at an alternate work site.

Please fill out this form, get your coworkers to fill it out, and review it to see where the potential for major security problems lie.

NAME:

WORK LOCATION:
(IN BUILDING OR ALTERNATE WORK SITE)

Do either of these two conditions exist in your building or at your alternate work site?

    Work alone during working hours
    No notification given to anyone when you finish work

Are these conditions a problem? If so when, please describe. (For example, Mondays, evening, Daylight Savings Time)

Do you have any of the following complaints (that may be associated with causing an unsafe work site)?

(Check All That Apply)
    Does your workplace have a written policy to follow for addressing general problems?
    Does your workplace have a written policy on how to handle a violent client?
    When and how to request the assistance of a coworker

PHYSICAL PROTECTION

> When and how to request the assistance of police
> What to do about a verbal threat
> What to do about a threat of violence
> What to do about harassment
> Working alone
> Alarm system(s)
> Security in and out of building
> Security in parking lot
> Have you been assaulted by a coworker?
>
> To your knowledge, have incidents of violence ever occurred between your coworkers?
>
> Are violence-related incidents worse during shift work, on the road, or in other situations?
> (Please Specify)
>
> Where in the building or work site would a violence-related incident be most likely to occur (specify)?
>
> Have you ever noticed a situation that could lead to a violent incident?
>
> Have you missed work because of a potential violent act(s) committed during your course of employment?
>
> Do you receive workplace violence–related training or assistance of any kind?
>
> Has anything happened recently at your work site that could have led to violence?
>
> Can you comment about the situation?
>
> Has the number of violent clients increased?

CONCLUSION

Many employers believe it can't happen to them. New policies to prevent violence were adopted after a fatal 1998 shooting at the state transportation department office in Greeley, Colorado. Accountant Robert "Scott" Helfer, 50, had a history of arguments with supervisors. As a meeting was held to discuss employee complaints against him, he shot and killed equal employment representative

Sharlene Nail and wounded Karla Harding, a regional transportation director. Helfer was later shot and killed by a state trooper in the parking lot.

Having effective policies and an aggressive program to track behavior if anything seems abnormal is essential. You never want to have to utter the following words to a reporter: "My mistake was I never saw him as a time bomb that would explode."

See also Contributors to Workplace Aggression; Employee Background Screening and Drug Testing; and Workplace Aggression.

NOTES

1. Bureau of Labor Statistics (BLS), U.S. Department of Labor, for the National Institute for Occupational Safety and Health (NIOSH), Centers for Disease Control and Prevention; "Survey of Workplace Violence Prevention, 2005," 2005, online at http://www.bls.gov/iif/osh_wpvs.htm (accessed 23 April 2007).

2. Stephanie Armour, "Managers Not Prepared for Workplace Violence," *USA TODAY*, 15 July 2004, online at http://www.usatoday.com/money/workplace/2004-07-15-workplace-violence2_x.htm (accessed 23 April 2007).

3. Bureau of Labor Statistics, "Survey of Workplace Violence Prevention, 2005."

4. Matti Huuhtanen, "Nokia CEO Laments 'An Era of Selfishness,'" *Moscow Times*, 25 Jan. 2007, online at www.themoscowtimes.com/stories/2005/01/25/259.html (accessed 23 April 2007).

5. Armour, "Managers Not Prepared for Workplace Violence."

6. Geary W. Sikich, *Integrated Business Continuity: Maintaining Resilience in Times of Uncertainty* (Tulsa, OK: PennWell Publishing, 2003).

7. Bureau of Labor Statistics, "Survey of Workplace Violence Prevention, 2005."

WORKPLACE AGGRESSION

W. Timothy Coombs

Workplace aggression, an important concern in the workplace, can be defined as "efforts by individuals to harm others with whom they work, or have worked, or the organization in which they are currently, or were previously, employed."[1] A wide range of behaviors in the workplace classified as aggressive can contribute to the creation of a toxic and harmful work environment. Most of these behaviors are not the media-hyped coworker murders or assaults. Instead, they can include the following: dirty looks, obscene gestures, theft, sabotage, defacing property, hiding needed resources, showing up late for meetings, intentional work slowdowns, refusing to provide needed resources, yelling, delaying work to make others look bad, insults and sarcasm, spreading rumors, failing to transmit needed information, failing to warn people of a problem, failing to return phone calls, and giving people the silent treatment.

IDENTIFYING TYPES OF AGGRESSION: BUSS'S TYPOLOGY OF AGGRESSION

Aggression expert Arnold Buss developed a typology for classifying aggression. Buss proposed three dimensions for classifying aggressive acts: (1) physical/verbal, (2) active/passive, and (3) direct/indirect. The physical aspect refers to physical actions such as hitting or shoving the target, whereas the verbal aspect indicates using verbal communication to harm the target. The active aspect refers to how harm is inflicted by performing an action (e.g., insulting another), whereas the passive aspect indicates how harm is caused by withholding some action (e.g., failing to provide information, refusing to provide assistance). The direct aspect reflects how aggression may be aimed directly at the target (e.g., yelling at the target, refusing to answer a question from the target). The indirect aspect indicates how aggression may be expressed through an intermediary or by attacking something valued by the target (e.g., saying negative things about someone to coworkers, spreading rumors, or sabotaging equipment or files). These three dimensions can be used to create eight categories of aggression that have provided the foundation of much of the research on workplace aggression.[2]

Although the distinction between aggression and violence is an important one, it often is blurred in media reports of workplace violence. Violence typically is conceptualized to refer to intense cases of aggression, such as physically attacking one's supervisor and shooting a coworker. These actions are physical, active, and direct. In contrast, aggression typically refers to all behavior intended to do harm and includes a much wider range of behaviors than those that garner dramatic media coverage or are reported in national statistics on workplace violence. Because these less intense forms of aggression occur with more frequency, they typically are the focus of research. Moreover, these lesser forms of aggression can negatively affect the targets of the aggression, those who witness the aggression, and the organization itself.[3]

Studies employing the typology based on Buss's work usually investigate the more common, less intense forms of aggression, which seem to represent aggression that has "gone underground." The perpetrators may recognize that their intent to harm another will be perceived as deviant and inconsistent with the norms for appropriate workplace behavior. Hence, they channel their aggression into forms that are more difficult to detect and identify as purposefully harmful. They also may rely on aggressive strategies that seem prevalent and unpunished in their organizations. The perpetrators act strategically (and rationally) when they seek methods of inflicting damage that are not covered clearly in company policies. By going underground, their tactics protect themselves while accomplishing their objective of harming a target.

To investigate the nature and pervasiveness of workplace aggression, researchers often assess the frequencies with which employees have engaged in or witnessed these categories of behaviors in response to a variety of situations (e.g., negative performance feedback, workplace changes, perceptions of injustice, etc.). This research supports the idea that perpetrators go underground with their aggressive acts. These studies typically demonstrate that most aggression is less intense and falls into categories reflecting more covert, indirect, and passive forms of aggression.

Examples of this include the research conducted by workplace aggression experts Deanna Geddes and Robert Baron. Geddes and Baron's findings are consistent with the assumption that perpetrators prefer covert, indirect, passive forms over physical, active, and direct forms of aggression because the former types are harder to detect and punish. Covert forms of aggression (verbal, passive, indirect) often are more frequently reported than overt forms of aggression (physical, active, direct), presumably because hostile intent is more difficult to prove. For example, much workplace aggression includes behaviors such as talking behind someone's back and creating rumors about a target. However, some studies demonstrate that direct and indirect forms of aggression occur with about the same frequency or that direct forms occur more frequently.[4] Along these lines, Coombs and Holladay reported that people viewed verbal and passive forms as the most acceptable types of aggression. They found no differences in perceived acceptability between indirect and direct forms of aggression.[5] At first blush, the similarity in occurrence of direct and indirect acts seems contrary to the theory that perpetrators go underground. However, upon closer examination a reason for the lack of difference becomes apparent, as the effects for indirect and direct aggression can be explained by the effect/danger ratio.

CALCULATING THE CONSEQUENCES OF AGGRESSIVE ACTS: THE EFFECT/DANGER RATIO

The *effect/danger ratio* explains that covert rather than overt actions are more likely to be perceived as desirable. Covert actions are likely to be seen as "safer" because they would be less likely to incur punishment. Noted social psychologist Albert Bandura observed that direct attacks on others carry a high risk of retaliation. When attacks are difficult to interpret and it is not easy to assign blame or intent to the aggressor, aggressors are likely to pursue these tactics because they will be protected from counterattacks. The anticipation of punishment and/or retaliation reflects the danger component of the effect/danger ratio. The effect component indicates the anticipation of creating the desired result. Aggressors prefer behaviors that are effective in harming the targets while, simultaneously, incurring as little danger to themselves as possible. The effect/danger ratio reflects the subjective estimates of these two components.

Robert Baron and Joel Neuman speculate that verbal and passive behaviors are effective in maximizing the effect/danger ratio whereas indirect tactics may be less effective than direct tactics. This would lead perpetrators to prefer forms of direct aggression over indirect aggression. Although indirect methods are more likely to go unidentified, they may not bring about the desired result. This interpretation stemming from the effect/danger ratio would account for the similarities in frequencies reported in a few studies.[6]

In sum, motivated perpetrators will select aggressive acts that accomplish their goals while minimizing their chances of being caught and punished. The fear of punishment is tied to workplace policies against aggression. Perpetrators go underground in ways that circumvent those policies.

FACILITATORS AND INHIBITORS OF AGGRESSION IN THE WORKPLACE

What leads workers to respond with aggressive acts to organizational events such as negative performance feedback, employee monitoring, downsizing, and budget cuts? Researchers following Bandura's social learning theory note that situational cues in the workplace play a major role in influencing aggressive behaviors. They suggest it may be shortsighted to look at individual-level traits when searching for an explanation for aggressive behavior. They argue that aggressive behavior is learned by observing others engaging in aggressive acts (modeling) as well as through direct experience with aggressive acts (as the target and/or as the perpetrator). Reinforcers in the workplace can either encourage or discourage aggression. These reinforcers may take the form of (1) seeing others go unpunished for performing aggressive acts and/or personally not being caught or punished when engaging in aggressive actions (positive reinforcers) or (2) experiencing or seeing a coworker experience reprimands or punishments for aggression (negative reinforcers). Furthermore, if incentives in the organizational environment reward aggressive behavior (e.g., getting a promotion for being aggressive, receiving a choice assignment over another candidate because of belittling the opponent, pitting coworkers against one another in bidding for assignments, receiving a budget increase because you were the "squeakiest" and most obnoxious wheel), it is no wonder that employees engage in aggressive acts. A culture that tolerates and rewards aggression is likely to sustain aggression.

Individual traits (believing that aggression will lead to desired outcomes) as well as organizational characteristics (presence of aggressive models, aversive treatment, incentives for aggressive behavior) can contribute to a culture that views aggressive behavior as acceptable. Among the notable individual traits are Type A personality, low self-monitoring, and hostile attributional bias. The entry Contributors to Workplace Aggression provides details on these individual traits.

ASSESSING WORKPLACE AGGRESSION

The Workplace Aggression Tolerance Questionnaire (WATQ) is a 28-item instrument based on Buss's eight categories of aggressive behavior. Sherry Holladay and Timothy Coombs conducted studies to establish the reliability and validity of the instrument; the overall reliabilities (Cronbach's alpha) were .95 (study 1) and .97 (study 2). The instrument is unidimensional; items seem to be tapping into a similar aggression construct. In tests of convergent and discriminant validity, the WATQ demonstrated expected relationships with measures of verbal aggression and vengeance, and was not contaminated by a social desirability bias. The instrument also shows face validity, as it was derived by Buss's typology. Overall, the psychometric properties suggest that although Buss's eight categories of aggressive behavior are reflected in the 28 items, it makes sense to view the responses as an aggregate evaluation of the appropriateness of aggressive behavior.

When using the WATQ, researchers present respondents with a stimulus situation depicting a manager conducting a performance review with a subordinate. The subordinate believes the manager has been unfairly critical of his or her job performance and explains to the manager that the negative comments are inaccurate. However, the manager refuses to make any changes to the evaluation. The subordinate is aware that this negative review could prevent a pay raise and/or promotion. After reading the scenario, respondents evaluate the twenty-eight items describing actions the subordinate might take in response to the meeting. They rate the actions on a five-point scale ranging from "very inappropriate" to "very appropriate." Refer to the Guidance Appendix for a complete copy of the WATQ.

The performance review scenario was selected because previous research by Geddes demonstrated a link between (1) negative feedback situations and perceptions of injustice and (2) aggression. The scenario has strong ecological validity because its elements have been shown to be associated with aggressive actions toward the target (manager). The scenario also does not request the respondents to report on their own likely actions or to imagine they are the subordinates. Rather, the WATQ instrument asks them to report on the extent to which they would perceive a range of subordinate responses as appropriate.

Application of the WATQ

As noted above, most workplace aggression policies target overt, not covert, behaviors. The WATQ provides a reading of the tolerance people have for a wide array of aggressive behaviors. Tolerance for aggressive behaviors can be viewed as a risk, something that could develop into a larger problem. Forward-thinking companies try to identify and minimize risk. Companies conduct a number of different annual audits to assess various risks related to insurance and regulatory compliance. The risk of workplace aggression should be no different because it is as much organizational as it is individual. If an aggressive behavior is tolerated, people are more likely to engage it and to model that behavior for others. The risk becomes manifested into a problem. Using the WATQ can map tolerance in the organization and benchmark for training.

The organizational culture can serve to facilitate workplace aggression. Is aggressiveness mistakenly rewarded? An organizational culture can also be dysfunctional; however, we know that it is not monolithic but actually composed of a variety of subcultures. Consider a university. Cultural differences exist between faculty in different disciplines, maintenance staff, support staff, and health services personnel. By mapping subcultures with the WATQ, managers can determine whether certain areas of the organization are (1) more tolerant of aggressive behavior and (2) in need of additional training to address that risk. Though unidimensional, a very high score on a particular type of aggressive behavior suggests training should focus on that type of aggressive behavior. The WATQ can help to locate subculture-based risks and to inform workplace aggression training needs.

Organizations have been investing heavily in training efforts designed to reduce workplace aggression. The training ranges from awareness of the problem to strategies for defusing tense situations. At its base, training represents the need for people to realize a wide array of aggressive behaviors is problematic, should be deemed workplace aggression, and should not be tolerated in the workplace. The WATQ provides a means of determining whether the training is changing people's views of workplace aggression. Tolerance of workplace aggression can be benchmarked prior to awareness-oriented training. A few months after the training, the WATQ can be used to determine whether the training had any effect on tolerance of workplace aggression. Periodic checks could be made to determine whether the level of tolerance is decreasing, increasing, or remaining the same. Additional training can be used if the scores suggest an increase in tolerance.[7]

CONCLUSION

The dangers presented by workplace aggression extend far beyond workplace violence, its most visible form. In fact, managers find themselves dealing most often with the less visible forms of workplace aggression. Many types of workplace aggression that are given a foothold in the organizational culture can grow and transform from a risk to a problem. The WATQ helps to identify potential problem spots in the organization as well as problematic subcultures. Some areas of the organization may be more tolerant of workplace aggression and be in need of training. Yet training does not guarantee results. It must be assessed. It is crucial to know whether all forms of workplace aggression are deemed less tolerable because of training. Moreover, regular assessment with the WATQ provides evidence of longer-term success or failure of workplace aggression initiatives.

See also Contributors to Workplace Aggression; and Workplace Violence Prevention and Policies. See the Guidance Appendix for the Workplace Aggression Tolerance Questionnaire.

NOTES

1. Joel N. Neuman and Robert A. Baron, "Aggression in the Workplace," in *Antisocial Behavior in Organizations*, ed. Robert A. Giacalone and Jerald Greenberg (Thousand Oaks, CA: Sage, 1997), p. 38.
2. Neuman and Baron, "Aggression in the Workplace," pp. 39–41.
3. Neuman and Baron, "Aggression in the Workplace," pp. 39–41.
4. Deanna Geddes and Robert A. Baron, "Workplace Aggression as a Consequence of Negative Performance Feedback," *Management Communication Quarterly*, 10 (1997), 433–454.
5. W. Timothy Coombs and Sherry J. Holladay, "Understanding the Aggressive Workplace: Development of the Workplace Aggression Tolerance Questionnaire," *Communication Studies*, 55 (2004), 481–497.
6. Robert A. Baron and Joel N. Neuman, "Workplace Violence and Workplace Aggression: Evidence on Their Relative Frequency and Potential Causes," *Aggressive Behavior*, 22 (1996), 161–173.
7. Coombs and Holladay, "Understanding the Aggressive Workplace," pp. 481–497.

CONTRIBUTORS TO WORKPLACE AGGRESSION

W. Timothy Coombs

People in organizations often encounter adverse events or actions, which include financial pressures, increased stress, betrayal, lack of control, verbal threats, taking credit for other people's work, and criticism. These adverse events or actions can create anger, resentment, and frustration, which may result in workplace aggression. But an adverse event or action does not always result in aggression. Researchers studying workplace aggression have found two factors that help in determining when an unpleasant event or action will result in aggression: perceptions of justice and personal characteristics. Workplace aggression is a combination of conditions in the organization and personality traits. Understanding the factors that contribute to workplace aggression can help management develop mechanisms designed to reduce it.

PERCEPTIONS OF JUSTICE: ORGANIZATIONAL FACTORS

Adverse events or actions can be viewed as violations of justice. In organizations, the three variations of justice are distributive, procedural, and interactional justice. First, distributive justice involves perceptions of the fairness of outcomes. Employees determine whether the effort they put into their jobs results in equal rewards from their jobs. This evaluation is part of equity theory. Employees try to balance effort and reward in their jobs.

Second, procedural justice involves perceptions of the fairness of procedures. Are procedures implemented consistently, without bias; do they consider the interests of all parties; and do they provide a means for correcting errors? An example would be a performance appraisal. Are all employees evaluated using the same criteria? Is there a mechanism for employees to challenge inaccurate evaluations? Third, interactional justice involves perceptions of the quality of interpersonal treatment during procedural justice episodes. Did the people in charge of the procedural justice treat those involved with dignity and respect?

Violations of the three forms of justice can result in different forms of workplace aggression. If an adverse event violated only distributive justice, no workplace violence is likely. The perceptions of procedural and interactional justice should moderate the frustration and prevent workplace aggression. The employee is unhappy with the outcome but feels the process is fair and that he or she was treated well.

When the adverse event violates only procedural justice, an employee is likely to engage in verbal types of workplace aggression that target the organization. The employee is upset with the organization, so the aggression targets the organization. When the adverse event violates only interactional justice, an employee is likely to engage in verbal types of workplace aggression that target the offending person. The other person created the injustice, so the workplace aggression targets that individual.

When two or more forms of justice are violated, employees are more apt to move beyond words to workplace violence. The multiple violations intensify the frustration and anger created by the adverse event. The heightened "upset" can result in more serious workplace aggression, such as violence.

INDIVIDUAL TRAITS

Researchers noted that individual traits (believing that aggression will lead to desired outcomes) as well as organizational characteristics (presence of aggressive models, aversive treatment, incentives for aggressive behavior) can contribute to a culture in which aggressive behavior is viewed as acceptable. Among the notable individual traits are Type A personality, low self-monitoring, and hostile attributional bias. Type A people are frequently irritable and impatient and try to control the situation when they work with others. Type A personalities have been linked to aggression. Low self-monitors are not skilled at being socially sensitive and do not fit their words and actions to the situation. Researchers have found a link between low self-monitoring and obstructionism. Hostile attributional bias is when people perceive others as having hostile intentions even when such intentions are not present. These individuals are more likely to respond to events in an aggressive manner. These individual traits, which can be assessed, can contribute to workplace aggression.[1]

CONCLUSION

Employees will encounter adverse situations in the workplace that could trigger aggression. Whether or not they engage in workplace aggression is a function of their individual traits and organizational factors. A number of traits have been linked to aggression such as hostile attribution bias. However, these aggression-related traits are not proof a person will become aggressive. Employees who are screened and found to have these traits will benefit, as will all employees, from training on conflict resolution or other skills for decreasing aggression. The key organizational factors relate to perceptions of justice. Employees are less likely to be aggressive when they see the organization as just. Perceptions of justice include employees feeling that the procedures in the organization are fair and that managers are treating them fairly when applying the procedures. Actively working to maintain a sense of justice in the organization will help to reduce the likelihood of an adverse event resulting in workplace aggression.

See also Workplace Aggression; and Workplace Violence Prevention and Policies.

NOTE

1. Robert A. Baron and Joel N. Neuman, "Workplace Violence and Workplace Aggression: Evidence on Their Relative Frequency and Potential Causes," *Aggressive Behavior*, 22 (1996), 161—173; and Joel N. Neuman and Robert A. Baron, "Workplace Violence and Workplace Aggression: Evidence Concerning Specific Forms, Potential Causes, and Preferred Target," *Journal of Management*, 24 (1998), 391–419.

EMPLOYEE BACKGROUND SCREENING AND DRUG TESTING

W. Timothy Coombs

An employee background check or investigation is an inquiry into a person's character, personal characteristics, or general reputation. A number of different sources can be checked including criminal records, financial records, and driving records. A background check may even include a psychological evaluation,

PHYSICAL PROTECTION

drug testing, or a physical. The type of background screening depends in part on the job qualifications. Driving records are critical if jobs include driving, whereas financial records are relevant to jobs involving finances. Employee background screening must protect consumer rights, comply with federal and state hiring standards, ensure a safe workplace, be part of homeland security, and avoid legal exposure.

Employee background screening is used for new hires and existing workers. Job applications can be checked for their backgrounds and drug tests. Current employees can be screened as well. Rescreening typically involves drug testing, random or follow-up, and periodic background rechecks. Background rechecks ensure that an employee's record has not changed. If there are changes that could place the company at risk, actions must be taken or else a negligent retention lawsuit could occur if an incident occurred involving that employee.

TYPES OF BACKGROUND SCREENING

Background screening can cover a wide array of information. The following is a list of thirteen common types of searches or screening activities.

1. *Social Security number trace.* This is done to verify the applicant's Social Security number. The trace determines whether the Social Security number belongs to someone who is dead and lists all the names and addresses associated with that Social Security number.
2. *Preemployment evaluation report credit report, or PEER.* A PEER uses the national credit bureau database to provide the applicant's national credit history. The report will reveal any bankruptcies, tax liens, foreclosures, or repossessions.
3. *County criminal record check.* This search, conducted through records of the clerk of county courts, is specific to that county only. This typically involves an on-site, manual search of records that cover the past seven years or longer as well as felony and misdemeanor filings.
4. *Criminal history search.* A criminal history search uses the Social Security number to construct a comprehensive criminal record search. A search is made of county records, as well as federal court records, for each address connected to the Social Security number.
5. *Statewide criminal record search.* Some states permit searches of their law enforcement or criminal records. The charges listed would include felonies, misdemeanors, and traffic violations.
6. *Federal criminal record search.* Searches can be conducted at any of the ninety-one federal district courthouses nationwide. The crimes covered in these searches would be violations of federal laws.
7. *Motor vehicle records check.* This search provides information on an applicant's driving history. The data would include speeding or moving violations,

chargeable accidents, DUIs/DWIs, suspensions or revocations, and accumulation of points.
8. *Employer reference check.* This search verifies past employment by checking with the human resource departments of previous employers. The search can check dates of employment and job titles. Applicants must grant approval for this contact.
9. *Education verification.* This search verifies all degrees claimed by the applicant. The information obtained includes the name of the institution, date of graduation, dates of attendance, degree obtained, and type or field of study. Applicants who are currently students can be checked for degree progress, field of study, and planned graduation date.
10. *Professional license verification.* This search verifies an applicant's professional license or certificate through the accrediting agency or professional association. This information includes the license number, expiration date, type of license or certificate issued, date of issue, and whether there have been any disciplinary actions or sanctions against the license or certificate holder.
11. *Military record verification.* This search contacts military branches to confirm dates served and type of discharge.
12. *Workers' compensation record search.* This is a search of an applicant's accident dates and nature or type of injuries involved. Such searches can be done only in states that permit the dissemination of workers' compensation claim history and must be used in compliance with the Americans with Disabilities Act guidelines. The State Workers' Compensation Commission of Industrial Relations Board will provide this data.
13. *Suspected terrorist watch list search.* This search determines whether the applicant is on a suspected terrorist watch list. See the Terrorist Watch List box.

BACKGROUND SCREENING AND LEGAL EXPOSURE

Background screening is a defense against negligent hiring and retention lawsuits. An employer can be held liable when an employee's actions harm someone. This can include attacks on coworkers or customers and some accidents. Employers have a duty to provide a safe workplace, and hiring and retention are a part of that. Employers should not hire or retain workers who pose a risk to others. This now includes intentional acts of violence. If an employee has a history of violence and attacks a coworker, the employer can be held responsible for the act. No employer is immune from negligent hiring lawsuits. However, a thorough background screening is a strong defense. The background screening should identify any red flags for the employer. If the applicant manages to hide the warning signs, the employer has shown good faith in trying to find the problems.

Negligent hiring occurs when a company fails to properly screen employees and the hiring results in injuries. Negligent retention takes place when a company keeps an employee on staff after learning the employee is an unsuitable worker and injuries occur. An employer must take corrective action such as retraining, reassignment, or firing when it discovers an employee is unfit for duty. There is also negligent supervision, which involves a failure to provide the proper oversight to ensure that employees perform their jobs properly. A taxicab company was held liable for hiring a driver with a history of criminal violence who later attacked a passenger. In another case, a security guard helped accomplices steal $200,000 in gold certificates. The security company was found liable for negligent hiring and supervision.[1] See the Minnesota Supreme Court's ruling in the Negligent Hiring box.

Suspected Terrorist Watch List

The following agencies or resources all can be contacted for suspected terrorist watch list searches:

- Office of Foreign Asset Control's SDN and Blocked Persons
- Federal Bureau of Investigation Alleged Suspects
- Suspected Terrorist List, National Counterterrorism Center
- Designated Foreign Terrorist Organization List U.S. Department of State
- Federal Bureau of Investigation's Most Wanted
- Office of the Superintendent of Financial Institutions—Individual Terrorism
- Office of the Superintendent of Financial Institutions—Entities of Concern to the Business Community
- Bank of England
- United Nations Sanctions List
- European Union List
- Sanctioned Countries Department of the U.S. Treasury
- Denied Persons List U.S. Department of Commerce Bureau of Industry and Security
- Unverified List U.S. Department of Commerce Bureau of Industry and Security
- List of Debarred Parties U.S. Department of State, Director of Defense Trade Controls
- Entity List U.S. Department of Commerce Bureau of Industry and Security
- World Bank Listing of Ineligible Firms

BACKGROUND CHECKS AND SECURITY

The concern over terrorism in the United States intensifies the need to verify employees. More importantly, background checks help to create a safer and more productive workplace. Screening may reduce theft and workplace violence, two important concerns for business security. Security is improved because the human risks are reduced through careful screening and evaluation of employees.

BACKGROUND SCREENING AND COST SAVINGS

In addition to avoiding the costs from liabilities for negligent hiring practices, background screening can help to lessen other operating costs as well. Background checks can reduce turnover by verifying that a person really is qualified for the job.

RESUME FRAUD

Resume fraud is rampant in the United States. The FBI estimates about half a million people in the United States make false claims about having a college

---

Negligent Hiring

The Minnesota Supreme Court has refined the central aspects of negligent hiring:

1. *Duty of care.* Employers must exercise reasonable care when hiring people whose jobs require that they have contact with the public. This contact places the public at risk for injury.
2. *Foreseeability.* The employer must anticipate how the employee's past could affect future actions. A bad driving record indicates an employee could have a vehicular accident. A history of violence demonstrates an employee could engage in a variety of violent acts in the future.
3. *Reasonable investigation.* Employers should investigate criminal backgrounds for jobs involving contact with the public. This is a reasonable precaution. Employers should also execute broader background investigation when the employee's application appears suspicious.
4. *Cause.* The hiring needs to determined to be the cause of the injuries. The employer placed others at risk by hiring the employee.

PHYSICAL PROTECTION

degree. The Society of Human Resource Managers believes that 53 percent of all applicant resumes have false information. Of the false information on resumes, 44 percent lie about work experience, 23 percent lie about credentials or licenses, and 41 percent lie about education. Applicants can buy diplomas from online "diploma mills" for under $500. These lies can create liabilities that cost companies money. The Association of Certified Fraud Examiners estimates that companies lose around $600 billion a year due to resume fraud.[2]

During interviews, behavioral interview techniques are an additional tool for checking work experience and skills. These techniques try to link the applicant's work experience to the current job. Based on the knowledge, skill, and abilities required for the job, interviewers ask applicants how they would handle a particular situation, such as "Describe an unpopular decision you had to make and how you implemented that decision." Or, "Tell me about a situation in which you influenced people positively through a presentation." Such questions provide insights into how applicants will perform, as well as expose job skills applicants might have embellished on their resumes.

Resume verification becomes an important step in the background check. The critical points on the resume to verify are education, work experience (especially job titles), previous employers (did the person really work there), specific job skills, and references (be wary of any glaring omissions in the reference list). Resumes should not be taken at face value.

DRUG SCREENING

Drug screening or testing is often a part of the background screening. Drug screening can help to prevent on-the-job accidents and reduce other business costs. Drug screening helps to lessen the costs of wages paid for absenteeism, wages paid for temporary staff to cover jobs, and costs of replacing damaged equipment; reduce sick leave; increase productivity; and reduce losses from theft. Around 80 percent of major U.S. companies conduct some form of drug testing. Drug tests can use urine, oral fluid, or hair as the samples. Drug testing minimizes negligent hiring by requiring all applicants to pass a drug test before being hired. Drug testing minimizes negligent retention by regularly testing employees and being prepared to intervene when a worker is identified as having a problem.

The legal situation for companies and drug testing involves two concerns. First, a company can be legally responsible if it does not drug test. If employees harm others while under the influence, the company can be held liable. Second, even well-intentioned drug testing polices can be challenged in court. Most of the challenges come in the form of right to privacy, freedom from unreasonable searches, and due process. Overall, the legal risk of not having a drug testing policy is greater than having one. A company can take actions to prevent successful legal challenges to a drug testing policy. A drug policy tells employees that they cannot be at work with "any detectable trace amount of drugs or alcohol in their system," not "under the influence" or "impaired." Drug tests identify the presence of drugs, not impairment from drugs. Drug testing should be kept confidential. Follow the testing standards set by the U.S. Department of Health and Human Services. See the Drug Testing Guidance box for information from the U.S. Drug Enforcement Administration (DEA).

WHY COMPANIES DO NOT USE BACKGROUND CHECKS

The two reasons for not using background checks are time and money. Some companies use so-called instant background checks. Though they are fast, they are not very accurate. It entails a computer search of a rather limited database. Thorough background checks are done by hand. The hand checks take longer and cost more money, but the client is getting more accurate and usable background checks. Poorly executed background checks are not a viable defense against negligent hiring lawsuits. Given the amount of money associated with the risks of hiring and retention, companies should be willing to pay for comprehensive background checks.

Drug Testing Guidance

Here are recommendations from the U.S. Drug Enforcement Administration (DEA) for steps to follow when implementing and maintaining a drug- and alcohol-free workplace program.

- Keep written records that objectively document suspect employee performance. These can be used as a basis for referral for testing.
- Know your employees. Become familiar with each one's skills, abilities, and normal performance and personality.
- Become familiar with common symptoms of drug use.
- Document job performance regularly, objectively, and consistently for all employees.
- Take action whenever job performance fails, regardless of whether drug or alcohol use is suspected.
- Know the exact steps to take when an employee has a problem and is ready to go for help.

Communicate immediately with your supervisor when you suspect a problem, and have a witness to your action when confronting an employee.[3]

Companies have a number of background screening companies to choose from. Keep in mind that background screening is becoming more professional. There's now even a professional organization, the National Association of Professional Background Screeners (NAPBS). The NAPBS, founded in 2003, has the following mission:

> The National Association of Professional Background Screeners exists to promote ethical business practices, promote compliance with the Fair Credit Reporting Act, and foster awareness of issues related to consumer protection and privacy rights within the background screening industry.
>
> The Association provides relevant programs and training aimed at empowering members to better serve clients and to maintain standards of excellence in the background screening industry.
>
> The Association is active in public affairs and provides a unified voice on behalf of members to local, state and national lawmakers about issues impacting the background screening industry.[4]

CONCLUSION

When hiring, always conduct a thorough background check that includes a criminal records search, employment verification, education verification, driving records, Social Security confirmation, suspected terrorist watch list check, and credit check. If you conduct the search yourself, be sure to follow all relevant laws, such as the Fair Credit Reporting Act. If possible, hire professionals to conduct the background check.

See also Security Guards/Officers; and Supply Chain Security.

NOTES

1. "A Look At Negligent Hiring Law Suits," online at http://www.verires.com/nhiring.htm (accessed 7 Feb 2007).

2. Mike Aamodt, "How Common Is Resume Fraud?" online at http://www.runet.edu/~maamodt/Research%20-%20IO/2003-Feb-Resume%20fraud.pdf (accessed 7 March 2007).

3. U.S. Drug Enforcement Administration, *Guidelines for a Drug-Free Workplace*, 4th ed., Summer 2003, online at http://www.usdoj.gov/dea/demand/dfmanual/index.html (accessed 27 Jan. 2007).

4. National Association of Professional Background Screeners, "Background Screeening: Past, Present and Future," online at http://www.napbs.com/images/pdf/HistoryBackgroundScreening.pdf (accessed 15 March 2007).

VIDEO SURVEILLANCE

W. Timothy Coombs

Technology allows organizations to watch their employees, customers, visitors, and would-be intruders through video surveillance. Organizations can identify intruders, discover employee theft or other misbehavior, or prevent shoplifting. Video surveillance helps security to see what is going on in and around a facility.

EVOLUTION OF VIDEO SURVEILLANCE

Video surveillance is in a transition from analog to digital. Old video surveillance consists of the stand-alone CCTV systems that record to videotape, which are slowly fading away. Tape coding and storage is problematic, locating particular footage is time consuming, and tapes degrade over time and need replacing. Many companies are transitioning to the digital age with hybrid digital-analog systems. These systems connect cameras to digital video recorders, much like a DVR or TiVo.

Fully digital video surveillance is networked Internet protocol (IP) based. The video surveillance is a component of the IT network. Each camera has an IP address and is controlled centrally through a software application. The video surveillance transition to digital is another illustration of the convergence of physical security and IT security. (Refer to the Integrating Physical and Information Security entry for a further discussion of the topic.) The move to digital video surveillance is costly, requires additional training of security officers, and demands integration with IT.

BENEFITS OF DIGITAL IP SURVEILLANCE

The benefits of digital IP surveillance justify the costs. Digital IP surveillance provides superior visual data. The cameras have good lenses that can record in low light and even do thermal imaging. One digital camera can cover a larger area than an analog camera and zoom in for fine detail. The digital IP surveillance is integrated into the IT infrastructure. It can be on the regular server and enjoy the benefits of IT security and backup systems.

Digital IP surveillance is easy to centralize for monitoring and automation. Multiple facilities in geographically diverse locations can be connected to one monitoring center instead of having multiple monitoring centers at each site. The digital video is easier to archive and to retrieve or access. Retrieval takes a few mouse clicks rather than searching through racks of videotapes. Software helps to automate part of the video monitoring process and serves as an extra set of eyes to

PHYSICAL PROTECTION 289

look for signs of trouble. An example is automated alarming that sends a warning signal when predefined signs are detected. The organization creates rules that help to identify a warning sign from normal activities. An example would be indications of human movement in an area where there should be no people or movement in an area during a specific time when there should be no people. These systems can differentiate between human and animal movements to prevent false alarms. The rules create a filter for detecting problems. Another example would be movement alerting a camera to switch to high-resolution mode and track the object.

Finally, digital IP surveillance has applications beyond security. Marketing departments can use the video to track the movement of customers through the store. They can determine how displays affect that traffic flow. The systems can also be used to work with safety systems and call for the fire department when needed.

Digital IP surveillance can be used for employee training and monitoring, sometimes called remote video auditing (RVA). RVA is much more than trying to prevent employee theft. Digital cameras are placed in locations to observe specific employee behaviors such as customer service or areas that involve regulatory compliance activities. The video is reviewed and critiqued by auditors, who can be consultants or in-house personnel. They then provide reports to management and employees at regular weekly or monthly intervals. RVA is currently being used in a variety of settings including health care, manufacturing, retail, food processing, and restaurants.

Feedback is an important component of goal setting and positive reward motivation systems. Both of these motivation systems have a high success rate across industries. However, neither works unless employees are given accurate feedback on their performance, which RVA does. Employees can see for themselves when they are acting properly or improperly. The video is used to show both good and bad examples to help employees learn how to improve their performance. RVA has assisted organizations in improving compliance scores, safety records, and customer satisfaction.

Plumrose USA is a meat-processing company. According to general manager Mike Rozzano, its facility was receiving poor performance safety reviews prior to using RVA. By employing an RVA system, Plumrose was able to identify areas that needed improvement and reward those areas with good performance. Safety review and employee morale improved with the RVA system. Employees could see their unsafe practices and examples of safe employee behavior. Clearly, digital IP surveillance has applications beyond safety and security.

PROBLEMS WITH VIDEO SURVEILLANCE

Some companies use fake cameras or hidden cameras as part of video surveillance. Security and legal experts agree that neither is a good idea. Fake cameras are designed to trick would-be troublemakers. However, safety manager Douglas Durden notes that fake cameras create a false sense of security for employees, customers, and visitors. Walter Palmer, a loss prevention specialist, adds that fake cameras can create a liability for negligent security when an organization has a responsibility to provide a certain level of security. In general, fake cameras do more harm than good.[1]

The same holds true for hidden cameras. Hidden cameras can violate surveillance policies and employees' right to privacy. Every company should have a clearly stated video surveillance policy that lets employees know where and how they will be observed. That policy should be well known and employees need to be regularly reminded of its reach. Lawsuits have reinforced the need for organizations to develop and publicize their video surveillance programs. Organizations do not have the right to use video surveillance just anywhere on their property; employees do have a right to privacy. Hidden cameras can create legal liability and angry employees. Before installing a surveillance system, then, you might want to consult with a local employment attorney to help draft a policy and set parameters for using the system.

CONCLUSION

Video surveillance has become much more than a person watching a television monitor and storing videotapes. Video surveillance is going digital and is becoming integrated with IT and IT security. The move to digital involves cost outlays in equipment. However, in most cases the many benefits of digital IP surveillance warrant the expense.

See also Crisis Sensing Mechanism; Integrating Physical and Information Security; Physical Security; and Supply Chain Security.

NOTE

1. Todd Datz, "The Hidden Camera," Sept. 2005, online at http:// www.csoonline.com/read/090105/hiddencamera_3824.html (accessed 20 March 2007).

COUNTERSURVEILLANCE
W. Timothy Coombs

Terrorism is a risk faced by most organizations. As the Terrorism entries note, terrorism can occur in the United States and in overseas facilities. Keep in mind that domestic terrorists in the United States include radical activist groups such as the Earth Liberation Front. (Refer to the entry Ecoterrorism for additional information on this topic.) Terrorist attacks are well-planned, not spur-of-the-moment, actions. Terrorists case a facility by visiting and collecting information many times before an attack. Organizations need to begin thinking like counterterrorism units. Security experts use countersurveillance to identify and prevent possible attacks. Among those advocating countersurveillance by organizations is Fred Burton, a former counterterrorism agent for the U.S. State Department. He is now a consultant and believes countersurveillance is the only means of identifying a potential terrorist attack.[1]

Countersurveillance is a process of collecting and analyzing data, something organizations regularly do. Countersurveillance seeks specific data on potential attacks. Security personnel begin by developing a list of potential suspicious activities, such as repeated visits by a person or vehicle. Suspicious activities narrow down the data to be collected and analyzed, the next step in countersurveillance. The final step is to take countermeasures.

Analysis is the most difficult part of the process. Companies, such as Abraxas Corporation, can provide software and training that facilitate the data analysis process. Abraxas has a software program known as TrapWire, which analyzes security information by looking for patterns of suspicious activities. Abraxas first examines an organization's physical vulnerabilities. The vulnerability analysis helps to determine what areas need to be monitored. It then installs TrapWire in the client's system. When security people see a suspicious vehicle or person, they open a TrapWire menu and enter information about it. The information can include physical appearance, types of vehicles, and license plate numbers. TrapWire then reviews the organization's security database (the security video) to construct a "PeoplePrint" or "VehiclePrint" profile for the suspicious activity. Based upon the security data, TrapWire creates a threat rating ranging from 1 to 100. The security personnel then take the necessary actions based on the threat level.

Other vendors that provide countersurveillance equipment and consulting help include Quest Consultants International, ADT, and Brink's Business Security.

Not every organization has the money to buy expensive computer-based countersurveillance systems. However, any organization can take a few simple steps to improve countersurveillance. First, train your employees to be observant. This means explaining that they should look for such things as unfamiliar people in an area, people without proper credentials, or unfamiliar vehicles parked nearby. Your people will be more effective if they know what they are looking for and why. Second, clarify whom employees should contact when they see something that causes concern. Employees should know whom to report specific suspicious behaviors to and the best way to contact those people. More than anything, countersurveillance is about people paying attention and reporting suspicious activities.

CONCLUSION

It is vital to remember that any countersurveillance is only as good as the people conducting it. Even advanced computer-based systems depend on security personnel noticing something suspicious and initiating the analysis. Relevant staff members must be trained in countersurveillance activities. This includes any staff that is in a position to spot suspicious activities. A vigilant staff is the key to effective countersurveillance. Successful countersurveillance, in turn, can save lives and protect organizational assets.

See also Agroterrorism; Corporate or Industrial Espionage; Crisis Sensing Mechanism; Ecoterrorism; and Terrorism.

NOTE

1. Joseph Straw, "Countersurveillance Foils Attacks," *Security Management*, Feb. 2007, pp. 24–26.

RADIO FREQUENCY IDENTITY

W. Timothy Coombs

Radio frequency identity (RFID) may be the most controversial of any business security tool. RFID provides a means of tracking shipments through a supply chain, products in a store, and eventually paper money through the economy. This entry focuses on the security uses of RFID and concludes with some of the privacy issues that make it such a contentious security tool.

BASICS OF RFID

RFID is a system of technology components used to track "things." Its primary use is to follow the movement of items through the supply chain. As such it can be an important component in supply chain security. (Refer to the entry Supply Chain Security for more information on this topic.) The central component of the RFID is a wireless radio frequency device known as a tag. The tags are small devices that have a transponder and an antenna. They transmit data signals when powered/queried by an RFID on the tag's frequency.

RFID tags can have multiple frequencies depending upon their purpose, distance, and cost. Low-frequency tags cost a few cents and have a range limited to a few centimeters. High-frequency tags cost around fifty cents with a range of one meter. Ultra-high frequency tags are just over fifty cents with a range of seven meters. Microwave technologies cost much more and have a range of ten meters. The type of RFID tag an organization uses depends on the cost of the items being tracked and the nature of their transportation and storage. For instance, the transportation and storage may not permit close scanning so an ultra-high frequency or microwave tag may be required.

Unlike bar codes, RFID tags do not require a line of sight, do offer high-speed reads, and can make multiple reads. No power source is needed for passive RFID tags. Active RFID tags have batteries and attributes similar to wireless communication and sensing devices. Passive RFID tags are powered by the RFID readers or RFID printers. The reader sends an RF signal to the tag for power. The tag then transmits data to the reader. An RFID reader can write information to read/write tags in addition to collecting data. RFID printers can print information received from the tag and write some tags as well. RFID is governed by the electronic product code (EPC) standards. This ensures each tag has a specific and unique identity as well as providing a standard way to identify and exchange information between readers and tags.[1]

RFID PRIVACY CONCERNS

Privacy advocates are concerned that RFID could invade people's privacy. There is a web site devoted to the topic, http://www.stoprfid.com, and the charge is being lead by Consumers Against Supermarket Privacy Invasion and Numbering (CASPIAN). The concern is over the use of RFID tags on individual products, not on shipping containers. Each RFID tag has a unique signature. Retailers can use the RFID tag to link purchases directly to a person and track that individual's purchases over time. This is already done on a smaller scale through consumer loyalty cards. Retailers can create a more detailed picture of each consumer's buying habits. Retailers believe RFID tags provide an effective means of inventory control to identify out-of-stock items or excess inventory. Wal-Mart has been systematically adding RFID technology to its stores and requiring that its major suppliers implement the technology.

Clothing maker and retailer Benetton, for instance, had a plan to embed RFID tags in all of its retail items. The RFID tags would have allowed Benetton to track individuals and inventory their belongings by linking a buyer's name and credit card information with the serial number in an item of clothing. Benetton agreed to drop the plan after complaints from privacy groups.

Some privacy advocates see a much more serious breach of individual rights. The passive RFID tags remain active. When a customer enters a store, a retailer could read any item on or being carried by that customer that had an RFID tag and then use the RFID tags to track a customer's movement through the store. Some even fear an Orwellian world in which the range of detection improves and people can be located anytime and anywhere by RFID tags. Current technology does not allow such tracking, but privacy advocates fear it could be developed and limit people's ability to evade the technology.

Privacy advocates do provide advice on evading current RFID tags, how to search products for RFID tags, and how to disable RFID tags. They recommend puncturing or crushing the RFID tags, but not microwaving them. Microwaving will destroy an RFID tag but it can also cause the tag to catch fire.

CONCLUSION

RFID tags are an effective way to follow the movement of material and products through the supply chain (shipping) and to track inventory of individual products (retail). The RFID tags are low cost, are fairly easy to read, and allow readers to scan multiple items. RFID tags can be an important element in supply chain security. They let organizations know where a shipment is at all times, thus reducing loss through theft or misplaced items. They also let an organization know it is receiving the authentic items. The RFID tags have critics in privacy advocates. Their concerns are often overstated, but an organization must be ready to reply to criticism when it decides to utilize RFID technology as part of its security efforts.

See also Supply Chain Security.

NOTE

1. Craig Dighero, James Kellso, Debbie Merizon, Mary Murphy-Hoye, and Richard Tyo, "RFID: The Real and Integrated Story," 2005, online at http://developer.intel.com/technology/itj/2005/volume09issue03/art09_rfid/p01_abstract.htm (accessed 5 March 2007).

BIOMETRICS

W. Timothy Coombs

A common question in security is "How do we know who someone is?" We need to know who people are to determine whether they should have access to physical locations (access to a building) and information assets (access to a computer system). Biometrics is one way of identifying who a person is as a prelude to access. Biometrics utilize body parts (physiological characteristics) and actions taken by a person (behavioral characteristics) to determine or verify identity. Biometrics can be used to permit entry to buildings, computers, e-mail accounts, Internet access, intranet access, and record the time of access and location of employees.

THE BASICS

Biometrics can use a variety of different physiological and behavioral characteristics to identify or verify a person. The physiological characteristics include fingerprints, iris recognition, retina recognition, facial recognition, and hand geometry. Behavioral characteristics include voice recognition, signature verification, and keystroke dynamics. A biometric system converts user information into a template. A user is then compared to that template for later matching. When a match occurs, the user is permitted access to the desired system or location.

A user "enrolls" by providing a biometric sample, which is assessed, processed, and stored. Later, a user provides a "submission," another sample, to be judged against the template. Biometric systems vary in reliability (number of false positives or negatives) and how acceptable the technology is to users.

We are most familiar with fingerprints. A simple reader is used to assess the fingerprint. Each fingerprint is unique. Many laptops now have fingerprint scanners built in. Fingerprinting is one of the most reliable and accepted technologies. Handprint geometry works in a similar fashion. Each handprint is fairly unique and can be read with a scanner. Currently, hand scanners cost a little more than fingerprint scanners. Facial, iris, and retina recognition are based on the unique features of these body parts. These scans are more invasive than fingerprint scans using cameras and other scanning technologies. Retina scans, for instance, require low-intensity infrared light, whereas facial and iris recognition require special cameras. These more invasive systems are resisted by users and are not as accurate as fingerprints. Retina scans are the most problematic because disease can change a person's retina.

The behavioral characteristics of biometrics are more susceptible to fraud than their physiological counterparts. Signature verification, sometimes called dynamic signature verification, examines a signature to see whether it matches the template using a signature tablet. You have probably used a signature tablet if you have ever charged an item in a store to a credit or debit card. Voice recognition examines voice samples. Keystroke dynamics, sometimes called typing rhythms, examines the way a person types on a keyboard. The ability to commit fraud with behavioral characteristics is not easy, but the risk for fraud is too high for many organizations that demand tight security.

CONCLUSION

Biometrics have a wide range of options and applications for business security. They are also a further illustration of the integration of physical security and information security. The identity of a person can now be verified by computers rather than individuals. The type of biometrics your organization uses depends upon the level of security that is needed, the amount of money you have to spend, and what types of biometrics your users are willing to accept.

See also Information Security; Integrating Physical and Information Security; Physical Security; and Portable Device Security.

SHELTER-IN-PLACE
W. Timothy Coombs

In an emergency, companies have two basic response options: evacuation or shelter-in-place. People are less familiar with the idea of shelter-in-place than evacuation. Shelter-in-place means that people should stay inside of buildings and seal the buildings off from outside air. For some reason, the outside air presents a danger. Shelter-in-place requests can be driven by chemical, biological, or radiological contamination in the air. At a company, shelter-in-place can affect employees, visitors, and customers. Anyone in the building must be notified of the situation, accounted for, and informed of what to do. Remember, you cannot force employees, visitors, or customers to shelter-in-place unless a government order has been issued. People have the right to leave if they choose to do so. An important part of preparation for your employees is to give them compelling reasons to stay inside.

HOW TO SHELTER-IN-PLACE

The basic steps to sheltering-in-place for a company include the following:

1. Close the business and post signs warning those outside the building.
2. Inform all those in the building, including visitors and customers, of the situation and the reason to shelter-in-place.
3. Shut and lock all windows and doors.
4. Have people assemble in designated rooms and bring disaster supplies to these locations. (Disaster supplies include nonperishable food, water, battery-powered radios, first aid supplies, batteries, flashlights, phones/phone access, garbage bags, duct tape, and plastic.) The National Institute for Occupational Safety and Health (NIOSH) recommends a variety of respirators that can be used during an evacuation to protect people from smoke inhalation. See http://www.cdc.gov/niosh/.
5. Have people call their emergency contacts to inform them of what is happening.
6. Turn off all heating, ventilation, and/or air conditioning. An employee who has been trained in the proper way to shut down these mechanical systems should do this.
7. Seal the windows and vents with sheets of plastic and duct tape. Ideally, the pieces of plastic are precut to fit the openings and employees are preassigned and trained in how to seal the windows and vents.

PHYSICAL PROTECTION

8. Seal around doors with plastic and duct tape.
9. Turn on the radio or television and listen for further instructions.
10. Once the all clear is given, remove plastic, restart the mechanical system, and go outside of the building until the old, possibly contaminated air in the system has been replaced with fresh, clean air.

From this list, we can identify three critical factors that must be considered before and during a shelter-in-place episode. First is the physical location of the shelter. The locations for shelters need be selected and designated, and people must be moved to those locations. Second is an accountability system. You'll need to take a head count, form a roster, post signs indicating shelter-in-place is in effect, and provide notification if a person leaves the shelter. Third is preassigning and training employees to shelter-in-place. Employees need to be assigned and know how to perform the task of sealing the building. It is essential to have a system in place so those sheltering can contact family members. The inability to contact family is the main reason people break with a shelter-in-place order. Security personnel can play a vital role during a shelter-in-place. Security personnel can help to ensure the facility is secure, provide direction to shelter areas, inspect the sealing-off efforts, and supervise the shelter areas.

When the need to shelter-in-place is a result of your company's actions, such as a chemical release, the responsibility for the shelter-in-place extends to the community around your facility. Community members need to understand that the warning signal means they must shelter-in-place, and they need to know what to do when they shelter-in-place in their homes. Companies work with area emergency personnel to inform residents about shelter-in-place and to warn them when such an action is needed. It is not uncommon in the United States for shelter-in-place warnings to be issued following an accident involving hazardous chemicals. According to Environmental Protection Agency statistics, there were 32 incidents requiring community members to shelter-in-place between 1989 and 1999.[2]

CONCLUSION

Some emergency situations or crises require people to shelter-in-place. Organizations must understand the equipment that is needed and the procedures to follow if there is a shelter-in-place order. As with other preparation efforts, organizations should conduct drills to test employee readiness for sheltering-in-place.

See also Benefits of Emergency Management; Disaster Recovery Management; Emergency Preparedness and Response Component; Exercise and Training Basics; Evacuation in Large and Multiple Tenant Buildings; and Emergency Response Training and Testing: Filling the Gap.

NOTES

1. American Red Cross, "Shelter-in-Place in an Emergency," online at http://www.redcross.org/services/prepare/0,1082,0_258_,00.html (accessed 5 May 2006).
2. "Toxic Release Inventory Program," online at http://www.epa.gov/tri/ (accessed 11 Feb. 2007).

EVACUATION IN LARGE AND MULTIPLE TENANT BUILDINGS
W. Timothy Coombs

One common security action all employees face is the evacuation of a building. It seems very simple—just leave the building. However, a successful evacuation requires careful planning and preparation. This entry reviews the key points for a successful evacuation, security's role in the process, and the challenges posed by multiple tenant buildings.

An evacuation is built around an emergency notification system and an active evacuation team. These two points constitute the evacuation plan. How do people know whether to evacuate a building? Again, this sounds deceptively simple. Recently, a university had trouble evacuating during an actual fire because people assumed it was a drill and kept working. People must know that when the warning to evacuate is given, everyone one must leave the building. Drills should be treated like actual emergencies. Management needs to convey the importance of always exiting a building when the alarms sound. Make sure emergency signage is visible, in working order, and in multiple languages if non-English-speaking people are ever in the building.

The evacuation team helps to ensure a safe and orderly egress from the building. Common evacuation team members include floor wardens, searchers, elevator monitors, stairway monitors, and ADA assistors. The floor warden, who runs the evacuation team, is in charge of a certain area during an evacuation. Searchers look in specified areas for people and mark areas as searched. This helps to make sure everyone is out of a particular area so that emergency personnel do not have to search those areas. Typical areas to search would include the copy room, offices, conference room, and bathrooms. Because people tend to congregate in these areas, they may not hear the evacuation notification.

Elevator monitors stand near the elevators to keep people from using them. Most elevators will automatically go to the ground floor and remain closed once an emergency notification is sounded. However, older systems or malfunctions could create a hazard. The elevator monitor prevents people from becoming victims of those hazards. Stair monitors in the stairwell tell people to go single file on the right side of the stairs. A single file on the right makes it easier for emergency responders to go up the stairs. During an emergency, people can forget and clog the stairs, so the stair monitor has an important job. Both monitors stay at their positions until the floor warden tells them to leave.

The ADA assistors are there to help anyone with disabilities to evacuate. An organization should talk with its employees ahead of time to find out who might need assistance and in what form. For instance, it may take two assistors to help an individual with a wheelchair. Organizations should consider purchasing special evacuation chairs to make the process even safer. It is a good idea to have extra assistors in case there are visitors needing extra assistance. The evacuation includes anyone who may be visiting your organization. These people require special guidance because they have not been a part of your drills. Everyone should be directed to the secure muster area.

Security is invaluable in an evacuation. It can help to identify who has left the building and is in the muster area. Some organizations use portable card readers to determine who is out of the building and in the muster area. Those records can be compared to security's records of who has logged in that day. Security is an excellent liaison with emergency responders. Security personnel can coordinate by supplying floor plans, locations of possible victims, evacuation status, and ADA employee locations. Security should secure the site as well, which includes establishing a perimeter and access points and providing access control to the site.

Frequently, organizations find themselves as one of many tenants in an office building. Multiple tenants add complexity to the emergency management process and can be illustrated with "simple" evacuations. Emergency management expert Patricia Bennett has found that the lack of integration between facility managers and the tenants' emergency managers/continuity managers is the main problem for evacuations of office buildings. The facility managers often do not communicate with tenants about evacuation and have little understanding of evacuations themselves. Bennett identified four common mistakes related to facility evacuations: (1) lack of communication about roles during an emergency (e.g., not knowing who will do what); (2) not knowing the real risks (e.g., failure to examine risks); (3) not knowing the audience (e.g., having signage in the right languages); and (4) not having a proper muster area (e.g., everyone trying to muster in the same location). These problems can be solved if residents and facility managers coordinate their evacuation plans and even have joint drills.[1]

Occupants of multiple tenant facilities should work together to develop and test evacuation plans. The same holds true for shelter-in-place plans. Evacuation and shelter-in-place are the two basic options in emergencies. Every organization should have its own emergency notification system and evacuation team in place before coordinating with others. Once the plans are developed, they must be tested in drills. Refer to the entry Exercise and Training Basics for the value of drills and the problems of failing to drill.

CONCLUSION

All organizations should create an evacuation team and train people for their various roles on that team. The organization should also hold regular evacuation exercises that involve everyone in the organization. For organizations in a multiple tenant building, be sure to coordinate your evacuation plans and exercise with the other tenants.

See also Exercise and Training Basics; Emergency Preparedness and Response Component; Shelter-in-Place; and Emergency Response Training and Testing: Filling the Gap.

NOTE

1. Patricia L. Bennett, "Evacuation Planning: Four Mistakes Managers Make," *Disaster Resource Guide Quarterly*, Fall 2006, pp. 24–26.

INTEGRATING PHYSICAL AND INFORMATION SECURITY

W. Timothy Coombs

The heavy reliance on technology as part of the physical security system is the driving force behind the need to integrate physical security and information security. Technology on the physical side can track such activities as use of identification cards, biometric scans, facial recognition software, parking management, and the storage of closed circuit television (CCTV) video or digital video. Consider the CCTV system. The old analog version of tapes is being replaced with computer/web-based digital video. Add to this the computerization of fire systems and HVAC systems, two systems frequently linked to physical security, and we can see the trend.

Integration occurs on two levels. The first level is the integration of the various physical security systems. An organization needs to assimilate such physical security systems as parking, employee access control, CCTV, and visitor monitoring systems. The multiple physical security systems must work together and communicate with one another. Here are a few examples. An employee uses an identification card to move through various parts of the building. CCTV cameras are operating as well. Security officers can use a simple mouse click to create a report about that employee's access along with the accompanying video. A visitor signs in to meet with an employee. The visitor's

PHYSICAL PROTECTION

driver's license is scanned into the computer and entered into the visitor system database including the license photograph. The employee is sent a text message that the visitor has arrived. The systems work in concert to make physical security more efficient and effective. For instance, if an employee's access requirements change, that can be added to the badge and access scanners. If an employee loses a badge, the reports from the access scans and input from the CCTV video can identify where the badge was used and capture the user on video.

The second level is to integrate the physical security system with other computer systems in the organization. This level includes integrating the physical security systems with the IT security systems. The IT security systems help to ensure the integrity of more modern, technologically advanced physical security systems. The second level also means using old-fashioned physical security to protect the IT hardware. One way to defeat IT security is to have direct access to the equipment. As security expert Salvatore D'Agostino notes, a combined system provides better protection from both physical and IT threats.[1] When an employee leaves an organization, the integration aids with security. The human resource system informs the security systems that the employee is no longer a part of the organization. The security systems immediately revoke that employee's access to physical and computer resources. The security systems can be used to monitor the employee's final exit from the facility.

INTEGRATION CHALLENGES

Integration sounds like a magical solution with all those different systems working in harmony. Historically, security systems have been proprietary and designed to be closed systems. They were difficult if not impossible to integrate with other systems. In others words, the security systems would not play well with others. More recently, security systems have improved integration by offering more open, standard systems and software kits to facilitate interfacing with other programs. Still, integration is not easy. The IT and security departments, like their system counterparts, will have to work together to make the integration work. Moreover, organizations may need to update the job descriptions and training for security officers. Security officers will need to be able to interact effectively with the technology that is part of today's advanced physical security systems.

CONCLUSION

Physical security is evolving, and that evolution is leading to the increased use of technology. To maximize the benefits of technologically enhanced physical security, an organization needs to integrate the systems with one another as well as other computer systems in the organization. Integration results in faster and more effective use of the physical security technology.

See also Information Security; Physical Security; and Video Surveillance.

NOTE

1. "Security Integration: Making It Work," March 2007, online at http://www.facilitiesnet.com/bom/article.asp?id=6273&keywords=security+integration (accessed 27 March 2007).

SECURITY ON A GLOBAL SCALE

Most people believe that technology continues to shrink the world. The size of the world has not changed, but it is easier for people to interact with others across vast geographic distances. Management may find its security and safety measures are complicated by its international connections. In particular, security on a global scale raises issues related to pandemics and supply chains.

PANDEMICS

Robert C. Chandler

A pandemic is a global disease outbreak. It occurs when a new contagion or communicable virus emerges for which there is little or no immunity in the human population, begins to cause serious illness, and then spreads easily person to person worldwide. Infected individuals experience the symptoms of the disease, including potentially serious ill effects, health consequences, or even death. Such a widespread illness strains the health care system and can create significant economic consequences for businesses, including disrupting travel, shipping, and transportation; impacting international commodities markets; high levels of employee absenteeism; lost productivity; disruptions in civic and infrastructure services; loss of key personnel and institutional knowledge; and a cascade of negative impacts spanning from vendors and suppliers to end-user customers and clients. It is impossible to predict precisely the time and severity of the next pandemic. Whenever and wherever a pandemic starts, everyone around the world is at risk. Countries might, through measures such as border closures and travel restrictions, delay the arrival of the virus but they cannot stop it. A pandemic is one of the most serious potential threats to business continuity, public health and safety, national security, and contemporary human civilization.

PANDEMIC PERIODS

The World Health Organization (WHO) and the U.S. Centers for Disease Control (CDC) and Prevention categorize pandemic status as having four distinct periods: (1) the interpandemic period, or between pandemics, (2) the pandemic alert period, (3) the (acute) pandemic period, and (4) the postpandemic period. Within each period various specific phases or stages are recognized as markers for pandemic alert. During the interpandemic period, a period of relative calm, there are no new or identified threatening viruses. In the case of influenza, for example, no new influenza viruses have been identified in animals, birds, pigs, other animals, and there are no new or threatening subtypes of a virus identified in humans. The interpandemic period can last for an indefinite amount of time, because it is really the pause between waves of pandemics.

During the pandemic alert period, the second major period, there is awareness of a pandemic threat and the potential for threats. There is not only an identifiable threatening virus detected in animals but also incidences of humans who succumb to the virus after having been infected by animals. In this period, small clusters of animal-to-human or perhaps even human-to-human transmissions, in very specific geographic areas, can be identified. There may even begin to be clusters of human-to-human transmission in a specific geographic region or epidemics on a local level.

During the acute pandemic period, the third major period, evidence exists of sustained human-to-human transmission in the general population in an expanding geographic area, and the virus jumps from one region to multiple regions and eventually becomes a global pandemic. The acute pandemic period is the time of active alerts and requires health control measures such as travel restrictions and quarantines. This would also be a period in which we might see waves of people succumbing in different demographic groups. The pandemic might affect the elderly first and then go through schools. Next there might be waves of secondary transmissions as the schoolchildren bring it home, and it thereby travels to different population centers. A pandemic period usually lasts anywhere from four to seven months, depending on the severity and the scale of the outbreak.

Also during the acute pandemic period, a substantial percentage of the world's population requires some form of medical care. Health care facilities can be overwhelmed, creating a shortage of hospital staff, beds, ventilators, and other supplies. The need for vaccine is expected to outstrip supply, and the supply of antiviral drugs is also likely to be inadequate early in a pandemic. Difficult decisions need to be made regarding who gets antiviral drugs and vaccines. There are significant infection rates with substantial consequences. The bottom-line number of fatalities varies depending on four factors: (1) the number of people who become infected, (2) the virulence of the virus, (3) the underlying characteristics and vulnerability of affected populations, and (4) the availability and effectiveness of preventive measures. The consequences of a pandemic transcend the numbers of sick, dying, and dead victims as businesses, schools, and many aspects of society are fundamentally disrupted during the acute phase of the pandemic.

Once the acute pandemic outbreak has subsided, we enter the fourth and last period, the postpandemic phase, which is a slow return to the interpandemic period with its preparations and threat monitoring for the next pandemic phase. The interpandemic period is the starting point of a new cycle. Potential pandemic viruses in animal populations during the interpandemic period may be seen, but have yet to be detected in humans. Because of the long history of pandemics, we recognize the basic cycle. This history also helps us take the appropriate preparedness measures and steps. Many organizational continuity planners may therefore want to tie the specific activation of their plans to the basic phases of a pandemic, and then communicate the period and actions required to the media as well.

PANDEMIC CONSEQUENCES

To begin to understand the scope of a potential pandemic impact, one need only look at the "routine" seasonal influenza outbreaks and project the impacts in geometric increments. The cold and flu season in a "normal" year is disruptive enough. It comes with tremendous loss in productivity, worker absenteeism, health costs, and lost opportunity cost. The seasonal flu itself is a health issue deserving some attention and preparedness in the event a sizable percentage of your workforce succumbs to it. For the public health sectors, the seasonal flu is a significant concern. However, the disruptions of the seasonal flu are only a minor sample of the disturbances that a fast-moving and widespread pandemic can unleash.

A pandemic brings with it substantial impacts and challenges. Each of its phases creates unique challenges, but the most significant impacts occur during the acute pandemic phase. The prime consequence to note is that pandemics happen quickly at the acute phase. A contagion can spread rapidly across the globe, and modern medicines and techniques can do little to stop it. For example, when an influenza virus emerges in even a remote part of the world, epidemiologists and world health authorities consider its global spread inevitable. The emergence of a new pandemic virus means that the entire world population could be susceptible. Individual countries might, through measures such as border closures, quarantines, and travel restrictions, delay its arrival or slow its spread, but such measures cannot stop a pandemic.

In the worst case, most people will have little or no immunity to a pandemic virus, and infection and illness rates will soar. A substantial percentage of the world's population will require some form of medical care. This suggests that the demand for health care access, at peak times, will be thousands of times greater than typical demand, which will stress and possibly overload health care systems. Death rates, moreover, are high, largely determined by four factors: the number of people who become infected, the virulence of the virus, the underlying characteristics and vulnerability of affected populations, and the effectiveness of preventive measures. Past pandemics have spread globally in two and sometimes three waves, creating an extended period of acute pandemic impacts that lasts for a number of months rather than merely weeks.

The projections include forecasts that medical supplies and health care system capacity will be woefully inadequate during the peak periods of the next pandemic. The existing supplies of general antiviral drugs are likely to be insufficient during the initial phases of an acute pandemic period. There are efforts currently underway to develop vaccines that would be effective against many strains of the existing pandemic possibility, the H5N1 influenza virus (the bird flu). Some of these vaccines require multiple injections (spaced up to one month apart) in order to provide even limited protection. In clinical trials, such vaccines have prompted immune responses only strong enough to prevent infection in about 45 percent of study participants who received the full-strength vaccines.[1] Even these levels might help mitigate the impact of infection for those receiving the vaccines. However, they are not yet available commercially. (Typically, some vaccines are stored in the Strategic National Stockpile and used at the discretion of the national government for key high-risk individuals deemed essential to national security. The U.S. government's goal is to stockpile enough vaccine to protect up to 20 million people.)

Any existing vaccines will be slow to arrive (if at all) to the general public or specific enterprises and possibly will not be available until a second or third peak of the pandemic. Even when available, demand for such a vaccine is likely to outstrip supply, creating not only shortages but also problematic "lifeboat" decisions about who should receive vaccinations and on what criteria to base such decisions. A pandemic can create a shortage of health care providers (whose ranks will also be reduced due to infection and absenteeism), as well as of emergency department space, local health care provider slots, hospital beds, ventilators, and other medical supplies.

The pandemic will create economic and social disruptions. An especially severe influenza pandemic can lead to high levels of illness, death, social disruption, and economic loss. Everyday life will be upset because so many people in so many places become seriously ill at the same time. Impacts can range from school and business closings to the interruption of basic services such as public transportation and food delivery. Travel bans, closings of schools and businesses, and cancellations of events could have a major impact on communities and citizens. Significant worker absenteeism will be the result of those who are sick and dying, as well as those who care for sick family members or simply fear exposure by coming to the workplace. Some studies have forecast worst-case scenarios of over 50 percent absenteeism.[2]

Although a communicable virus might move rapidly across the globe, a pandemic disaster is a comparatively "slow-motion" disaster compared with tornadoes, hurricanes, industrial accidents, terrorism or criminal events, earthquakes, wildfires, or other types of catastrophic events. A pandemic may unfold in a series of peaks (e.g., waves), at which there are extremely high levels of infected and symptomatic people, each wave of which can last for six to eight weeks.

PANDEMIC RISK

Although many different possible contagions could spur the next pandemic, the most publicized threat source in recent years is the (cyclical) risk of an influenza virus. Influenza pandemics are recurring and unpredictable calamities that have occurred during at least the last four centuries and probably for longer than those recorded outbreaks. Since 1900, three major pandemics and several "pandemic threats" have arisen.

Influenza Pandemics

The Spanish influenza pandemic is the catastrophe against which all modern pandemics are measured. An estimated 20 percent to 40 percent of the worldwide population became ill and over 50 million people died. Between September 1918 and April 1919, approximately 675,000 deaths from the flu occurred in the United States alone. Many people died from this flu very quickly. Some who felt well in the morning became sick by noon and were dead by nightfall. Those who did not succumb to the disease within the first few days often died of complications from the flu (such as pneumonia) caused by bacteria. One of the most unusual aspects of the Spanish flu was its ability to kill young adults, and the reasons for this remain uncertain. With the Spanish flu, mortality rates were high among healthy adults as well as the usual high-risk groups. The attack rate and mortality were highest among adults 20 to 50 years old. The severity of that virus has not been seen since.

In February 1957, the Asian influenza was first identified in the Far East. Immunity to this strain was rare in people under 65 years of age, and a pandemic was predicted. In preparation, vaccine production began in late May 1957, and health officials increased surveillance for flu outbreaks. Unlike the virus that caused the 1918 pandemic, the 1957 pandemic virus was quickly identified due to advances in scientific technology. Vaccine was available in limited supply by August 1957. The virus came to the United States quietly, with a series of small outbreaks over the summer of 1957. When U.S. children went back to school in the fall, they spread the disease in classrooms and brought it home to their families. Infection rates were highest among schoolchildren, young adults, and pregnant women in October 1957. Most influenza- and pneumonia-related deaths occurred between September 1957 and March 1958, with the highest death rates among the elderly. By December 1957, the worst seemed to be over. However, during January and February 1958, another wave of illness broke out among the elderly—an example of the potential "second wave" of infections that can develop during a pandemic. The disease infects one group of people first, and infections appear to decrease and then they increase in a different part of the population. Although the Asian flu pandemic was not as devastating as the Spanish flu, about 70,000 people in the United States alone died out of approximately 2 million worldwide.

In early 1968, the Hong Kong influenza pandemic was first detected in Hong Kong. The first cases in the United States were discovered as early as September of that year, but illness did not become widespread in the United States until December. Deaths from this virus peaked in December 1968 and January 1969. Those over the age of 65 were most likely to die. The same virus returned in 1970 and 1972. The number of deaths between September 1968 and March 1969 for this pandemic was 33,800, making it the mildest pandemic in the twentieth century.

There could be several reasons why fewer people in the United States died due to this virus. First, the Hong Kong flu virus was similar in some ways to the Asian flu virus that had circulated between 1957 and 1968. Earlier infections by the Asian flu virus might have provided some immunity against the Hong Kong flu virus, which might have helped to reduce the severity of illness during the Hong Kong pandemic. Second, instead of peaking in September or October, as the influenzas had in the previous two pandemics, this pandemic did not gain momentum until near the school holidays in December. Because children were at home and did not infect one another at school, the rate of influenza illness among schoolchildren and their families declined. Third, improved medical care and antibiotics for secondary bacterial infections were available for those who became ill.

"Bird Flu"

Avian (bird) flu is caused by influenza A viruses, which occur naturally among birds. Wild birds worldwide carry avian influenza viruses in their intestines but usually do not get sick from them. Avian influenza is extremely contagious among birds and can make some domesticated birds, including chickens, ducks, and turkeys, very sick and then kill them. Infected birds shed influenza virus in their saliva, nasal secretions, and feces. Domesticated birds may become infected with avian influenza virus through direct contact with infected waterfowl or other poultry, or through contact with surfaces (such as dirt or cages) or materials (such as water or feed) that have been contaminated with the virus. This virus, which has not yet been detected in the United States, had infected about 300 people (with over 50 percent of the infections proving to be fatal) by early 2007.

In the mid- to late 1990s, another two avian influenza flu viruses—A/H9N2 and A/H5N1—were identified. Each of these caused illnesses among people after exposure to infected birds. Although neither of these viruses has yet gone on to start a pandemic, their continued presence in birds, their ability to infect humans, their very significant contagion and mortality rates, and the ability of influenza viruses to change and become more transmissible among people are ongoing concerns in the early part of the twenty-first century.

The risk from avian influenza is generally low to most people, because the viruses do not usually infect humans. H5N1 is the most deadly of the few avian influenza viruses to have crossed the species barrier to infect humans. The H5N1 virus has raised concerns about a potential human pandemic because of the following: it is especially virulent, can be spread by migratory birds, has already been transmitted from birds to other animals including humans, and (like other viruses) continues to evolve. A number of cases of human infection with A/H5N1 have periodically been reported in Asia in the past decade, and more than half of those infected with that virus have died. As of April 2007 there had been no sustained human-to-human transmission of the disease, but the concern is that H5N1 will evolve into a virus capable of human-to-human transmission at some point in the future.

Human influenza virus usually refers to those subtypes that spread widely among humans. There are only three known "A" subtypes of influenza viruses (H1N1, H1N2, and H3N2) currently circulating among humans with human-to-human contagion patterns. It is likely that some genetic parts of current human influenza A viruses originally came from birds. Influenza A viruses are constantly changing, and other strains might adapt over time to infect and spread among humans. Researchers are concerned that the continued spread of the highly pathogenic A/H5N1 virus across eastern Asia and other countries represents a significant potential threat to human health worldwide.

Most cases of H5N1 influenza infection in humans have resulted from contact with infected poultry (e.g., domesticated chicken, ducks, and turkeys) or surfaces contaminated with secretion/excretions from infected birds. So far, the spread of H5N1 virus from person to person has been limited and has not continued beyond one person. Nonetheless, because all influenza viruses have the ability to change, scientists are concerned that one day H5N1 virus could infect humans and spread easily from one person to another. Symptoms of avian influenza in humans have ranged from typical human influenza-like symptoms (e.g., fever, cough, sore throat, and muscle aches) to eye infections, pneumonia, severe respiratory diseases (such as acute respiratory distress), and other severe and life-threatening complications. The symptoms of avian influenza may depend on which virus caused the infection. Because these viruses do not commonly infect humans, the human population has little or no immune protection against them. If H5N1 virus were to gain the capacity to spread easily from person to person, a pandemic (worldwide outbreak of disease) could begin. No one can predict when a pandemic might occur. However, experts from around the world are watching the H5N1 situation very closely and are preparing for the possibility that the virus may begin to spread more easily and widely from person to person. No commercially available vaccine is currently available to protect humans against the H5N1 virus that is being seen in Asia, Europe, and Africa. A pandemic vaccine cannot be produced until a new pandemic influenza virus emerges and is identified; there is always some delay in developing vaccines from the emergence of a new contagion virus.

CONSEQUENCES OF THE NEXT INFLUENZA PANDEMIC

In the twentieth century, three influenza pandemics caused high death rates and great social disruption. Although health care has improved since the last one, epidemiological models from the Centers for Disease Control and Prevention project that a pandemic is likely to result in from 2 million up to 7.4 million deaths globally. Losses to businesses are projected to range upward to over $100 billion worldwide for a moderate to severe pandemic due to transportation limitations, disruptions to the commodities markets, productivity losses due to absenteeism, business continuity disruptions, personnel losses, and unusual medical and health care costs.

According to the WHO and CDC, if an influenza pandemic were to appear, the following could be expected:

- Given the high level of global traffic, the pandemic virus may spread rapidly, leaving little or no time to prepare.
- Vaccines, antiviral agents, and antibiotics to treat secondary infections will be in short supply and unequally distributed. It will take several months before any vaccines become available.
- Medical facilities will be overwhelmed.
- Widespread illness may result in sudden and potentially significant shortages of personnel to provide essential community services.
- The effect of influenza on individual communities will be relatively prolonged when compared to other natural disasters, as it is expected that outbreaks will reoccur in a series of peaks or waves.

PREPARING FOR THE NEXT PANDEMIC

Central to preparedness planning is an estimate of the deadliness of the next pandemic. Experts' answers to this fundamental question have ranged from 2 million to over 50 million. The disruptions to society, families, businesses, and the national economy may be substantial. Although pandemic preparedness planning has dramatically increased, a large number of businesses and organizations are still underprepared for the threat of a pandemic.

Pandemic preparedness includes the necessity to allocate resources (money, time, personnel, and efforts) for creating a pandemic plan and a pandemic communication plan. Look beyond the boundaries of your business, and consider whether your suppliers, contractors, vendors, and distributing channels are also prepared for pandemic risks. The threat of a pandemic should be monitored.

It is also vital that your plans establish policies to adjust to and accommodate for the disruptions that a pandemic will create. Consider revising sick leave policies, benefits (bereavement and health care), hygiene and social distancing protocols, cleaning procedures, alternative work (off-site or telework), travel policies, and contingency (backup) plans for all aspects of your business operations. It may also be necessary to stock emergency supplies such as masks, gloves, and supplies for those who may be isolated at home or work, or even medical stockpiles (possibly including antiviral medications). Prepare your people to minimize their risk of becoming infected, and know how to cover for the anticipated personnel shortages. Explain pandemic sick leave policies or alternative work arrangements. Have a specific response plan in place for the first warnings of a pandemic outbreak. One of the most critical aspects of surviving a pandemic is to have a comprehensive pandemic communication plan.

Steps Toward Pandemic Preparedness

The following are key steps generally acknowledged as essential for moving toward pandemic preparedness.

1. Secure a senior management commitment to pandemic preparedness.
2. Identify a pandemic preparedness coordinator.
3. Conduct a pandemic business impact analysis (BIA).
4. Establish pandemic HR policies.
5. Create a pandemic communication plan.
6. Create a pandemic response plan (including activation criteria).
7. Establish a threat monitoring system.
8. Work with insurers, vendors, and customers to ensure a coordinated response.
9. Train and prepare your workforce, customers, and constituents for pandemics.
10. Test, practice, and revise your pandemic plan.

CONCLUSION

It is impossible to predict exactly when the next global pandemic will strike. However, the scientific consensus makes one point absolutely clear: pandemics are recurring events and another pandemic will occur. Disaster management and recovery plans have long sought to ensure business continuity by seeking to protect and have alternatives to losses of information technology, data, buildings, and equipment. The new reality is that no disaster management or recovery plan can be considered comprehensive or complete if it does not address the inevitable next public health disaster—a global pandemic.

See also Business Continuity; Crisis Communication: External and Internal; Crisis Management, and Pandemic Communication.

NOTES

1. Lisa Schirring and Robert Roos, "FDA Approves First H5N1 Vaccine," 17 April 2007, online at http://www.cidrap.umn.edu/cidrap/content/influenza/avianflu/news/apr1707 vaccine.html (accessed 20 April 2007).

2. "Pandemic Planning Assumptions," 13 Sept. 2006, online at http://www.pandemicflu.gov/plan/pandplan.html (accessed 20 April 2007).

PANDEMIC COMMUNICATION
Robert C. Chandler

A pandemic is a global disease outbreak. It will strain the health care system and can create significant economic consequences for businesses including disrupting travel, shipping, and transportation; impacting international commodities markets; high levels of employee absenteeism; lost productivity; disruptions in civic and infrastructure services; loss of key personnel and institutional knowledge; and a cascade of negative impacts spanning from vendors and suppliers to end-user customers and clients. It is impossible to predict precisely the time and severity of the next pandemic. The threat of a pandemic is one of the most serious potential threats to business continuity, public health and safety, national security, and contemporary human civilization.

According to the World Health Organization (WHO) and the U.S. Centers for Disease Control (CDC), even in the best-case scenarios of the next pandemic, 2 million to 7 million people would die and tens of millions would require medical attention. If the next pandemic virus were a very virulent strain, deaths could be dramatically higher. The global spread of a pandemic cannot be stopped but preparedness can reduce its impact. It is of central importance that businesses take the necessary steps to develop their own preparedness plans. Some have already implemented structures and processes to counter this threat; however, the plans of others are far from complete and many organizations have yet to begin.

WHO believes the appearance of H5N1, now widely entrenched in Asia, signals that the world has moved closer to the next pandemic. Although it is impossible to forecast the magnitude of the next pandemic accurately, it is clear that much of the world and many businesses and organizations are unprepared for a pandemic of any size.

A pandemic will significantly disrupt your business operations. Consider the following: key personnel may be unavailable for substantial periods of time or may be lost permanently; organizations will need to communicate to and notify different key target audiences representing various demographics; the crisis will create pressures, constraints, and stress, which in turn will negatively affect access, comprehension, and compliance with messages; misunderstandings and rumors will create havoc; key people will hunger for accurate and useful information; and updated two-way communication will be of the utmost importance to the functionality of your operations and care for your people. Every organization needs a pandemic communication plan to mitigate, survive, and recover from a pandemic disaster.

Some projections estimate that, at peak periods, approximately 40 percent (or more) of the workforce may be absent from their duties during a pandemic. Operational, as well as personal, financial, and safety, messages critical to your operations will need to be disseminated and confirmed. These communication challenges will occur when people are too distracted, preoccupied, and overstressed with the pressures of the crisis to devote all of their attention to trying to understand and comply with such messages. This creates a very difficult communication problem. Prudent companies should thoroughly prepare and have various communication contingency plans in place to sustain coordination and communication with all key constituents before, during, and after a pandemic crisis. The pandemic communication planning process is a key priority for overall pandemic preparedness, because in the end, your pandemic communication plan must be prepared to overcome these challenges.

PANDEMIC COMMUNICATION PLANNING

All businesses face common communication challenges during crises, including receiving inaccurate, incomplete, and contradictory information, especially early in the critical events; rapidly changing circumstances; and a variety of sensitive HR and personnel information issues. By their very nature, crises are usually beyond the power of "routine" processes and procedures. The methods, processes, procedures, and even people that handle the day-to-day events cannot be consistently relied upon to successfully manage crisis events and ensure the survival of both your people and your business. You must anticipate the particular communication needs for your organization during a pandemic outbreak. Be prepared to take the initiative to communicate effectively despite the disruptions and people out of their usual position. Assume that there will be breakdowns of "routine" systems and technologies during the peak periods of a pandemic. Your target audiences (both individuals and teams) will experience high levels of stress, and all these factors present an additional level of communication challenges to prepare for. Although pandemics are "slow-moving" disasters compared with many other types of threats, rapidly occurring events/changing information on a local level will challenge your communication plans. People will aggressively demand information during a pandemic. All of the decisions about when to communicate, how, and to whom will be subject to the critical analysis of your stakeholders, your constituents, the news media, and the general public.

Prepandemic education and outreach are critical to both preparing your audience for the pandemic as well as prepositioning knowledge with your audiences that will aid your messages' effectiveness during an outbreak. Understanding what a pandemic is, what needs to be done at all levels to prepare for pandemic influenza, and what could happen during a pandemic helps in making informed decisions both as individuals and as a nation. During the pandemic, the public must be able to depend on its government to provide scientifically sound public health information quickly, openly, and dependably. The capacity to assess risk and employ effective mechanisms to mitigate and manage risk has advanced far, but one key factor to the success of any risk management is risk communication.

Pandemic communication planning begins with self-assessment. First you need to determine the basic objectives of your communication plan. Determine to whom and with whom you need to communicate before, during, and after the pandemic. Clarify how you intend to reach these key audiences, what your alternative channels (modalities) are for connecting with them, when and how often you will need to communicate, and whether the capability exists for two-way/interactive communication. Then create specific messages to have available for use (to provide information, warnings, notification, and requests for behavioral compliance). Finally, test, revise, and continue enhancing your communication plans to be ready when the pandemic unfolds.

A pandemic communication plan should detail how your organization plans to communicate (and who is responsible for each action) with the following potential target audiences during and following a disaster: employees/families, customers, suppliers/vendors/partners, local community, emergency responders, government authorities, and the news media.

RISK COMMUNICATION

Risk communication, as described by the U.S. Department of Health and Human Services, is an interactive process of exchanging information and opinion among individuals, groups, and institutions. It often involves multiple messages about the nature of risk or expresses concerns, opinions, or reactions to risk messages or to legal and institutional arrangements for risk management. Any written, verbal, or visual risk message contains information about risk that may or may not include advice about risk reduction behavior. A formal risk message is a structured written, verbal, or visual package developed with the express purpose of presenting information about risk.

Effective risk communication requires a proactive plan. Merely disseminating information without regard for communicating the complexities and uncertainties of risk does not necessarily ensure effective risk communication. Systematic, thorough, and validated testing efforts will help guarantee that your messages are constructively formulated, transmitted, and received, and that they result in meaningful actions. Successful risk communication messages can assist in preventing ineffective, fear-driven, and potentially damaging responses to pandemic crises. Appropriate risk communication messages can create trust and credibility, which are vital in a pandemic.

Here are some general guidelines for effective risk communication, which can prevent making the crisis worse with inadequate messages or failure to communicate effectively:

- Have clear goals and objectives for your communication plan.
- Develop "message maps" in advance for all key audiences and phases of the pandemic. (There's more to come on message maps.)
- Give your target audience specific behavioral guidance on what to do/how to respond.
- Test your communication plan and messages—assess and evaluate. During the pandemic, consistently stay on message.
- Plan your communication to ensure timely and accurate information. Your messages should disclose real risks and be honest, frank, and transparent. When in doubt, acknowledge uncertainty. Your risk communication should create and sustain your credibility.
- Avoid unnecessary communication of complex, medical, technical, or scientific information.

Accurate, timely, and appropriate information reassure your audiences.

INFORMATION DISSEMINATION

As a general rule, more information (rather than less) is usually better during crises. However, the nature of communication during pandemics must consider specific and more complicated questions of timing, load, and source credibility, which make the decisions about how much information to include difficult. Sometimes, too much information is just as problematic as too little information. Getting the appropriate amount of information to the right people at the right time is vital. Underloaded messages serve no purpose at all. Overloaded messages, containing too many details and too much information, tend to overwhelm the receiver, and important parts of the message are "lost."

Determining how best to get the word out is another aspect of pandemic communication planning. You must decide what information is crucial to convey in messages during a pandemic in order to secure comprehension, understanding, and also prompt appropriate audience responses. Likewise, it is important to decide what messages should be delivered prior to, during, and after a pandemic and via which channels or modalities. Also, you must anticipate the possible breakdowns and obstacles to effective communication and how to minimize these weak points. Although every organization is unique, consider the following factors in your information dissemination and emergency notification communication planning: evacuation or shelter instructions; notification or activation of emergency operational procedures; warnings; rumor control information; resumption-of-operations information; assembly-of-security information; emergency operations center (EOC) functions as well as "command, control, and coordination" (C3) messages; response to phone line jams and loss of Internet services; response to media intrusion, inaccurate reports, and mistakes; and how to stay in touch with your people as they are out of place or scattered during the pandemic.

Information must arrive to the appropriate recipient at the optimal time via the best channel—misrouted, tardy, or misdirected information delivery is a significant problem during a crisis such as a pandemic. Those who receive the information must also be able to recognize and understand it. Review your communication technology and anticipate the impact of breakdowns in communication systems. What are your backup modalities if phone lines are down or Internet servers are off-line? What vulnerabilities exist due to outmoded or inadequate notification technology (e.g., is your company still using a traditional "manual calling tree" system for emergency notification)? Is your contact information database outdated or unreliable? Do you have the capacity for multiple channels of notification or automated notification? Is there multiple two-way communication capability?

MESSAGE CHARACTERISTICS

Much could be written about the quality of pandemic communication messages. Obviously, emergency warning or notification messages should be precise, transparent, oriented toward actions, compassionate, and framed in personal terms relevant to the audience. The risk communication messages you create before, during, and after a pandemic should be prioritized (state the most important facts at the beginning of the message) and factual (do not downplay risks or exaggerate facts). Avoid any unnecessary speculation. Remember that word choice is critically important both for the messages and their implications as well as for the "tone" of the message.

META-MESSAGES AND FRAMING

Meta-messages are literally messages about messages. More generally, these are (usually implicit) interpretative clues that provide the audience with cues to guide their understanding of messages. Meta-messages are the deductions that one gains by "reading between the lines" in an interpretive process to discern the tone, urgency, or implicit (but unstated) meanings of a message. Some meta-messages are deliberate (e.g., staging, context, putting points first or last in a message, etc.). Others may not be intentional, but nonetheless audiences "decode" them believing that they have "understood" or "gotten the message," even though such understanding originates from their own perceptions and interpretations rather than any deliberate construction of the message.

Perhaps the most basic meta-message can be illustrated with the language function of irony. Irony is revealed by context and/or nonverbal behavior that alerts the audience that the message is something other than/more than what the words themselves mean. Irony can transform a salutation of "good morning" into a belligerent confrontational challenge. Meta-messages are more generally the messages between the lines that guide audiences' interpretation of what they are hearing or reading. Meta-messages usually exist as either intentionally positive or unintentionally negative. Meta-messages may include the setting (location, context, etc.), nonverbal communication behaviors (tone, timing, appearance, movement, etc.), presence/absence of specific individuals, topics/issues discussed (or not discussed), or any other aspect that can be taken as a hint of something more than what is literally being said.

Obviously negative meta-messages can undermine your message strategy and perhaps even have the reverse persuasive effect on your audience than what you intended. Thus, on one hand, it is essential to exercise caution about potential negative meta-messages that can undermine your communication goals. Critically evaluate how your audience could (mis)understand and (mis)interpret elements of the message or the medium. On the other hand, consider the strategic use of meta-messages to bolster the effectiveness of your communication. Think about how you can embed meta-messages that are consistent with your message, and be attentive to the setting and timing of your communication.

Framing refers to the process by which selective words, narratives, or terminology subtly influences how audiences interpret or evaluate otherwise objective information. Framing includes labels and embedded words as well as how information is packaged. In political contexts this is related to the idea of "spin" and "spin control." Words function as form of label that defines the reality for the audience. The classic illustration is the maxim that one side's *freedom fighter* is the other side's *terrorist*. In this case, specific word choice determines the persuasive interpretation of all other elements presented and carries with it the "implications" of guilt embedded in concepts associated with one word or the other.

Framing also sets the parameters of defining the issue for the audience. A common illustration of this principle is in the social controversy over abortion rights. It is significant that each side of this debate self-describes its position with a one-sided polarized term, which in turn implicitly (negatively) defines its opposition. Therefore, the "pro-life" position is juxtaposed by implication with the "anti-life" position, as is the "pro-choice" side with the "anti-choice" position.

Let's examine one dramatic example to demonstrate the concept of framing in the context of communicating in the midst of a pandemic. One could use different words (labels) to describe those infected with a virus, calling them *infected, victims, patients, ill, contagious, sick, sufferers, carriers, symptomatic*, or in the end *survivors*. Perhaps some of these are extreme examples, but I use them to illustrate the point. Each term carries with it a set of implications that implicitly frames the nature of the act under question. The reality is that many, if not all, of these terms are correct at some level for describing the act/person/situation in question. Nonetheless, when thinking about these word choices in the relative calm of the prepandemic period, referring to someone as *infected* or a *carrier* of the virus brings with it a number of depersonalizing implications and spins our perception of that person in a dehumanizing (and threatening) direction. It is even possible that such word choices could help sustain the "logic" of mistreating or discriminating against someone during a pandemic.

Some of these words unnecessarily play on our fears and prejudices. Other words instead call upon the spirit of our "better angels" and point us in a more positive direction. It is imperative, however, to recognize and understand the framing implications of these word choices. In the past pandemics, hysteria and fear were fueled by some of the language used, the meta-message (between the lines) implications of those words, and the way that the crises were framed. Careful consideration of communicated frames and the role that they play to facilitate the interpretative process and guide the audience toward particular conclusions is very important.

In its most basic manifestation, framing can influence the agenda of the expected communication topics and information. In its strongest expression, framing can serve as a subtle aspect of persuasion capable of changing opinion, shifting attitudes, and motivating specific action among the audience.

RUMORS, MISINFORMATION, AND ERRORS

Rumors will happen during a pandemic. Rumors have always thrived during public health emergencies and tend to emerge in almost every category of disaster or crisis. This is due, in part, to the state of heightened emotional responses, the inherently limited availability of accurate and timely information, and the tendencies for narrative weaving during these events. This combination of elements gives rise to rumors, gossip, speculation, assumptions, inferences, and no small number of conspiracy theories. Obviously, in this modern age of instantaneous communication, the Internet, and personal weblogs (blogs), the unstable local information environment extends electronically. It is important that your pandemic crisis communication plan include specific strategies to respond to substantive rumors, speculation, and misinformation as they circulate.

When responding to a rumor or widespread misinformation, consider how the rumor was initially generated and how it might further evolve as well as spread to new audiences. This may be helpful in designing a communication response that "nips the problem in the bud." As you plan, note that it is important to move quickly to correct these examples of dysfunctional communication; that you will need to keep the level of your response appropriate to the level of the problem; that overreacting to an isolated statement of error might attract attention to the very inaccurate "fact" you are trying to correct; and that underreacting to widely reported erroneous information might allow for a compounding of the error. If a significant rumor is confined to a small audience, then focus your corrective communication to those connected with that group. On the other hand, if misinformation or a rumor is widely known and spreading, you may need to communicate to the broadest possible audience or even consider mass media options.

MESSAGE MAPS

Message mapping is an important process for developing clear and concise message templates that will expedite your communication during a crisis. Message maps are navigational charts or road maps for plotting the risk communication messages that you will need to utilize during and after a critical event. Message maps, developed in advance of a crisis, are detailed, systematically organized, "prepositioned" responses to anticipated message needs, informational issues, questions, or concerns in your various target audiences. Message maps should be constructed according to the principles of effective communication, audience perception and information processing tendencies, and the goals and objectives for your pandemic crisis management. Message mapping and the generation of specific message map templates are crucial to ensuring that an organization has a central repository of effective, consistent, and easy-to-use messages at each stage of the pandemic.

Message maps can be very useful in helping an organization reach its communication goals before, during, and after a pandemic. Message maps should be created well in advance of the pandemic outbreak, linked to different periods/phases of the pandemic, and adjusted for maximum effectiveness with specific target audiences. Every organization needs to have a set of message maps for its particular situation. Message maps should contain adjustments for different demographic groups—including language, education, life situation, and occupation or work role—and cover all of the information agenda questions of most importance to each category. The process of mapping messages for pandemic communication takes time and energy. It should involve a number of different perspectives across the organization.

As you develop maps, consider people affected, walk through every possible outbreak scenario for each period and phase, make choices about types of messages, determine channels of communication, and hone and refine actual message maps. It is also useful to test messages to have a sense of how various audiences will perceive, interpret, understand, comprehend, and respond to these messages.

Message maps can help you connect with your key audiences. Achieving and sustaining effective communication with your target audiences depends (in part) on selecting channels (modalities) of communication that reach them and allow them to reach back to you. Consider both your messages and your target audiences in selecting the most appropriate communication notification systems.

THE ROLE OF AUTOMATED MASS NOTIFICATION IN PANDEMIC COMMUNICATION

Automated mass notification systems can help address common pandemic challenges. These tools enable communicating quickly, easily, and efficiently with large numbers of people or targeted groups of people in minutes, not hours. They allow you to utilize all contact paths, including the preference list of communication devices provided by the targeted audience. These capabilities are especially important when regional or local communication infrastructure is damaged or not working. These emergency notification systems can help ensure two-way communication before, during, and after the pandemic as well as reduce miscommunications and squelch rumors with accurate, consistent messages. By automating manual, time-intensive, error-prone processes of initiating communication during the pandemic, automated systems can free up key personnel at the periods of high absenteeism so they can perform other critical tasks. These systems can improve communication effectiveness by eliminating any single point of failure. And they can be designed and maintained "in-house" or outsourced to an external provider. Typically these systems are easier to use and require only a modest investment given their capacity to enhance your communication during a pandemic crisis. There are many different vendors and providers for the automated notification technology and services; however, one of these companies—3n (National Notification Network)—also provides a pandemic communication message/content solution foundational product, which I have developed, that you might consider to aid your own planning. You can find more information about 3n in the Resource Appendix under Vendors.

CONCLUSION: PANDEMIC COMMUNICATION READINESS

Your pandemic communication planning goals should ensure that you can fulfill all of your critical communication goals and objectives before, during, and after a pandemic outbreak. You should be able to demonstrate the capability of reaching your target audiences (key people with whom you need to communicate) with valid messages in a reliable, confirmable, and efficient way.

When testing and validating your pandemic communication plan, be sure to assess whether your plan is reliable; includes communication technology that addresses potential communication failures; has redundancy and overlapping message paths; sustains two-way and interactive communication; has high usability; is flexible and provides mobility capability; is verifiable; and enables you to communicate quickly, easily, and efficiently with large numbers of people in minutes, not hours.

See also Crisis Communication: External and Internal; Pandemics; and Risk Communication.

OUTSOURCING AND SECURITY
W. Timothy Coombs

Outsourcing is when management shifts noncore operations from internal production to an external entity that specializes in that operation. An outside vendor is contracted to handle what was once done internally at the organization. Organizations can outsource both manual and intellectual labor. Examples would include hiring a cleaning company rather than having a janitorial staff and having data entry performed by a vendor rather than an in-house team. Companies often outsource physical security by hiring private security firms. (Refer to the Security Guards/Officers entry for more information on this topic.) The main reason for outsourcing is cost savings.

A major downside to outsourcing is loss of control over those noncore operations. Part of the lost control is security. Outsiders now have access to information that could be sensitive, such as customer data. In 2005, Citibank had a case in which call center workers in Pune, India, stole $350,000 from customers. The workers secured the customers' passwords and transferred the money to fake accounts. The customers noticed the problem and brought it to Citibank's attention. Similarly, outsourcing payroll information can expose Social Security numbers so that they could be used in identity theft or other fraud.

The first concern in outsourcing security is to protect sensitive information. Security expert Bar Biszick-Lockwood recommends that organizations classify information and reconsider any outsourcing activities that would place sensitive information at risk. Sensitive information includes mission critical data, private customer information, private employee information, proprietary information, data used to calculate the organization's financial performance, and any data that would damage an organization if it were exposed. If you decide to outsource activities that involve this information, make sure the outside vendor can guarantee security.

The security issue on outsourcing is complicated further by governmental regulations such as the Sarbanes-Oxley Act. Such laws hold organizations accountable for actions that could be compromised by insecure information. Therefore, organizations must carefully vet their vendors for security. Generally, going offshore increases the risk in outsourcing deals because international law is weak at protecting intellectual property rights. For sensitive information, the recommendation is to keep the outsourcing within your national borders.

Make it very clear to vendors that security is a top priority and research their security efforts. It is appropriate to request an external security audit of the vendor. In addition, include a contract clause prohibiting the vendor from subcontracting or re-outsourcing the project, check contact information for other clients (references), review details on any recent security breaches and countermeasures taken to prevent their repetition, and ask the vendor to agree to U.S. jurisdiction.[1]

CONCLUSION

Outsourcing can result in cost savings for an organization. However, security should be a higher priority than cost savings. Organizations must evaluate each outsourcing project to determine the sensitivity level of the information associated with it. An organization should reconsider outsourcing any function that would involve highly sensitive information. Moreover, avoid outsourcing offshore if you do decide to outsource a project or function that includes sensitive information. Be sure to vet the vendor thoroughly for security concerns. Monitor the security risk throughout the entire outsourcing process. Ultimately, your organization will suffer the consequences if there is a security breach through your outsourcing. Customers blamed Citibank, not the call center, when money in their accounts was compromised.

See also Customs-Trade Partnership Against Terrorism; Information Security; Security Guards/Officers; and Supply Chain Security.

NOTE

1. Bar Biszick-Lockwood, "4 Steps to Secure IT Outsourcing," online at http://www.sourcingmag.com/content/c050824a.asp (accessed 11 March 2007); Stephen Reed, "Managing Risk in Outsourcing," online at http://www.sourcingmag.com/content/c051017a.asp (accessed 10 March 2007).

SUPPLY CHAIN SECURITY

W. Timothy Coombs

Supply chain security ties together ideas from many other entries in this work. In fact, this entry's "see also" list is one of the longest. A supply chain is the series of steps from the extraction of raw materials to getting finished products in the hands of consumers. These connections require a coordinated system of organizations, people, activities, information, and resources to move a product or service from supplier to customer. Every organization that comes into contact with a product is part of its supply chain. In a manufacturing organization, the typical supply chain includes organizations that manufacture the parts, assemble the parts, deliver the parts and products, and sell the products.

Today's business climate demands global supply chains to remain competitive. Organizations need to be able to access parts of their supply chain from anywhere in the world to help reduce costs. This is related to the market forces behind outsourcing. (See the Outsourcing and Security entry for security issues related to this practice.) Supply chains are multicountry and multivendor, and this trend is unlikely to change.[1] With long multivendor supply chains, it is critical that organizations collaborate for supply chain security. Supply chain security is complex, covering the areas of physical security, access control, personnel security, education and training awareness, procedural security, documentation processing security, information security, incident management and investigation, trading partner security, and conveyance security. The exact security concerns within these various areas depend upon the nature of your supply chain. This entry provides a generic overview of the security concerns involved in supply chains.

PHYSICAL SECURITY

Materials pass through various physical locations as they move through the supply chain. The physical security of each location needs be considered. There must be measures in place to monitor and control the exterior and interior perimeters of the various locations. Some key physical security concerns to consider are perimeter fencing, lighting of the perimeter, locking devices for internal and external doors, clear identification of restricted areas, restricted access to cargo areas, and oversight of all trash removal. Refer to the Physical Security entry for a more detailed discussion of the topic.

ACCESS CONTROL

Access control is often considered part of physical security as it covers access to facilities, conveyances, vessels, aircraft, shipping, loading docks, and cargo areas. The goal is to prevent unauthorized personnel from gaining access to materials as they move through the supply chain. Some key access concerns include being able to limit the access of people and vehicles, inspecting all vehicles entering and exiting access areas, enacting procedures for challenging unauthorized people, having people sign in and out of high-risk areas, restricting access to cargo storage areas, recording loading areas with CCTV, and installing alarms that can be sounded if unauthorized people enter a restricted area.

PERSONNEL SECURITY

Organizations must make sure the authorized people with access to materials in the supply are trustworthy. Personnel security involves the screening of employees and potential employees as well as being able to monitor employee access to cargo areas. Background and drug tests prior to and periodically after hiring are recommended. Color-coded identity cards and uniforms can be used to designate where an employee should have access. Biometrics can be implemented to limit access to sensitive areas. Refer to the entries Employee Background Screening and Drug Testing and Biometrics for further information on this topic.

EDUCATION AND TRAINING AWARENESS

Many entries have repeated the mantra that security is every employee's responsibility. To be effective, employees need to know and to understand security policies that are relevant to them. Education and training programs help employees with security policies, awareness of deviations from the policies, and knowing what actions to take when a violation does occur. The security policies and standards must be communicated to all employees. Incentives can be offered to employees reporting suspicious activities.

PROCEDURAL SECURITY

Procedural security involves efforts to record and verify the introduction and removal of materials from the supply chain. Supply chain partners must know when something enters or exits the supply chain. Written verification security procedures should be in place and shared with supply chain partners. Methods need to be established for identifying and verifying authorized carriers and authorized cargo. Guidelines are required for affixing, replacing, recording, tracking, and verifying devices used to ensure the cargo is authentic and has not been adulterated in some fashion. Such devices include seals, serialized tape, or radio frequency identity tags.

Procedural security also covers the proper storage of cargo, correct storage of empty containers to prevent unauthorized access, and checking whether empty containers received for storage have been altered. Employees should guard against unauthorized materials from entering the supply chain as well. All personnel and packages that can come into contact with the cargo should be searched, and the security devices such as seals stored securely.

DOCUMENTATION PROCESSING SECURITY

Documentation processing security tries to ensure that all information is legible and safe from exchange, loss, or falsification. All cargo should be recorded, including packing condition, unit type, and security devices. The people logging in cargo must provide printed names and signatures. Investigate any deviation from the reporting process. Record the entrance and exit time of people receiving and delivering goods. Put special control processes in place for emergency or last-minute shipments. The documentation processing security must be responsive to the special needs of emergency shipments.

INFORMATION SECURITY

Information security seeks to protect the integrity of information about the supply chain against loss, exchange, and introduction of false information. Supply chain information should be available only to those who need to know and procedures in place to keep the information away from unauthorized personnel. This includes physical and information security for computer systems. Refer to the Information Security entry for more information on this subject.

INCIDENT MANAGEMENT AND INVESTIGATION

Incident management and investigation cover the tracking and information coordination capability of an organization with the purpose of timely reporting missing or lost cargo. The point of supply chain security is to make sure cargo arrives when and where it is intended. Supply chain partners need to know when cargo has gone missing and how the loss will be investigated.

TRADING PARTNER SECURITY

An organization must be sure the concern for supply chain security extends to and is shared by its suppliers and customers. Trading partners should agree to security procedures, even to the point of writing them into contracts. Key activities include requesting supply chain partners to assess supply chain–related security, using seals or other security devices, and documenting supply chain security policies. Coordination is central to trading partner security. Trading partners must share security-related information, share educational and training materials, and work together to identify, prioritize, and address supply chain security concerns.

CONVEYANCE SECURITY

Conveyance security seeks to protect against the introduction of unauthorized personnel or materials into the supply chain. Security devices and procedures and access control all help to protect conveyance security. High-risk cargo should be given special consideration. For instance, using two drivers, escort services, driver security training, and varied routes are examples of special considerations. Clearly conveyance security overlaps all of the other elements of supply chain security in this entry.[2]

CONCLUSION

Organizations will find a continuing and growing need to address supply chain security because the trend is for supply chains to have more links and links that have greater geographic diversity. Supply chains challenge supply partners to create a secure environment for the storage and transportation of cargo. A number of security concerns and organizations must converge to craft an effective supply chain security program.

See also Biometrics; Supply Chain Continuity; Customs-Trade Partnership Against Terrorism; Employee Background Screening and Drug Testing; Information Security; Outsourcing and Security; Physical Security; Radio Frequency Identity; and Terrorism.

NOTES

1. Kenneth Karel Boyer, Markham T. Frohlich, and G. Tomas M. Hult, *Extending the Supply Chain: How Cutting-Edge Companies Bridge the Critical Last Mile into Customers' Homes* (New York: AMACOM, 2005), pp. 1–23.

2. "C-TPAT Security Guidelines," 2006, online at http://www.cbp.gov/xp/cgov/import/commercial_enforcement/ctpat/security_guideline/ (accessed 5 March 2007).

SUPPLY CHAIN CONTINUITY

Geary Sikich

Cargo security is a problem for industry. Annual losses due to cargo theft run between $10 billion and $15 billion. According to the National Cargo Security Council (NCSC), motor carriers are the victims in 85 percent of all cargo theft, the majority taking place at terminals, transfer facilities, and cargo consolidation areas. However, these statistics may not reflect the actual situation. The FBI reports that only 40 percent of businesses or individuals actually report theft. If indirect costs are factored in, total losses are estimated to be between $20 billion and $60 billion a year.[1] Cargo safety is but one of the problems that can befall a global supply. Business continuity planning must extend beyond a business to the members of its supply chain.

No one company can deliver products and/or services end to end in today's complex business environment. Your company, like other companies, is most likely dependent on vendors of various types (manufacturing, professional services, software, transportation, etc.) to meet customer expectations.

All the people and organizations involved in getting product to market make up a supply chain. Because a supply chain is typically a mix of competencies, it poses a risk to an organization. That's why developing a process for ensuring business continuity throughout the supply chain is important. An interruption in your supply chain could mean a major disruption to your company.

Effective business continuity strategies, such as supply chain assurance, need to be designed. This entry discusses the process for building business continuity capabilities into the supply chain. The first section considers how to examine the supply chain to ensure business continuity in the face of trouble. Figure 1, Supply Chain Business Conituity, provides a visual summary of business

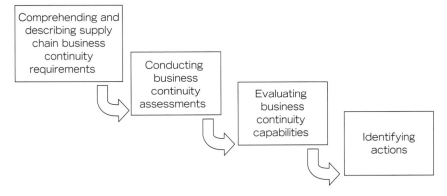

Figure 1 Supply chain business continuity elements.

continuity and the supply chain. The second section discusses how to integrate business continuity into the supply chain using the procurement process as a guide. The third and final section reviews the need for early identification of problems through incident management.

ASSESSING A SUPPLY CHAIN FOR BUSINESS CONTINUITY CONCERNS

Developing and implementing criteria for assessing the business continuity capabilities of vendors operating within your supply chain can be a daunting task. The scope of activities involves assessing current procurement processes, first by collecting and analyzing data and then by making the necessary changes to the procurement planning. This may include determining a vendor's ability to maintain supply chain continuity in the face of disaster, economic trouble, or other challenges; developing assessment processes; defining contract terms and conditions; developing sustainability procedures; and making business continuity an integral component of your procurement process.

It is helpful at this point to review the risks that can impact an organization's supply chain. These risks can be articulated as either internal or external, as depicted in Figure 2, Internal and External Vulnerability Drivers.

Figure 2   Internal and external vulnerability drivers.

These drivers and the ability to manage them—by putting contingency measures into place—often are interconnected. Understanding this potential interconnectedness helps in assessing vendor business continuity capabilities. Internal and external vulnerability drivers can materialize in a variety of ways. Risk can be context sensitive, as risk elements interact differently depending on the situation. Understanding the potential interaction of risk factors facilitates measuring business continuity capabilities and planning for actions that can be implemented should a disruptive event occur.[2]

Vendor Continuity Capability Questionnaire

The first action to take when assessing the supply chain for business continuity is evaluating current practices of vendors in the supply chain. Developing a vendor continuity capability questionnaire is a good first step but needs to be carefully thought through. You are, in essence, creating a legal document that could contain sensitive information that must be protected. You are also creating a potential liability document for yourself. With the type of information collected to assess vendor continuity capabilities, your organization could be held liable under the concepts of negligence (foreseeability), constructive notice, and/or constructive knowledge, for *not* taking action to mitigate potential losses.

The structure of a typical eight-part vendor continuity capability questionnaire is summarized below:

Part 1: Governance Provisions and Management Commitment
    The purpose of this part of the questionnaire is to establish that the vendor has a formal governance program in place that has management commitment. You also want to ascertain whether the vendor's program is integrated into the way it does business or is an adjunct to the business that it is in.

Part 2: Business Continuity Strategies: Developing and Implementing BCP
    The purpose of this part of the questionnaire is to gain an understanding of the vendor's strategy for continuity in the face of trouble and how it will implement the business continuity plan (BCP).

Part 3: Business Impact Analysis, Risk Evaluation, and Control Mechanisms
    Part 3 of the questionnaire seeks to gain an understanding of the extent to which the vendor assesses possible negative impacts and how often; identifies and evaluates risks; and institutes control mechanisms to address risk and mitigate threats, hazards, and overall vulnerability.

Part 4: Maintaining Continuity: Training, Awareness, Exercises, and Business Continuity Plan Updates
    This part of the questionnaire seeks to gain an understanding of the extent and adequacy of the vendor's continuity program training and whether it updates its plans regularly to take advantage of new situations and information.

Part 5: Incident Response Operations
> Part 5 looks at the tactical level of business continuity with the goal of understanding how the vendor identifies, responds to, and communicates information on disruptive events.

Part 6: Crisis Communications
> Part 6 focuses on the vendor's internal and external communications relating to policies, information flow, and the management of crisis communications.

Part 7: Coordination (External Entities)
> Part 7 of the questionnaire is designed to assess vendor coordination with external parties, including the government, customers, and its own vendors. It also assesses key components of the coordination process.

Part 8: Vendor Certification
> The final section asks the vendor to certify the accuracy of the answers on the questionnaire.

The exact questions and the length of the vendor questionnaire will vary depending on the industry and the depth of analysis that you want to perform. The questionnaires that I have developed for clients contained approximately fifty questions that require the vendor to provide quantifiable answers. Should you determine the need for further analysis, assemble a formal audit team to resolve the concern over vendor continuity capability.

Vendor Continuity Capability Assessment

During the course of assessment, data will be collected, analyzed, and developed into findings and recommendations. The data should be organized by essential element of analysis (EEA) criteria that the organization establishes and uses to conduct data collection, analysis, and evaluation. Examples of typical EEA are provided in the Essential Elements of Analysis box.

Essential Elements of Analysis

Organization

This refers to the current procurement process, vendor roles/responsibilities and deliverables during the procurement process life cycle, and current criteria for the organization's business continuity programs and plans. Evaluation issues include, but are not limited to, determining the following:

- Whether continuity is a strategic consideration for the vendor that results in specific ways of doing business or is an adjunct to the vendor's business

- Where in the vendor organization ownership of the business continuity program and plans resides
- To what extent the vendor's senior management team gets involved in the vendor continuity program

Vulnerability Identification and Control

This refers to establishing minimum acceptable criteria for vendor vulnerability identification and control methodologies. The purpose is to maintain continuity programs and allow the vendor to integrate its methodologies on a sustainable basis with the client's business continuity management strategy. Topics assessed include the following:

- Vendor methods for identifying and assessing threats, hazards, vulnerabilities, and risks
- Vendor methods for determining risk acceptance levels
- Vendor validation methods for assessing response, management, and recovery capabilities
- Vendor methods for integrating methodologies with current business impact assessment efforts

Continuity Strategy and Approach

This refers to the measures developed and used to verify vendor integration of the business continuity management program and plans with the client's business continuity management strategy. Assessment includes the following:

- Vendor methods for determining critical business functions and requirements
- Vendor compliance with minimum acceptable criteria for the business continuity management program and plans
- Vendor monitoring and enforcement of compliance across the business continuity management program and plans

Documentation

This refers to the documentation of vendor business continuity management program and plan capabilities. Key issues include the following:

- Policy, plan, and supporting documentation
- Measures

(continued)

- "Value chain" integration
- Communications (internal, external, including media and stakeholder handling)
- Vendor verification of integrated response, management, and recovery capabilities and documentation

Resource Management and Development

This refers to the metrics for vendor validation of staffing (business continuity staffing) and associated vendor integration of continuity planning, resource development, and awareness of continuity. Areas audited include the following:

- Vendor personnel, funding, expert knowledge, and future requirements to support the program and when they will be needed
- The functional roles and responsibilities of vendor organizations, departments, and individuals in support of business continuity management
- Vendor program, protocols, and effectiveness
- Vendor knowledge transfer and integration
- Vendor training and simulation activities
- Vendor facilities and equipment program status

Continuity Maintenance

This refers to the procedures used to ensure resilience of the vendor continuity process. Key areas audited include the following:

- Vendor maintenance protocols and effectiveness
- Vendor change control protocols

PROCUREMENT PLANNING CONSIDERATIONS

The overall objective of integrating business continuity criteria into the supply chain is to make the procurement process more efficient and safer from disruption. Once you determine what is needed, the focus shifts to addressing weaknesses through procurement planning. With any large-scale project, such as the integration of vendor business continuity criteria into the procurement process, attempting to implement on a grand scale can lead to chaotic results. A phased approach to implementation and integration is a better idea and should generally consist of five phases:

- *Phase 1:* Assessment and vendor continuity questionnaire—deliverable: letter report with executive summary that will include discussion and recommendations based on the results of the review of essential elements of analysis (report)
- *Phase 2:* Procurement integration (vertical/horizontal)—deliverables: procurement management system, vendor business continuity management program, and plan integration criteria guide (tools), and training program materials for each (knowledge transfer)
- *Phase 3:* Monitoring and enforcement—deliverable: procurement management system, vendor business continuity management program, and continuity plan integration criteria guide maintenance criteria (sustainability)
- *Phase 4:* Sustainability—deliverable: periodic metrics, event response reports
- *Phase 5:* Maturity model evaluation—deliverable: metrics for maintaining the process, change management procedures

PROCUREMENT INCIDENT MANAGEMENT CONSIDERATIONS

A vendor can complete the vetting process and still experience a disruption affecting your company's ability to meet customer demand. Problems will occur, and the sooner these are detected, the sooner they can be resolved and the better an organization can protect its assets. Having an incident management system as a component of the procurement process can allow your company to respond, recover, and restore supply chain operations with less potential for massive disruption. Incident management can range from assessing and classifying a vendor incident to implementing response actions, such as sending your personnel to vendor facilities to assist in incident mitigation processes.

Contingency alternatives can range from having backup response plans to alternative sources of supply. Once the risks are identified and evaluated, actions to address them throughout the procurement process can be taken. Identifying risk themes across a number of risk dimensions can help to determine where your company should place effort to mitigate the risk exposure.

Disruptive events as they occur need to be classified by their level of severity in order to determine their potential impact. Figure 3, Disruptive Events, offers an example of a classification system for disruptions. A classification system can provide a consistent framework for evaluation; enhance the communication process between internal and external groups; and facilitate response, management, recovery, and restoration efforts.

Consider creating an event assessment form to use in conjunction with the event classification system for determining the event classification level and for

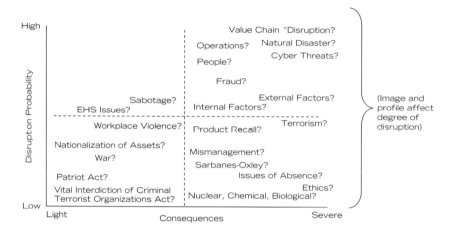

Figure 3 Disruptive events.

facilitating discussion within your company and with the affected vendor(s). The less prepared an organization is for service disruption, the longer it takes to recover operations and restore service levels. Having a classification system can help in identifying potentially disruptive situations early and determining how to respond effectively to minimize the level of service impacts.

The procurement process represents the first line of direct contact with vendors, suppliers, and so on. The ability of procurement personnel at any stage of the procurement cycle to detect and classify potential disruption and severity allows your company to implement its business continuity plan and coordinate with affected vendors to assure continuity of operations. Early detection, classification, and response can also help keep service at a high level, reduce potential chaos associated with a disruptive event, and lead to shorter recovery and restoration time frames.

CONCLUSION

Assuring supplier continuity capabilities is of paramount concern today. Realizing that most business processes extend beyond the boundaries of a single entity, businesspeople are more aware of critical supply chain interdependencies. Simply having profiles of potential high-risk suppliers, while extremely important, is by itself not enough. You need to develop capabilities to assess and monitor vendors to identify potential problems before they occur.

Business leaders have the responsibility to protect their organizations by facilitating continuity planning and preparedness efforts. Senior management and board members can and must deliver the message that survivability depends on being able to find the opportunity within the crisis. Today, we cannot merely think about the plannable or plan for the unthinkable, but we must learn to think about the unplannable. Market research indicates that only a small portion (5 percent) of businesses today has a viable business continuity plan, but virtually 100 percent now realize they are at risk. Seizing the initiative and getting involved in all the phases of crisis management can mitigate or prevent major losses.[3]

See also Business Continuity; Customs-Trade Partnership Against Terrorism; and Supply Chain Security.

NOTES

1. Sean Kilcarr, "Cargo Theft a Growing Threat," 10 Oct. 2002, online at http://driversmag.com/ar/fleet_cargo_theft_growing/index.html (accessed 23 April 2007).

2. Lord Levine, "Changing Risk Environment for Global Business," 8 April 2003, online at http://www.lloyds.com/NR/rdonlyres/E3A3A19D-F459-4D76-8677-CD3E89EE43F5/0/LordLeveneSpeechcostoflitigation2003.pdf (accessed 10 April 2007).

3. Geary W. Sikich, *Integrated Business Continuity: Maintaining Resilience in Times of Uncertainty* (Tulsa, OK: PennWell Publishing, 2003).

TRAVEL OVERSEAS

W. Timothy Coombs

When businesspeople travel, especially overseas, they must address some of their own security concerns. Companies should provide training to make employees aware of key security issues when they travel. Security measures must be taken to ensure personal safety as well as security of sensitive or proprietary company information in their possession. Overseas, personnel are at risk of criminal and terrorist acts. (The Terrorism entry outlines the basic terror groups and their geographic range.) This entry is divided by the elements of the travel process and emphasizes the need to be aware of one's environment.

TRAVEL PREPARATION

Research or be briefed on the country you will be traveling to. Check with the U.S. State Department for any travel warnings for that country or region of the world. Do not pack any sensitive or proprietary information; keep that with you in your carry-on luggage. For extra security, double envelope the material. Select luggage tags that have a cover and do not use laminated business cards for luggage tags. The objective is not to advertise your country of origin or company's name. Avoid exchanging currency in public areas of the airport that are targets for criminals. It is advisable to exchange some currency before you leave or to exchange currency before leaving the secure area of the airport. Make sure you will be flying through secure airports, even on layovers. Again, the U.S. State Department has information on the level of security at various international airports. Wide-body planes are less attractive targets for terrorists, although aisle seats place you closer to the action and at greater risk if there is a hijacking. Dress in casual clothes to draw less attention to yourself.

AT THE AIRPORT

Try to avoid long lines at the airport by arriving early for check-in. Pack for security ahead of time; do not repack your bags at the airport. Check with the Federal Aviation Authority (FAA) for the latest guidelines on carry-on luggage. This will help you to avoid problems at security checkpoints. Keep in mind that regulations are not the same in every country. As always, keep your luggage in sight at all times. When possible, use a hotel vehicle for transportation to and from the airport. If that is not possible, use "official" transportation. Always be vigilant for suspicious activity and people. Avoid open public areas, especially those with a lot of glass, which are inviting targets for bomb attacks.

AT THE HOTEL

Before you book your stay, check on the security level of your hotel. One way to do this is to contact the regional security officer at the local U.S. embassy for a list of hotels that government officials use when visiting. The best option is to stay at a hotel that is part of a U.S.-based chain. It will have security standards consistent with those in the United States. Parking garages are the most dangerous places because they are difficult to secure. In the lobby, notice if anyone seems overly interested in your arrival. Have a hotel staff member take you to your room and make sure the room is empty. Inspect the security features, such as locks, before the hotel employee leaves. Review the safety features such as where the nearest fire exit is located as well as the nearest house phone. Check and verify when hotel staff requests entry to your room. Criminals can use pretexting, pretending to be an employee, to gain entry. Use the optical viewer to determine whether the person has the proper uniform, and call the main desk to confirm someone was sent to your room.

Never discuss sensitive or proprietary information while in your hotel room because it may not be secure. The same goes for using the telephone in your room. Keep your key and passport with you at all times. Do not jog or walk in cities or areas you do not know well. Vary the times you leave and return to the hotel. Keep valuables in a secure location such as a safety deposit box at the main desk.

SECURITY ON A GLOBAL SCALE

Exercise caution in and around public restrooms at the hotel, a prime location for criminal activity. Keep your hotel door closed and the deadbolt locked when in your room. Be sure the door closes fully and locks when you leave your room.

DRIVING A CAR

Many business travelers like to rent cars for local transportation. When you rent a car, make sure to choose a vehicle that is common in that location. And, if possible, ask that the rental company remove any markings signaling the vehicle is a rental. You should be trying to blend in with the other cars so that you do not become an obvious target. Air-conditioning is a good idea to prevent rolled-down windows, which are a safety risk.

Here are some other car safety points people should follow in any city: avoid driving at night, keep your doors locked at all times, wear a seat belt, do not leave valuables in the car, do not park on the street overnight, and do not pick up hitchhikers. Be smart about how and where you drive. This will serve to maximize your safety when you feel you must rent a car for transportation.

TRAIN TRAVEL

Train stations have open access and are targets for criminals and terrorists. Be very vigilant when using these facilities. Trains are inviting targets for bombings and sabotage as well. They are considered "soft" targets because they are easy to access. Trains travel over miles of tracks that are simple to target.

TARGETING RECOGNITION

We hate to think the worst of people, but that is a useful trait when it comes to personnel security. Be suspicious of people who act that way. Here are some examples of suspicious behavior:

- Repeated contact with the same person who is not a business contact
- A business contact who tries to push the relationship beyond business
- Accidental meetings whereby a person then tries to strike up a conversation by practicing English, asking about your employer, or wanting to buy you a drink

These may be innocent encounters or else someone is targeting you for theft or kidnapping. Be alert and suspicious, especially when traveling in a high-risk country or area. If you think you are under surveillance, watch what you say. Do not try to give someone the slip; do not search your room for listening devices.

These actions send the wrong signal to surveillants. Report any suspicions to the U.S. embassy and follow its recommendation.[1]

CONCLUSION

Most of personal security while traveling is common sense or points we have heard announced to us numerous times over the airport public address system. The key is planning, preparation, and common sense. For more detailed information on personal security overseas, read the U.S. State Department's "Personal Security Guidelines for the American Business Traveler Overseas," included in the Document Appendix.

See also Countersurveillance; and Terrorism.

NOTE

1. "A Safe Trip Abroad," 22 Jan. 2007, online at http://travel.state.gov/travel/tips/safety/safety_1747.html?css=print (accessed 3 Feb. 2007).

CORPORATE OR INDUSTRIAL ESPIONAGE
W. Timothy Coombs

The U.S. government does track attempts by foreign agents to acquire sensitive information. The focus is on military-related technology. However, government reports also document efforts to acquire economic-related technology that provides companies with a competitive edge. Annually, agents from nearly one hundred countries attempt to procure what is classified as sensitive U.S. technology. Companies should realize they may be targets of foreign agents seeking sensitive technology. Frequent targets are companies involved in semiconductor production processes, computer microprocessors, software, and chemical processes. The government has documented that U.S. companies that lose sensitive information experience a decline in investor confidence and stock price. This is one form of corporate or industrial espionage, defined as illegal or unethical efforts to collect information for commercial gain rather than national interests. Examples of corporate or industrial espionage include bribery, theft, blackmail, technological surveillance, and violence.

U.S companies are attractive targets for foreign agents because of the openness of the United States to visitors. Many companies and universities employ foreign workers, who often have the skills necessary to acquire technology illegally. Companies may host foreign visitors as well for short-term or long-term visits.

Visitors may try to secure information by engaging in conversation and asking about the technology, by sneaking into areas where they can view or access sensitive information, or by trying to circumvent security. Long-term visitors are a greater risk than short-term visitors. A long-term visitor has more of an opportunity to learn the security procedures and devise ways to circumvent them. Information security is at the greatest risk. A long-term visitor has time to acquire passwords and can learn where sensitive information is stored. Short-term visitors are easier for people to see and to observe, whereas long-term visitors begin to blend in with other workers and are given less scrutiny. Technology, such as small portable storage devices with large capacities and picture phones, is another factor making it easier to engage in industrial espionage.[1]

Electronic storage devices that employees take abroad, such as laptop computers, personal digital assistants (PDA), and cell phones, are another source of risk. A recent survey found that over 66 percent of U.S. businesspeople carry

---

Corporate Espionage Case Study: KPMG

KPMG is a consulting company that provides a variety of business services, including financial services. Guy Enright worked as an accountant for KMPG in its Bermuda office. In May 2005, he was contacted by a man claiming to be a British secret agent named Nick Hamilton. Agent Hamilton claimed KPMG had information vital to national security interests in Britain. Hamilton wanted financial audit information about IPOC, a client of KPMG's. Enright agreed to provide the information, and the plot of a spy novel ensued. Enright left documents for Hamilton in a number of clandestine locations, including the storage compartment of a moped.

The problem was that Nick Hamilton was not an agent for the British government. Instead, his real name was Nick Day and he was the cofounder of Diligence Inc., a private intelligence company located in Washington, DC. Its advisory board includes former major politicians from both the United States and the United Kingdom. Diligence Inc. was working for a company named the Alfa Group Consortium, hired through the powerful lobbying firm of Barbour Griffin & Rogers. The Alfa Group Consortium was in competition with IPOC for a lucrative stake in the Russian telecom company MegaFon.

Business Week printed a lengthy story about the espionage incident. Diligence Inc. labeled the operation project Yucca.

(continued)

> Memos from Diligence Inc. claimed there was virtually no chance of being detected and that their actions would maintain plausible deniability. Diligence Inc. first collected a list of names of the people in the KPMG Bermuda office. Enright was carefully watched to be sure he was not a plant or a corporate spy himself. Project Yucca ended when an unknown source left a collection of records and e-mails about the project at the New Jersey offices of KPMG. KPMG filed a lawsuit against Diligence Inc. in U.S. District Court in November 2005. Diligence Inc. settled in June 2006 for $1.7 million. IPOC has sued both Diligence Inc. and Barbour Griffin & Rogers.[2]
>
> The actions of Diligence Inc. crossed the boundary from competitive intelligence to corporate espionage. Nick Day's posing as a British secret agent was definitely unethical and possibly illegal. Blatantly lying about one's purpose and organization falls well outside the realm of competitive intelligence, which prides itself on using ethical and legal means to obtain information.

important company information on their PDAs. Electronic storage devices can be compromised or stolen from hotel rooms or at security checkpoints.[3] Some PDAs can even be compromised/accessed remotely. See the entries Portable Device Security and Travel Overseas for ideas on securing data on electronic devices and security while traveling. Travel for conventions, expositions, and trade shows creates special problems. If your company is an exhibitor, it is a target. Hotels can be searched for sensitive information including technical reference manuals. Be sure all such information is secure at all times in all locations. The KPMG case is reminder that corporate espionage is real.

Foreign agents do not have to rely on cloak-and-dagger techniques to gather sensitive information. The most common strategy is simply to ask for the information through e-mail, by fax, or in person. The foreign agent often builds a relationship through conversations with an individual at a company before making the request. One strategy is to pose as a graduate student conducting research. The "fake" graduate student converses with a scientist about his or her work. When the time is right, the request for sensitive information is made and many times the scientist provides it; the graduate student seems to be a friend who is simply interested in research. Employees should be reminded not to share sensitive information with anyone, period.

Another common access point is via the Internet, one more reason companies must be concerned about cyber security. Refer to the entry Information Security for recommendations on protecting a company from cyber attacks. No system can be made perfectly safe from cyber attacks, but security should be better than it is at most companies. Consider a recent survey that found nine out of ten companies had serious Internet vulnerabilities when the companies believed they were extremely secure.

Finally, the foreign agents are not just government operatives. Government statistics indicate the largest classification of foreign agents trying to gather sensitive information is private companies at 36 percent. Those working for a government represented 21 percent, with another 15 percent being affiliated with a foreign government. Keep in mind these statistics do not include domestic activities. Competitors in the United States can be trying to extract sensitive information from a company as well. The techniques for collecting the information remain the same; only the names of the actors seeking the information changes.[4]

CONCLUSION

If your organization handles sensitive technology, assume that you are a target for foreign agents trying to acquire that technology. Carefully monitor all visitors to your facilities whether they are short-term or long-term visits. Review basic cyber security and electronic storage device security with your employees. Emphasize the danger presented when traveling with laptops and PDAs. Finally, warn your employees about the common strategies foreign agents use to trick employees into divulging information. Having strong security policies and creating awareness among employees are critical to preventing corporate/industrial espionage.

See also Competitive Intelligence; Countersurveillance; Information Security; Portable Device Security; Social Engineering: Exploiting the Weakest Link; and Travel Overseas.

NOTES

1. Office of National Counterintelligence Executive (ONCIX), *Annual Report to Congress on Foreign Economic Collection and Industrial Espionage—2004*, April 2005, online at http://www.fas.org/irp/ops/ci/docs/2004.pdf (accessed 10 March 2007), pp. ix–xi.

2. Eamon Javers, "Spies, Lies & KPMG," *BusinessWeek*, 26 Feb. 2007, pp. 86–88.

3. "Confiscated Laptops," *Security Management*, Feb. 2007, p. 44.

4. ONICIX, *Annual Report to Congress on Foreign Economic Collection and Industrial Espionage—2004*, pp. 1–7.

ENHANCING THE HUMAN SIDE OF SECURITY AND SAFETY

A common theme through most discussions of security and safety is the central role of people. If people fail to follow policies or execute desired behaviors, security, safety, or both can be compromised. This section provides general advice on how to improve the human side of security and safety.

IMPROVING TEAM EFFECTIVENESS

W. Timothy Coombs

Teams are often a part of business continuity, security, crisis management, and disaster recovery efforts. A number of the entries in these two volumes mention the use of teams to create various plans and policies for organizations. Moreover, teams are expected to execute various response plans when trouble arrives. Although we put a lot of faith in teams to succeed, we often fail to provide an environment in which this can happen. Our own mistaken assumptions about teams create the negative environment for teams. By exposing these common mistaken beliefs, we gain insights into the training needed to make teams effective and help them to succeed.

PROBLEM ASSUMPTIONS ABOUT TEAMS

Assumption 1

If we put people together on a team, they will work as a team. People are not born knowing how to function as a team. In fact, in the United States, our culture works against teams because we value and praise individual effort. This is reinforced in the workplace with rewards systems that focus on individual rather than team performance. We are not "wired" to be effective team members, so training is needed to form teams that work.

Assumption 2

Teams are always better than people working alone. It is true that a "synergy" occurs when people form teams. The interaction of people creates something unique from the sum of individual efforts. However, unique does not necessarily mean better. Teams can still make mistakes and ineffective decisions. Effective decision making requires training; it does not happen by chance.

Assumption 3

Select the best employees and excellence will follow. The belief is that if we place the best and brightest employees on a team, the team will perform at a high level. However, these are the people most likely to value and have benefited from individual effort. The best individual performers are not necessarily the best team performers. The skills to do your job may not be the same skills needed to function as a team.

These mistaken assumptions combine to blind us to the need to train people to be team members. We cannot assume people know how to function in teams or that they learned it somewhere else. Even our top people must be trained on key team skills if we are to have effective teams. This leads to the second point, the training necessary to develop effective teams.

DEVELOPING EFFECTIVE TEAMS

A team is a group of individuals that works together to perform tasks. People on a team need to take some actions on their own and some in coordination with others. For teams to work, people need both individual-level and team-level competencies. People must be responsible for their own actions as well as the welfare of the team.

Individual-level competencies center on contributing to the team. A person is placed on a team to contribute. He or she has unique knowledge and skills that can help the team. The benefits a person can provide a team are lost if that person does not contribute to the team. Contributing means a person is willing to share his or her ideas and opinions by speaking up in a team setting. Training must encourage people to participate in the life of the team by reinforcing the

HUMAN SIDE OF SECURITY AND SAFETY 345

value of individual contributions. It should be noted that in rare cases some people have extreme phobias about communicating in a team. Those suffering from such anxieties should not be placed on teams. The average person may be a bit anxious about expressing him or herself, but training can help to reduce that apprehension.

Speaking up on a team is linked to listening. We all think we are good listeners. However, a number of factors work against effective listening. Our minds work quicker than people can speak. As result, we can distract ourselves when listening to others by thinking of something else. Our mind's extra processing time can result in us drifting off. We must learn to stay focused. Also, we tend to dismiss or ignore ideas that are different from our own. We must learn to listen to all perspectives. Part of effective listening is trying to understand the point of view of each person speaking.

Team-level skills center on the decision-making process. The central deliverable of a team is its decisions. In fact, we typically judge team effectiveness by its decisions. Have you ever been trained in team decision making? Again, it is dangerous to assume teams know how to make decisions. Teams are unsuccessful at decision making for three reasons: failure to analyze the problem, failure to develop decision criteria, and failure to evaluate decision options carefully.

A team starts a decision by determining the problem it faces. Ineffective teams assume they know what the problem is and jump to solutions. Team members may think the situation looks like a problem they've handled before or want it to be so. The result is the team solves the wrong problem! Start by reviewing the situation to determine the real problem. Ineffective teams may have no criteria for evaluating decision options. As a result, they pick the one that sounds best or is pushed by the most powerful person on the team. Finally, ineffective teams do not evaluate their decision options carefully. If you have not developed decision criteria, you cannot effectively evaluate decision options. However, even teams that develop criteria may not apply them. The team does not carefully assess each decision option against the decision criteria to see which one is the best fit. Again, the selection is based on some preference unrelated to the effectiveness of the decision options.

Any one or combination of the three decision errors can sink a team. Teams need to be trained on proper decision-making techniques. This does not have to be a major training effort; it can be done in a few hours. People learn the common problems and strategies for overcoming those problems and then apply those newfound skills by practicing team decision-making tasks. The end result will be teams that are less likely to make the three common decision errors.

Part of decision making and life on a team is conflict. Disagreements can arise over the definition of the problem, decision criteria, evaluation of decision options, or where to order lunch. Conflict can be positive for a team. Airing multiple viewpoints helps a team to see the problem better and to evaluate the decision options. However, conflict can be destructive if handled negatively. Conflict needs to center on ideas, not personalities. People should be trained in the basics of conflict management, including the different styles for approaching conflict and the differences between productive and injurious conflict. Proper conflict management is a skill that benefits people in their regular jobs as well as in team activities.

CONCLUSION

One factor that effective teams have is their belief that they can handle any task; this is known as team potency. Training feeds into team potency by setting the stage for team success. By training on individual-level and team-level competencies, the team gains the skills necessary to succeed and to build confidence in its abilities. We entrust teams with important tasks, such as developing and implementing business continuity plans. Avoid the dangerous assumptions that lead to the belief that people naturally work effectively together as a team. Spend the time to train your teams to maximize the likelihood of their success.

See also Business Continuity; Crisis Management; Crisis Management Team; Disaster Recovery Management; and Benefits of Emergency Management.

AVOIDING THE SILO EFFECT TO IMPROVE BUSINESS SECURITY

Betty A. Kildow

An all-too-common condition in both government and private organizations today is business units that seem to operate almost autonomously with little understanding of what other departments are doing and how they impact one another. If we consider the silos used in farming for more than 130 years, it is not difficult to see how the term *silo effect* came to be. Although silos can be located next to or near one another, silos are soundproof, and people inside silos are unable to communicate with those on the outside. It's dark in a silo, and its closed environment allows for little fresh air.

WHY SILOS EXIST

In business, silos almost exclusively value their own business functions, needs, and interests with little appreciation for the contributions of other business units. These silos may be formed by intent or default—without realizing that is what is happening (ignorance) or because "that's the way it is here" (accepting and at times expanding the status quo).

Silos exist for a variety of reasons: a perceived means of building or maintaining power, politics, turf protection, history, and specialization of tasks. Simply put, silos are self-contained departments or other business units that struggle when communicating or working with other departments or silos. Large, complex organizations are a natural building site for silos, and silos are also easily created whenever two or more organizations merge—by whatever process.

DANGERS OF SILOS

The existence of silos in organizations significantly debilitates effective communication and productivity. The duplication of effort due to interdepartmental tension is at times astounding. In some organizations, silos are deliberately built and encouraged by upper management to create what is seen as "healthy competition." However, the lack of effective interdepartmental communication has proven time and again to be the undoing of effectiveness, and the existence of strong silos will almost assuredly negatively impact the end results.

These disconnected units, silos, become multiple organizations within an organization, focusing inward. Silos often store information that can be accessed only by either those within the silo or those with control or credentials deemed to qualify the holder for need-to-know status. Communication tends to be vertical, making timely coordination and communication among business units difficult, if not impossible. True collaboration becomes increasingly difficult. There is no sharing of resources. In some organizations, getting work done cross-functionally can be difficult at best, impossible at worst.

This can lead to reduced productivity, possible duplication of effort, and even tension and mistrust within the organization.

The "right hand not knowing what the left hand is doing" can result in a lack of valuable information sharing between business units, in losing sight of the goals of the organization as a whole, and in extreme situations by creating an "us" versus "them" mentality or turf wars (covert or overt). In addition, silos do not stand up well against high winds, such as the quickly changing requirements of responding to threats to the organization's security or dealing with emergencies and disasters.

That being said, it is possible that silos have value in some environments, such as where security of highly sensitive information is important or in short-cycle projects with a narrow mission.

SILOS AND EFFORTS TO PROTECT ORGANIZATIONS

There continues to be a growing awareness of silo mentality and its harmful impact, and those of us involved in keeping our organizations safe, secure, and operational are not necessarily immune. In some large organizations, a multitude of business units, such as those involved in business continuity (BC) and disaster recovery (DR), emergency management, security, crisis management, safety, environmental health and safety, and perhaps others, has shared responsibility for addressing risks, possibly leading to both planning gaps and overlaps. We would do well to learn from the mistakes of others.

The disciplines charged with keeping the organization's people, operations, and facilities safe and functioning do span silos. For example, effective business continuity planning follows, examines, and tracks the entire organization's operations, not only internally but also externally (e.g., suppliers, contractors, regulatory agencies).

That is not to say that for multiple reasons we have not collectively already developed our own silos, or at least laid the foundation for them:

- As the "new kids on the block" getting varying levels of respect and acceptance, we circled the wagons and began silo construction.
- Our language is "foreign"—between both us and them and us and us (public vs. private sector, aforementioned functions such as safety, security, etc.).
- Many organizations are not quite sure where some of the newer disciplines—disaster recovery and business continuity, in particular—"fit" in the organization, and as a result new silos are established.
- Historically, security, business continuity, disaster recovery, emergency management, and so forth have been segregated in individual departments, creating isolated silos.
- Disaster recovery practitioners in particular bear the scars of Y2K, as those who look back and put Y2K into the technology hoax category say, "I told you so," rather than considering the possibility that the blood, sweat, and tears that went into Y2K planning is the reason that the arrival of the new millennium did not create significant problems.

Although BC/DR is an organization-wide issue, there is a tendency on the part of people not directly involved in the planning process to believe that business continuity and disaster recovery are separate from the rest of the organization, "something that Bob, Mary, or XYZ Department does." BC/DR professionals can be equally at fault, focusing on what "we need to accomplish," while forgetting that to be truly successful and effective, BC/DR requires enterprise-wide involvement. Even those of us with interrelated responsibilities may tend to develop a stand-alone approach to our responsibilities. Rather than accept this as "just the way it is," to be truly effective, we need to plan for and take proactive steps to make business continuity and disaster recovery an integral part of the overall organization, its culture, and day-to-day operations. The following section on Evaluating Whether You Have Silos can also be found in the Guidance Appendix.

EVALUATING WHETHER YOU HAVE SILOS

To provide a general indication of how entrenched silo mentality is in your organization... in your department... at your desk, choose the most accurate

HUMAN SIDE OF SECURITY AND SAFETY 349

response to each of the following ten statements. Be objective as you select the most accurate of three possible responses to each statement:

- True
- Partially true
- Not true

1. Employees have a good understanding of what all other business units do, how they operate, and how they all fit in the big picture.
2. Collaboration is a strong element of the organization's culture.
3. Your peers from other areas of the organization are open to collaborative efforts and joint projects.
4. You believe that silos are detrimental, and you model that belief on an ongoing basis.
5. You encourage your direct reports to collaborate with those outside your business unit.
6. Business units in your organization freely share resources and information with one another.
7. You receive consistent messages from different levels and/or functions within the organization.
8. Your organization has been successful in building commitment for security, business continuity, disaster recovery, and emergency management projects and programs.
9. A vehicle (e.g., work group) exists to facilitate ongoing communication among related functions within your organization (BC/DR/emergency management/security/safety).
10. ALL employees at all levels are aware of the BC, DR, security, and related programs within the organization.

SCORING:
Give yourself:
—10 points for each TRUE response
—5 points for each PARTIALLY TRUE response
—0 points for each NOT TRUE response
Maximum possible points = 100

AVOIDING SILOS

Avoid creating or further strengthening silos. BC, DR, security, and related disciplines demand an ability to solve complex problems that cuts across traditional organizational silos, and in many cases they also require effective integration of multidisciplinary and even multicultural workforces.

Be alert for areas in the organization where business continuity, disaster recovery, and security aren't adequately integrated. Eliminate damaging silo

Score	Assessment
90–100	Outstanding. Congratulations to you and your organization. Continue to avoid silo formation.
70–80	Very good. Look for additional improvement opportunities, and avoid further entrenching existing silos.
50–60	Mediocre. Note which questions were answered "no," and concentrate efforts there.
30–40	Improvement needed. Further assess the situation and develop a plan for positive change.
0–20	HELP!! A great deal of work to be done. Get started now.

thinking whereby you and members of your team focus only on your group's goals to the detriment of other business units and even the organization as a whole. Do everything possible to encourage other business units to be active participants in the BC/DR/security process on an ongoing basis.

Keep the big picture in front of you. Remember that business continuity, disaster recovery, emergency management, safety, and security all have a common mission—to protect the organization and its people, physical assets, and operational capability, and to meet the needs of all stakeholders.

Start by developing and maintaining executive management's commitment. Its support of and involvement in the planning process, training, and testing are invaluable in demonstrating the organization-wide importance of the BC/DR and related programs. Work toward having all functions with a related focus be centrally managed and report to the same executive. For example, consider establishing a department that includes emergency management, security, safety, business continuity, disaster recovery, and risk management and is headed by a person who reports directly to the CEO, CFO, COO, or other position at the upper executive level. The resulting improved communication and coordination will help to ensure that the organization is better prepared to face and recover from disasters.

All employees from the mail room to the boardroom are critical to the overall success of your programs, and each employee has a role to play, be it large or small. Continually educating everyone in the organization about the importance of the BC/DR programs and each employee's roles and responsibilities in carrying out the related strategies and plans is essential. Beginning with new employee orientation, provide the appropriate level of training needed for all employees. Include regularly scheduled reviews and updates.

If you require that business unit managers provide information, develop work-around procedures for their departments, or make employees available to participate in training or a test, ensure they understand why their contribution is important, how it will be used, and where it fits in the big picture. Focus on commonalities, the value to the organization as a whole. Don't stress what's in it for *me* (WIIFM), that it is something *you* need them to do for *your project*. Make certain they are aware of what's in it for *them* (WIIFT)—the value to *their* business units and the organization. Develop partnerships. Make sure there is an understanding that assistance is available if needed, and make the needed help readily available when requested.

If it's not possible to tear down existing silos, build bridges between them by establishing and maintaining working relationships with people in all business units throughout the organization. Get out of your office; avoid being only an e-mail signature or a voice over the telephone. Whenever possible have face-to-face meetings.

Each day when you make your to-do list, be sure to include at least two items that require you to interact with people outside your work group. Avoid the temptation to communicate only via phone and e-mail. Take the time to have lunch or coffee with people both directly and indirectly involved in BC/DR. If there are multiple locations involved in your BC/DR programs, make every effort periodically to visit those locations. Offer help by sharing resources with other departments, including information, equipment, and time. Introduce yourself and the organization's BC/DR programs to new department managers. Be available to acquaint them with the BC/DR programs, and bring them up to speed on their business units' involvement.

Beware of using jargon and terminology that is unique to your area of responsibility. As an example, if you're involved in business continuity planning, avoid communicating in "business continuity-ese" when interacting with those who are not directly involved in business continuity planning. Avoid using our favorite acronyms—such as BCP, DRP, BIA, RTO, and RPO. It is likely that these are not much more meaningful to most people than a bowl of alphabet soup and can lead to those not directly involved in business continuity planning feeling like "outsiders" being excluded from an exclusive club. In addition, make sure a common vocabulary of terms in used universally throughout your organization.

Establish a BC/DR planning group composed of representatives of all major business units. This representation gets people from multiple areas of the organization involved and invested in the process and enhances BC/DR visibility. Beyond BC/DR benefits, representation from multiple areas assists in building better understanding of how other departments function and a greater awareness of the interdependencies between business units. To the extent possible, the planning team should be composed of those who will be carrying out the developed strategies and plans when a disaster occurs. This leads to plans that are considered the property of the end users (*our plans*), rather than plans that belong to those with primary responsibility for BC/DR planning (*your plans*).

Be open to suggestions. Ask for and truly listen to feedback. Remember that BC/DR is still relatively uncharted territory and that none of us knows everything or the best way to do anything. Realize that we don't yet have all the answers. Those not directly involved in BC/DR who are looking at strategies and procedures with a fresh set of eyes may have ideas for new approaches that are of tremendous value. This openness and the involvement of others in the planning process help make our programs and plans better and something that belongs to the entire organization, not just those directly involved in BC/DR.

In today's complex world that constantly delivers new challenges for those charged with keeping organizations operational in the face of disasters, there can no longer be independent silos. One silo cannot successfully manage risks. If we cannot tear down the silos, we must make sure that all efforts related to disaster management are interdependent and coordinated.

Develop a working network among those with similar and interrelated responsibilities, for example, safety coordinator, security director, disaster recovery manager, business continuity coordinator. Representatives from human resources, facilities, or engineering might also be included if they have responsibilities related to responding to emergencies and disasters. Take it upon yourself to establish and foster effective lines of communication and working relationships with these important partners. Go out of your way to say hello and touch base with these allies at times when you are not asking them for something.

Once established, maintain your relationships with others through mutual trust and respect. Walk your talk. Mistrust and disrespect allow silos to flourish.

Another subgroup of silos that can form is the various business continuity–related functions including emergency preparedness and response, business continuity, disaster recovery, security, safety, and so on.

To help avoid polarization among these functions, consider forming a peer group that includes all BC/DR-related functions. The key here is that it be a true "peer" group. When bosses are in the room, the consequence may be a reduced comfort level, with more vulnerability and more motivation to look good rather than openly and honestly collaborate and share information.

Focus on individual or shared work issues and challenges. These peers may have some excellent suggestions for a project you're tackling, and you may discover some opportunities to work together on projects. When the issues are compelling and the work results in tangible benefits on the job, the benefits of meeting as a group become even more evident.

Commit to meeting on an ongoing, regular basis to help ensure that group discussions translate into action. It takes time to overcome the effects of a silo history. After several meetings, relationship and trust are strengthened. Continuing to meet over time results in moving beyond the exchange of surface-level information and office gossip to addressing deeper challenges and mutual problem solving.

Collaboration with these fellow employees will result in not only a safer and more secure working environment for everyone but also a greater organizational capability to continue or more rapidly resume operations following a major emergency or disaster. A synergy of functions including security, business continuity, contingency planning, and disaster recovery, and—as appropriate based on roles and responsibilities—human resources, safety and health, purchasing, facilities management and real estate, and other key areas creates programs that are better integrated into the organization's culture and operations and, therefore, much more effective.

Never assume that help, assistance, and cooperation are not available or that employees from other departments are not interested in what you are doing. Ask for involvement. Don't wait until after a training, test, or other activity only to hear a department manager or other colleague say, "If I had known you needed help, I would have volunteered," or "I would have welcomed a chance to be involved in or to learn more about. . . ."

Articles in organization newsletters and on an intranet and announcements at departmental and other work group meetings can help familiarize all employees with BC/DR. Use break room and cafeteria bulletin boards to promote your programs. Be available to speak at internal meetings and training sessions.

Don't forget the basics. Even when time is at a premium and a project task deadline is looming, *make* the time to recognize and thank those who have made a contribution. Send an e-mail to a BIA survey respondent; make a phone call to the person who developed a business unit's continuity plan; or pay a visit to a department manager who made employees available to participate in a hot site test. These seemingly simple, yet often overlooked, steps both provide appropriate recognition and build the working relationships that help tear down silo walls.

CONCLUSION

Continually look for ways to prevent silos from being built, to establish and maintain bridges between existing silos, and to avoid the damage that can be caused by silos. Although this takes time and effort, this time is well spent and will lead to programs that involve and benefit the entire organization, rather than becoming another silo.

See also Managing Organizational Culture and Change Acceptance.

WINNING ACCEPTANCE FOR SECURITY OR OTHER NEW PROGRAMS

W. Timothy Coombs

It seems reasonable that an organization would embrace ideas such as information security and business continuity. There is clear evidence that both benefit an organization. The problem is that organizations are just a collection of people. People regularly do things that are bad for them, such as driving over the speed limit and eating fatty foods. Like the people who populate them, companies can be irrational and avoid doing things that are good for them. This means it is naïve to assume people will just embrace and support efforts to introduce a new program. The key to winning acceptance of a new program is to anticipate resistance and devise a strategy for overcoming it. Here are some tips for winning acceptance of new programs that can be applied to security and business continuity.

STARTING POINT

The starting point is to develop your arguments for a new program—your central message. Begin by outlining why the program is needed and how such a program will benefit the organization. These are your basic talking points. Presenting bulleted lists can be boring, so you need a good story. Craft a vision around the new program. A vision is story—a picture of what the organization will be like with the new program. Paint a picture of how the new program will assist an organization in surviving common business problems faced in your industry. You are helping people to visualize the new program, and visualization is a powerful persuasive tool. Have a short and a long version of your message, because sometimes you have a minute or two to make an informal pitch and other times ten to fifteen minutes for a formal presentation. All of your later messages designed to win support will come from this central message.

Your early efforts will concentrate on developing a core group of supporters for the introduction of the new program. The first step is to identify the people in management who make the real decisions. This typically includes top managers. Without their support, you stand no chance of introducing an effective new program. The second step is to consult with employees who will be involved with the new program once it begins. These people must become a part of the process from the start. Ask these employees for their input on the project, thoroughly brief them on why a new program is being developed, and update them on the progress of the project. You need to secure buy-in from this core group.

RESISTANCE TO CHANGE

Now you are ready to tackle the organization as a whole. Keep in mind that introducing a new program is a form of organizational change, and in general, people do not like and will resist change. Organizational change creates anxiety because it is new and unknown, and inconvenient, as people must alter their schedules and learn new things. Be prepared for resistance. Negative reactions to change follow a pattern of denial, anger, bargaining, and acceptance. Develop message strategies that can counter any and all of these four negative reactions.

Denial should not be dismissed. Instead, acknowledge people's concerns about the new program. Review the positive and negative aspects of the change. Do not just push the positive parts, or you will sound like a used car salesperson. But return to your central message and reinforce why a new program will be worth the effort. Anger is unpleasant, but do not take it personally and do not escalate the anger with a like response. Let people vent their feelings. Most people will feel better after they complain, and it shows you respect their feelings. You need thick skin to deal with angry employees.

Bargaining means you can be flexible on inconsequential matters. For instance, you can negotiate the exact date the plan takes effect, the colors of the binders, and the name of the plan. However, you must hold firm on the basic position that a new program will be part of the organization's future. Think about which points can and cannot be changed. Knowing this in advance makes the bargaining phase go more smoothly. Once people accept the business continuity program, praise them for their support. Resist the temptation to gloat or ridicule their reluctance. Praise will help ensure that the business continuity program functions smoothly.

CHANNELS AND VIVID EXAMPLES

There are two final points to think about once you have developed your central message and prepared to address the various negative reactions to change: the communication channels and vivid examples. The communication channels you use are a critical decision. Common communication channels used by managers include meetings, e-mails, memos, and reports. It is important to use a mix of channels to reinforce your message. Also, the various channels differ in terms of being dynamic. Dynamic channels allow the receiver of the message to give feedback to the sender. Feedback will help you, the sender, understand what type of negative reaction you are encountering. The most dynamic channels involve face-to-face communication such as meetings. Make sure meetings are used in the mix of communication channels to provide employees a chance to give feedback and to ask questions. They also give you a chance to assess the types of negative reactions and the level of resistance.

Finally, develop some vivid examples to illustrate key points. Statistics are important and will be the base for many of your messages. A cost-benefit analysis or statistics about the number and types of business disruptions in your industry are critical to have. However, anyone who has sat through a budget presentation knows that statistics alone can be boring and easy to forget. Vivid examples, using colorful words and details, are needed to illustrate your statistics and facts and make the information more engaging and easier to remember. Consider the following example. Your data show the number of IT-related business problems in your industry and the total cost of those disruptions. Now, take one specific case and give the details. How and why did the disruptions occur? How much money did the company lose and why? How were the employees affected by the problem (lost wages, benefits, and/or stock price)? The details make the statistics real for people—they bring the data to life. Vividness connects back to the vision you created for the central message. The vision of a future with the new program should be vibrant. It can even include an alternate view of the future should the organization not adopt the new program.

CONCLUSION

Introducing a new program into an organization can be difficult. If you are fortunate, people will love the idea and embrace your change. But, it is important to prepare for the worst-case scenario. Create your central message complete with talking points and vision. Get your core group of supporters to buy into the business continuity program. Be prepared to respond to any and all of the four negative reactions to change. Finally, be sure the communication channel mix includes some face-to-face interchange, and use vivid examples to support your statistics. The better prepared you are for resistance, the greater the likelihood your organization will adopt a functional business continuity program.

See also Managing Organizational Culture and Change Acceptance.

MANAGING ORGANIZATIONAL CULTURE AND CHANGE ACCEPTANCE

W. Timothy Coombs

A recurring theme in many of the topics related to business security is the need for winning acceptance of changes. This can include the development of business continuity programs, use of computer security protocols, or acceptance of ethics and compliance codes. New policies and programs are a form of organizational change. One thing we do know is that change is never easy in an organization. This entry examines the role of organizational culture in managing change.

People naturally resist change because it creates something new and different. People in an organization have to learn new behaviors with a change, which requires investing their time. There is also the possibility that people will not be successful in learning the new behaviors. Any or all of these factors can lead people to resist change efforts in an organization.

WHAT IS ORGANIZATIONAL CULTURE?

In general, organizational culture is the way things are done in an organization. Organizational culture represents the beliefs, values, norms, and practices that are taken for granted in an organization and guide actions in the organization. Some people make the mistake of assuming the culture is the same as the mission statement or vision statement. That may not be the case if people in the organization do not live by the mission or vision. The organizational culture is a manifestation of what is important in the organization and what directs people's actions and thinking. Cultures evolve over time and are difficult to control.

HUMAN SIDE OF SECURITY AND SAFETY

ASSESSING ORGANIZATIONAL CULTURE

Before trying to change a culture, management needs to know what the culture is. A variety of methods can be used to assess the current culture, including surveys, focus groups, interviews, observations, and an analysis of organizational documents such as web sites, policies, newsletters, and manuals. Utilizing these research tools enables understanding of what is important in the organization—its core values. Focus on what employees believe management pays attention to, what actions are rewarded or punished, and what beliefs and values seem to dominate messages in the organization. Because we have a hard time seeing our own culture, it may be helpful to bring in a consultant to facilitate assessing the organization's culture.

SKEPTICISM IN THE ORGANIZATIONAL CULTURE

An analysis of the organizational culture can determine whether there is skepticism in your organization about change. Skepticism is when employees doubt a change will succeed. This intensifies employee resistance to change because they see the effort as a waste of time. Experts in change management refer to this phenomenon as organizational cynicism. A culture plagued by organizational cynicism reflects a lack of faith in the leadership to make successful changes. Typically, organizational cynicism is a result of past change failures. Employees are pessimistic that any change will work and blame management for past failures. Organizational cynicism is reflected in comments such as the following:

- "Most of the programs that are supposed to solve problems around here will not do much good."
- "Attempts to make things better will not produce good results."
- "Suggestions on how to solve problems will not produce much real change."
- "Plans for future improvements will not amount to much."
- "The people responsible for solving problems around here do not try hard enough to solve them."
- "The people responsible for making things better do not care enough about their jobs."
- "The people responsible for making improvements do not know enough about what they are doing."
- "The people responsible for making changes do not have the skills needed to do their jobs."
- "The people responsible for solving problems around here cannot really be blamed if things do not improve."
- "The people responsible for solving problems around here are overloaded with too many job responsibilities."

- "The people responsible for fixing problems around here do not have the resources they need to get the job done."
- "The people responsible for making changes around here do not get the cooperation they need from others."

Organizational cynicism can be overcome through strategic change efforts. Management must be willing to confront the past failures that have bred the organizational cynicism. Management must acknowledge and explain past failures, which includes owning up to its own contributions to those failures. Be sure to talk about past success as well so that employees remember change can work.

STRATEGIC APPROACH TO CULTURE MANAGEMENT

Change needs to be planned carefully. Management must be strategic in its approach to change management if it wants the change to take root and flourish. There are six steps in the strategic approach to culture management: (1) change evaluation, (2) cadre construction, (3) message development, (4) organizational buzz, (5) cultural reification, and (6) progress monitoring.

Change evaluation seeks to decide whether the change is consistent with the existing culture. Management must determine whether the change is like something the organization already does or values. Management needs to find a connection to current culture because this connection makes the change less threatening. Management must also determine what new "features" need to exist in the culture to support the change. It will also need to develop a framework to support the desired change.

Cadre construction organizes a group of people who support the change. This serves as a foundation for change acceptance. The cadre construction should include what management researchers call prime movers—powerful figures in the organization that others look to for guidance or rely on as important sources of information. Involving a wide number of people in the cadre helps to create greater buy-in to the change.

Message development involves creating a vision for the change and talking points for the change cadre. The vision indicates what the future of the organization can be when the change is in place. It is a picture of what the organization can become. The vision creates a sense of purpose but must be realistic. Management should avoid overpromising and setting up the change for failure. A talking point is a short rationale for the change. Talking points can include current vulnerabilities, the problems with the status quo, and how the organization will benefit from the change. The talking points also need to create a sense of urgency.

Organizational buzz attempts to get people in the organization talking about the change. This is a time when past change failures must be addressed. Management must send messages about the change through multiple communication channels, including an intranet, newsletters, e-mails, position papers, and town hall meetings. Management must also solicit feedback from people in the organization. In this way management answers questions people might have about the change.

Cultural reification happens when management integrates the change into the day-to-day operations of the organization. The policies and behaviors become linked to employee evaluations. Some organizational rewards or punishment will be tied to the change. Finally, management must monitor the change to determine whether it is successfully taking effect. Are new policies being followed? Are new appraisal items being used? Are rewards and/or punishments being delivered consistently? If the change is failing, the change management effort should be adjusted.

CONCLUSION

As organizations look to implement or to improve policies related to business security, management will have to grapple with organizational change. Culture is an essential factor in getting change to become an effective part of the organization. People must become comfortable when change is introduced into an organization. Because change can create uncertainty, stress, and resistance, communication is essential to helping people accept a change and its becoming integrated in the organizational culture. A carefully constructed plan is needed to help an organization manage its change.

See also Crisis Management; Avoiding the Silo Effect to Improve Business Security. and Winning Acceptance for Security and Other New Programs.

CULTURE OF INTEGRITY

W. Timothy Coombs

Millie Kresevich, a retail loss-prevention specialist, believes one mechanism for reducing misconduct in organizations, including employee theft and fraud, is to develop a culture of integrity. Employees must learn to talk about integrity, comply with ethics policies, and deal with problem situations.[1] A culture of integrity is something that management must craft. The following four principles serve as a foundation for a culture of integrity:

1. Have a clear statement of the organization's ethical policies, and explain them to every employee.
2. Model ethical behavior for all employees.

3. Encourage everyone in the organization to talk about ethics.
4. Make adhering to the policy on ethics part of employee evaluation. Ethical behavior must be rewarded, and unethical behavior punished.

As noted in other entries related to ethics, management conduct is critical to an effective ethics program. Employees will follow the examples set by management. Managers can lead by example with integrity. Management must learn to trust others, become a good listener, clarify expectations about integrity, and help to create a climate of honesty.

Kresevich has developed a training program designed to facilitate a culture of integrity. The program teaches employees communication skills, how to comply with ethics policies, and how to address specific ethical situations. The training centers on a twenty-four-page workbook containing ethics exercises. It places employees in ethical dilemmas so they can work through the problems with other employees. The exercises emphasize the issues that helped to create the situation rather then just the behavior or feelings. A series of questions guides employees through the scenarios, which provide insights beyond the specific problems. Trainees learn why people do not commit to ethical policies and how to react to unfair situations. One sample scenario involves a top-producing employee coming in late regularly. Participants are asked the following questions: "What would you do and why?" "What would you say to the employee that was late?" "What is the benefit of addressing the situation?" and "What is the impact if you do not address this?" The questions allow trainees to dig deeper into the issues and gain richer insights.[2]

CONCLUSION

Training in integrity can help to reduce theft, improve morale, and increase profitability. Although we want to believe employees naturally bring integrity to the workplace, this is not always the case. Employees simply might not know what is an ethical violation and how they should behave in situations that could compromise ethics. Having a clear policy and training employees according to that policy help to reduce uncertainty about ethics and promote an ethical workplace. Organizations can benefit from ethical policies and training employees about those policies and the idea of integrity in general.

See also Corruption as a Business Security Concern; Ethics as a Business Security Concern; and Ethical Conduct Audit.

NOTES

1. Millie Kresevich, "Using Culture to Cure Theft," *Security Management*, Feb. 2007, pp. 47–51.
2. Kresevich, "Using Culture to Cure Theft," pp. 49–51.

EXERCISE AND TRAINING BASICS
W. Timothy Coombs

Every response plan developed by an organization should be practiced and tested. This includes emergency management plans, business continuity plans, and crisis management plans. Plans are practiced and tested through exercises. People practice putting the plans into action and the plans are examined for any weaknesses. An exercise is a focused activity that puts organizational personnel in a simulated situation requiring them to act as they are expected to in a real event. Exercises involving emergency plans will be used for illustrative purposes in this entry, but the concepts and ideas can be applied to any type of response plan.

WHY EXERCISE?

A plan is an idea that should work. An actual event is not the time to discover the plan does not work or that individuals charged with enacting the plan cannot do so. The two main reasons for exercises are (1) for individual training as people experience their roles and (2) for system improvement as plans and personnel are adjusted. Exercises test and evaluate the plans, policies, and procedures. The organization finds out whether the plan can work. A weakness in the plan or lack of resources is revealed in a low-risk environment. Exercises clarify roles and responsibilities, help personnel improve their individual performances, and develop coordination and communication between units and people in the organization.

Exercises are used to spot and correct problems before an actual event. The end result should be a more complete plan and a more effective response to an event. Some organizations are required by regulations and laws to engage in exercises. Airports, health care facilities, and nuclear power plants are obligated to hold regular exercises. The Occupational Safety and Health Administration (OSHA) requires employers to develop emergency plans. Organizations in which chemicals are produced, used, or stored are to conduct yearly exercises and evaluate their hazardous materials response and recovery plans.

PROGRESSIVE EXERCISING

An exercise program uses a variety of types of exercises. The idea is for an organization to perform a series of different types of exercises that build in complexity. Each exercise is designed to achieve particular goals and builds upon the results of the previous exercise. An exercise program demands careful planning and specific goals. The progression in complexity is designed to build confidence. People are able to master a set of skills before more demands are placed on them. The people involved in an exercise vary by the nature and size of the exercise. The progression also serves to increase the number of people involved in the exercise. By the final stage, your organization will be coordinating with local emergency responders as part of the exercise.

An exercise program cannot be developed overnight. Management must examine capabilities, costs, scheduling of tasks, and developing a long-term plan. A planning team should be developed with representatives from across the organization and perhaps even some community responders. The team can develop a plan and adjust that plan, as early exercises indicate how well the organization is or is not progressing.[1] The Exercise Program Rationale table reviews the reasons to conduct an exercise program.

Exercise Program Rationale

		Reasons to Conduct Exercise Program Activities[2]		
Orientation	Drill	Tabletop Exercise	Functional Exercise	Full-Scale Exercise
No previous exercise	Assess equipment capabilities	Practice group problem solving	Evaluate a function	Assess and improve information analysis
New plan	Test response time	Promote management familiarity with plans	Observe physical facility use	Assess and improve cooperation
New procedures	Personnel training	Assess plan coverage for a specific situation	Reinforce established policies and procedures	Test resource and personnel allocation
New staff or leadership	Verify resource and staffing capabilities	Assess plan coverage for a specific risk	Test seldom-used resources	Assess personnel and equipment locations
New industrial risk		Examine staffing contingencies Test group message interpretation Observe information sharing		

HUMAN SIDE OF SECURITY AND SAFETY

EXERCISE TYPES

A comprehensive exercise program is composed of five types of exercises: (1) orientation session, (2) drill, (3) tabletop exercise, (4) functional exercise, and (5) full-scale exercise. The Exercise Comparison table summarizes the value of each type of exercise.

Orientation Session

An orientation session introduces the plan and process. The purpose is to familiarize people with their roles, plans, procedures, and equipment. It can also help them understand how to coordinate activities and clarify their responsibilities. There are no set rules for conducting an orientation session. The facilitator must carefully plan the orientation session and keep it moving. This is not a time to make it up as you go along or to let people drift off on tangents. Be creative and try to involve the participants in the process. There should be interactivity in the session to keep people connected to the discussion.

Key characteristics:

- *Format.* It is a low-stress exercise that is usually presented in a group setting with rather informal but structured discussion. A variety of formats can be used including lecture, discussion, slide or video presentation, computer demonstration of software program, panel discussion, and guest lectures.
- *Applications.* The orientation session can address a number of purposes, including the group discussion of a problem or topic; introduce new ideas or concepts; inform new people about existing plans and procedures; discuss the progressive nature of exercises; or motivate participants to be involved in future exercises.
- *Leadership.* Orientation sessions are guided by a facilitator, who should organize the session, present the information, and guide the discussion.
- *Participants.* The participants are cross-functional, drawn from a variety of departments in an organization.
- *Facilities.* Orientation sessions can be conducted in conference rooms or training rooms. The location depends on the size and equipment needs of the session.
- *Time.* Orientation sessions should last from one to two hours.
- *Preparation.* Preparation time is fairly short, usually about two weeks. Participants are not required to have had previous training.

Drill

A drill is a supervised, coordinated activity used to test a specific operation of part of a response plan. The idea is to work on smaller parts of the response plan before integrating these components into a large drill that tests them all.

Exercise Comparison

Comparing Five Types of Exercises[3]

	Orientation	Drill	Tabletop Exercise	Functional Exercise	Full-Scale Exercise
Format	Informal discussion in group setting Variety of ways to present it	Actual facility response Actual equipment	Narrative presentation Problem statement and simulated messages Group discussion No time pressure	Interactive Players respond to messages given by simulators Realistic but equipment is not used Real time and stressful	Realistic event announcement People gather at assigned locations Actions on the scene are the input for the participants
Leaders	Facilitator	Manager or exercise designer	Facilitator	Controller	Controller(s) Evaluators
Participants	Cross-functional	People involved with the function being tested	Anyone who may be involved with the particular event	Players Simulators Evaluators	All people who might be involved in the event
Facilities	Conference or training room	Command center and the field	Conference or training room	Command center and other rooms	The entire facility
Time	1 to 2 hours	1/2 to 2 hours	1 to 4 hours	3 to 8 hours	1 hour to 2 days
Preparation	2 weeks	1 month	1 month	At least 6 months	About 1 year

HUMAN SIDE OF SECURITY AND SAFETY

An example would be an evacuation drill or a test of the organization's emergency notification system. Begin a drill by briefing participants on the scenario to be used and the function of the drill. Prepare for the drill by carefully reviewing the part of the response plan to be tested. During a drill, monitor its progression, making sure participants are taking the expected actions. If they are not, the drill designer needs to provide messages to prompt those actions. The drill tests specific tasks and skills, so you must make sure the participants engage in those tasks and use those skills.

Key characteristics:

- *Format*. Drills involve actual field responses. They should be as realistic as possible and use any equipment that would be needed during an actual event.
- *Applications*. Drills work on specific tasks and skills. They also provide training on new equipment, new policies or procedures, and maintenance of current skills. For instance, a chemical facility might work on evacuations, isolating a spill area, or valve system shutdowns.
- *Leadership*. The person in charge could be a manager or an exercise designer who must have a thorough understanding of the tasks and skills being tested.
- *Participants*. The people involved are those associated with the specific task or skill being tested.
- *Facilities*. Drills are conducted on-site where the task or skill would need to be performed.
- *Time*. Drills run from a half hour to two hours.
- *Preparation*. Drills are fairly easy to design. Preparation time is usually a month. Part of the preparation includes creating a short orientation for the drill.

Tabletop Exercise

Tabletop exercises are designed to be low-stress analyses of problematic situations an organization is likely to face. The tabletop, led by a facilitator, is not an attempt to simulate the event or use equipment, but is just an analysis of the situation. Participants improve their critical thinking skills by analyzing and resolving problems using plans and procedures. The key is to have participants identify and analyze the problem areas.

Key characteristics:

- *Format*. A tabletop begins with people reading a short narrative about the event. The facilitator then starts the discussion with a problem statement describing major events. Participants then talk about how they might respond. The facilitator also uses simulated messages to further the discussion. These messages are more detailed than the problem statement and add

more information to the discussions. The simulated messages mirror how teams learn more about an event as it unfolds. The discussion should focus on roles (how a person would respond), plans, coordination between units, the effects of decisions, and similar concerns.
- *Applications*. Tabletop applications include the following: low-stress discussions of plans and procedures, favorable conditions for problem solving, allowing team members to get to know one another better, and serving as preparation for a functional exercise.
- *Leadership*. The facilitator who leads the tabletop decides who gets messages, calls on particular people for comments, asks questions, and tries to keep the decision making on course.
- *Participants*. The organization should select participants who would be involved in that particular event so that they can practice their roles.
- *Facilities*. A large conference room or training room is used depending on the number of participants and the equipment requirements.
- *Time*. Tabletops generally run from one to four hours but can be longer. The facilitator wants to give people a chance to discuss topics fully, so there are no time pressures. The aim is not to get through all the material but to have arrived at decisions based on in-depth analysis.
- *Preparation*. Tabletops take about a month to prepare. Participants should have been through an orientation session and one or more drills prior to a tabletop.

Functional Exercise

The functional exercise tests the capabilities of the organization to respond to a simulated event. As a simulation, it is interactive and time pressured. Events and information unfold in real time. The functional exercise allows an organization to examine the coordination, interaction, and integration of its roles, policies, procedures, and responsibilities.

Key characteristics:

- *Format*. Functional exercises are interactive. They simulate events but do not involve moving resources or using equipment. Simulators provide messages to participants in a carefully planned sequence that mimics the actual event. These messages make information available and present problems that participants would encounter in the actual event. The idea is to see how well the participants or "players" can coordinate and use the plans and procedures to address the event. The exercise is high stress because it requires real-time decisions and actions. Players experience the consequences of their actions, and their decisions and actions influence the development of the exercise. Creating the specific messages and anticipating contingencies make a functional exercise difficult to construct.

- *Applications*. Specific response functions—actions or operations required during a response—are tested. Examples would be damage assessment and coordination of units. Testing functions is the last step before a full-scale exercise.
- *Leadership*. Functional exercises require a controller who manages and directs the exercises.
- *Participants*. Include players, simulators, and evaluators. Players are the organizational personnel who would be involved in the particular function in that specific situation. The simulators play external roles, such as the fire chief, and deliver the planned messages to the players. Evaluators only observe and assess performance. They do not interact with players.
- *Facilities*. The organization's command center is a designated area to be used in emergencies or crises. Players should gather where they would be during the event.
- *Time*. It takes three to eight hours for a functional exercise.
- *Preparation*. It takes organizations six to eighteen months to plan a functional exercise. Due to the complex nature and time consumption, organizations may consider hiring a vendor that specializes in running simulations. The vendor can customize the exercise to your organization's needs and supply the materials, controller, simulators, and evaluators. An additional benefit of vendors is the third-party evaluation of performance. Their critiques are not influenced by workplace relationships.

Full-Scale Exercise

A full-scale exercise simulates an actual event as closely as possible. Equipment is used, people mobilized, and simulated victims appear. The organization often coordinates the full-scale exercise with local emergency responders. These responders get practice, and the organization better understands how to coordinate responses with these units. Your people are on the scene moving and using equipment as they would in the actual event, as well as coping with the simulated victims. This is the most stressful of the exercises and is executed in real time.

Key characteristics:

- *Format*. People are given a description of the event just as they would be in a real event. This might mean telephone calls or a warning siren. Those who must go to the field take their places and cope with the simulated problems encountered there. The command center should be operational and directing the response. Those assigned to the command center take their places to engage in their roles.
- *Applications*. Full-scale exercises are the ultimate test to reveal whether people can do what they should during an event and how well the plans and procedures work in an event. The downside is the cost and time commitment.

- *Leadership.* There will be one or more controllers and multiple evaluators. The controllers will send the necessary messages to participants.
- *Participants.* Anyone who might be involved in the actual event is included. Evaluators are needed too.
- *Facilities.* The event unfolds across the organization's facility and must include the command center.
- *Time.* Full-scale exercises can be as short as two to four hours or last one or two days. It depends on the scope and nature of the event.
- *Preparation.* It can take a year to prepare a full-scale exercise. As with functional exercises, organizations should consider hiring a vendor to develop and run the full-scale exercise.

The exercise process encompasses a variety of tasks. FEMA organizes exercise tasks in a two-by-three matrix involving the exercise phases (preexercise, exercise, and postexercise) and the type of task (design or evaluation). The Task Categories table reviews the range of tasks involved with exercises.[4]

THE EXERCISE PROCESS

It is best to think of exercises as a process rather than an event. The exercise process is composed of five accomplishments: (1) establishing the base, (2) exercise design, (3) exercise conduct, (4) exercise critique and evaluation, and (5) exercise follow-up.

Establishing the Base

The purpose of any exercise is to get people to think or act the way they would in a real situation. Management begins by laying a foundation—known as establishing

Task Categories

	Preexercise Phase	Exercise Phase	Postexercise Phase[5]
Design	Review plan Assess capability Determine cost and liabilities Organize design teams Build support for an exercise Draft a schedule Design the exercise	Prepare facility Assemble props Brief participants Conduct exercise	
Evaluation	Select evaluation team Develop evaluation methodology Select and organize evaluation team Train evaluators Contract the evaluation	Observe assigned objectives Document actions	Assess achievements of objectives Participate in post-exercise meetings Prepare postexercise report Participate in follow-up activities

the base—for the exercise to be sure it gets the desired results. To prepare for an exercise, management must review the current plans, assess the organization's ability to conduct a drill, define the scope of the exercise (its limits), select the type of exercise based upon organizational needs, consider the costs and liabilities, develop a statement of purpose, and build interest and support for the exercise. Three of these points require elaboration here.

Current plans outline responsibilities for personnel during an event and include contact information. They are also a reference tool for decision making and problem solving. The review of a plan should consider the types of situations (hazards and risks) covered; the roles people are to play in various situations; current training level of team members; and a review of the personnel, resources, and procedures outlined in the plan. You are looking for areas that could use improvement or revision.

Management must determine whether the organization is ready for a particular exercise. Reviewing the last exercise helps establish what the team would be ready for next. Is it prepared to advance to the next level, or is there a need to work on existing skills? Does the desired exercise fit into the upcoming schedule of the personnel who need to be in the exercise? Why have an exercise if the participants will not have time to spare to be fully involved. Costs and liabilities are a related issue. Does the organization have the financial resources to cover the desired exercise? Exercises have costs including employee salaries, equipment, materials, and there are vendor fees, if an exercise is outsourced. An organization must also review its insurance to ensure coverage for possible injuries or equipment damage during an exercise.

Exercise Design

Exercise design is one of the most complicated stages. The design team must create the materials necessary to have an exercise. This exercise design process covers eight points: assess needs, define scope, purpose statement, objectives, narrative, major and detailed events, expected actions, and exercise messages. Needs assessment helps the design team in selecting an appropriate scenario for the exercise, one that fits with the hazards or risks an organization is most likely to encounter along with the actions and skills it has identified as needing to be addressed. A key part of needs assessment is the results of the last exercise. What still needs work, and which areas of the response are ready for more intensive testing?

The scope sets limits to the exercise. One exercise cannot test everything. The scope narrows the focus of the exercise and flows from the needs assessment. It should consider the type of event, the participants involved in that event, the location of the event, and the functions to be tested in the event. The scope is also shaped by the available budget and personnel.

The purpose statement—a broad articulation of the exercise goal—guides the entire exercise. The purpose statement limits the objectives and clarifies to participants why the exercise is being conducted. Objectives are more specific and describe the performance you expect from the participants. Objectives are critical to the entire design process. The exercise events need to capture the objectives, whereas the evaluation is based on the performance of the behaviors specified in the objective. Good objectives are concise and clear. An objective is written with an action phrased in observable terms, the conditions under which the action is to be performed, and the standards of performance. The objective states who should do what under what conditions according to what standards. A sample objective can found in the Objective and Expected Action box. FEMA recommends the SMART guidelines for objectives. An objective should be *s*imple, *m*easurable, *a*chievable, *r*ealistic, and *t*ask oriented.

The narrative briefly describes the events that occurred right before the exercise began. The narrative sets the mood for the exercise as well as the stage

---

Objective and Expected Action: FEMA Example for Airplane Crash

Function: Coordination and communication between the airport and the emergency responders.

Objective: Upon notification that a crash is imminent, responding unit will stage within three minutes according to the emergency plan.

Event: Landing of disabled aircraft is imminent.

Expected Actions:

Airport
Notify police, fire, and medical personnel.
Alert hospitals to potential mass-casualty incident.
Activate its response team.

Hospital
Notify other medical facilities as appropriate.

Crash/Fire Rescue
Initiate command center.
Notify dispatch of command center and staging areas.

Possible Messages:
Radio call from plant to tower.
Tower calls police, fire, and rescue.
Plane requests runway to be designated.
Call from hospital requesting information.
Calls to dispatch from media.
Degrading radio communications with plane.
Pilot feels major vibrations/noise on the plane.[6]

for later actions by providing the initial information players use to make choices and take actions. It is akin to the briefing people receive when an actual event begins. A good narrative is about five paragraphs long, is very specific, is phrased in the present tense, presents a chronology of events, is written in short sentences, and emphasizes the critical nature of the event. Points to consider for a narrative include what the event is, the danger surrounding the event, how the organization learned of the event, reported damage, sequence of events, where the event is, the weather conditions, and whether there was any advanced warning. The Sample Narrative box provides a sample exercise narrative from FEMA.

Major and detailed events serve to organize the development of the scenario. These are large or small events or occurrences that are created by the major event presented in the narrative. Major and detailed events supply a structure that links the simulated events to the actions people need to take and provide unity to the exercise as it unfolds. Major events are the big problems resulting from the event. Detailed events are specific problem situations that require a response from participants. Detailed events should prompt one or more expected actions from the players. The design team picks events that fit with the objectives and the actions it wants the players to take.

Here are sample major and detailed events FEMA provides for an airplane crash.

Sample Narrative

Here is a sample exercise narrative from FEMA:

A Boeing 747, en route from Panama to San Francisco, is experiencing in-flight engine problems and will have to make an emergency landing. Plans have been made to land at a large airport 200 miles north. However, the latest communication with the pilot indicates that the plane has lost engine power and is losing attitude too quickly to reach the large airport. Even though your city airport is too small to handle a 746, you are the only hope for the 350 passengers and 10 crew members.

Conditions at your airport are clear and the surrounding area is dry. A hot, dry wind is blowing from the north.

The main runway lies along a relatively unpopulated suburban area. However, the likelihood of the pilots being able to control the huge plane and stay within the landing space is slim. The approach passes over populated suburban housing developments. The airport control tower alerts its own crash/fire rescue units and requests that the local emergency services provide backup assistance in fire, police, medical, welfare, and search and rescue capabilities.[7]

Major events: Fuselage breaks apart and hits buildings on the ground, several homes in the area are ignited by jet fuel, survivors are believed trapped in the front section, a crowd gathers near the crash including family members of victims, bystanders are injured on the ground, and casualties are estimated at between 100 and 150.

Detailed events: Mortuary capacity is too small to accept the large number of remains, local hospitals do not have the specialized burn care needed by victims, and the American Red Cross has agreed to fund a family information center.

The expected actions are the actions or choices that you want players to carry out in order to demonstrate their competence. In short, it is what you want people to do so you can evaluate their proficiency. The expected actions shape the messages developed for the exercise. The messages must lead people in the direction of the desired actions. The expected actions are drawn from the performance expectations of the objective. The desired actions are essential to evaluation because they tell evaluators what to assess. There are four basic actions: verification, gather and verify information; considerations, consider information and discuss among players; deferral, defer action to later; and decision, deploy or deny resources. The Objective and Expected Action box contains sample actions.

Finally, the design team prepares the messages used to communicate details of unfolding events to the players. The messages that relate the major and detailed events to the participants can be delivered by telephone, e-mail, radio, fax, written notes, or in person. The messages should have a direct connection to expected actions. The Objective and Expected Action box also contains some sample messages. Each message must have the source (who sends the message), the transmission (how the message is sent), message content (what is in the message), and recipient (who is to receive the message). The Message Format box illustrates how FEMA recommends creating messages. Not surprisingly, exercises do not always go as planned, as players respond differently than anticipated. Spontaneous messages, used to address the off-script developments, need to be created quickly but with care. Controllers and simulators must remember the four message factors and try to keep the messages realistic and consistent with the exercise scenario.

In the end, a design team's master scenario of events is used to monitor the development of the exercise and help keep it on track. The information can be converted into a chart that provides guidance for the controllers and simulators.

Exercise Conduct

Exercise conduct involves the leaders and participants taking part in the exercise. A successful exercise depends on a number of factors. Participants must have a clear understanding of the rules for the exercise and what is expected from them. Controllers and simulators must provide a consistent flow of messages. The messages serve to sustain action by providing participants with something to react to. Participants must be encouraged to take the exercise

---

Message Format

This is the format FEMA recommends for preparing exercise messages.[8]

Emergency Exercise Message

To: Method: From:

Number: Time:

Content:

Actions Taken:

---

seriously and treat it as a real event, not as a mere game. They need to know there will be consequences if they do not take the exercise seriously. To aid the realism, a valid timeline for an event is required. If the exercise does not mirror reality, participants will have a difficult time treating it as a real event.

Exercise Evaluation and Critique

Exercise evaluation and critique determines how well the exercise achieved its objectives. This includes organizational-level and individual-level evaluations. Common concerns for evaluation consist of the following needs: to improve the plan, to improve the policies and procedures, to provide additional training, and to overcome staffing problems. A report should be drafted analyzing and critiquing the exercise effort. This report includes recommendations for correcting any problems identified in the report. An organization should not conduct an exercise if there will be no evaluation and critique.

Exercise Follow-up

Exercise follow-up is the most neglected area of the exercising. The lessons learned from the evaluation and critique should be put into action. People have spent time and money to create and execute an exercise, but that investment pays off only if recommendations for improvements are implemented. People should be assigned responsibility for executing each recommendation and given a schedule for when the changes need to be made. Management then monitors the situation to ensure these changes were made. Part of the next exercise should include efforts to test the changes to make sure they work.[9]

DESIGN TEAM

If exercises are done in-house rather than through a vendor, a team must be assembled to design the exercise. The team leader should be someone who has experience with the topic of the exercise, knows the plan to be used in the exercise, and has the time to devote to the design. Other members of the design team should represent the different areas of the organization to be involved in the exercise. The design team will be responsible for determining the exercise objectives, tailoring the scenario to the organization, developing the sequence of events and related messages, helping to create and distribute preexercise materials, and assisting in conducting the preexercise training sessions. Design teams need to have a clear goal and agree on a realistic schedule for preparing the exercise. The leader must schedule regular meetings and monitor the team's progress against its agreed-upon schedule.[10]

Exercise Documents

The design team creates four major documents: the exercise plan, the control plan, the evaluation plan, and the player handbook. The exercise plan is common body of knowledge for everyone involved in the exercise. It serves to guide the design team and helps participants to appreciate and understand the exercise.

Exercise Design Document

Exercise Plan	Control Plan	Evaluation Plan	Player Handbook
Exercise type and purpose	Exercise concept	Exercise concept	Exercise scope
Scenario narrative	Preexercise player activity	Preexercise player activity	Scenario narrative
Scope	Concept for management, control, and simulation	Concept of evaluation management	Player procedures and responsibilities
References		Evaluation team staffing	Safety and security
Objectives			Communications
Exercise management structure and responsibilities	Control team staffing	Evaluation team training	Reporting
	Control team training	Evaluation team staff responsibilities	Administrative system
	Control team staff responsibilities		Recommended preexercise training event
Safety and security	Control team procedures	Evaluation team responsibilities	Schedule of player exercise briefings
	Communications, logistics, and other support	Support for the evaluation team	Command center procedures

The control plan, which is not for players, sets the rules and explains the roles for the controllers and simulators. Most importantly, it defines the communications, logistics, and administrative structure of the exercise. In other words, the control plan details when events will happen and when certain messages should be delivered.

The evaluation plan provides guidance to evaluators for exercise evaluation procedures, responsibilities, and support. Controllers and simulators will have access to the evaluation plan as well. This plan explains the purpose of the exercise, establishes the basis for evaluation, and provides evaluation criteria that reflect the focus of the exercise. The player handbook details the information that participants need to be involved effectively in the simulation. It serves as the main source of the information present in the player briefing session. The Exercise Design Document table provides a summary of the four documents.[11]

CONCLUSION

Exercises and drills can be used to test and refine emergency management, crisis management, business continuity, and disaster recovery skills. You do not know how well a plan will work or how well your people will perform before you test them. It is better that your plans and people are tested in a nonthreatening exercise or drill than in a real negative event such as an emergency or crisis. Exercises and drills take a commitment of time and finances but are well worth the investment.

See also Benefits of Emergency Management; Business Continuity; Crisis Management; Evacuation in Large and Multiple Tenant Buildings; and Emergency Response Training and Testing: Filling the Gap.

NOTES

1. Federal Emergency Management Agency (FEMA), *Exercise Design* (Washington, DC: U.S. Government Documents Office, 2003), p. 2.1.
2. FEMA, *Exercise Design*, p. 2.17.
3. FEMA, *Exercise Design*, pp. 2.18–2.19.
4. FEMA, *Exercise Design*, pp. 2.5–2.16.
5. FEMA, *Exercise Design*, p. 3.3.
6. FEMA, *Exercise Design*, p. 4.34.
7. FEMA, *Exercise Design*, p. 4.28.
8. FEMA, *Exercise Design*, p. 4.42.
9. FEMA, *Exercise Design*, pp. 3.1–3.11.
10. FEMA, *Exercise Design*, pp. 3.23–3.27.
11. FEMA, *Exercise Design*, p. 3.22.

GUIDANCE APPENDIX

Evaluation Tools

Evaluating Whether You Have Silos	378
Evaluating an Emergency Preparedness and Response Program	379
Assessing an Organization's Reputation	381
Organizational Reputation Scale	381
Assessing the Climate for Workplace Aggression in an Organization	383
Workplace Aggression Tolerance Questionnaire (WATQ)	383
Pandemic Preparedness Evaluation	386

Guidelines, Document Recommendations, and Warning Signs

Information Security Policy Content Areas	391
Requesting an Exception to an Information Security Policy	392
Common Signs of Social Engineering	392
Steps for Constructing the Crisis Sensing Mechanism	393
Step 1: Identify Existing Crisis Sensing Activities	393
Step 2: Assess the Existing Crisis Sensing Activities	393
Step 3: Assess Information Gathering Techniques	393
Step 4: Develop Procedures for Funneling Information	393
Step 5: Establish Evaluative Criteria	394
Step 6: Test the Crisis Sensing Mechanism	394
How to Shelter-in-Place	394
Components of an Ethics and Compliance Program	395
Components of Anticorruption Policies	395
What Creates Outrage in Risk Communication	396
Spokesperson Tasks for Crisis Management	397
Basic Guides for Using Passwords	397
Spyware Symptoms	397
Information Security Basic Guidance	398

Sample Crisis Management Plan Elements 398
    Generic Components of a Crisis Management Plan 398
    Incident Report 399
    Crisis Management Team Strategy Worksheet 399
    Stakeholder Contact Worksheet 400
Initiating Emergency Notification Communication 400
    Initiating Emergency Notification 400
    Reaching the Right People with the Right Message at the
      Right Time 401
    Delivering and Receiving Communication 401
    Communication Efficiency 401
    Ease of Use 402

EVALUATING WHETHER YOU HAVE SILOS

Management can use the following self-survey to determine whether it has a "silo" problem.

To provide a general indication of how entrenched silo mentality is in your organization or department, choose the most accurate response to each of the following ten statements. Be objective as you select the most accurate of three possible responses to each statement:

- True
- Partially true
- Not True

1. Employees have a good understanding of what all other business units do, how they operate, and how they all fit in the big picture.
2. Collaboration is a strong element of the organization's culture.
3. Your peers from other areas of the organization are open to collaborative efforts and joint projects.
4. You believe that silos are detrimental, and you model that belief on an ongoing basis.
5. You encourage your direct reports to collaborate with those outside your business unit.
6. Business units in your organization freely share resources and information with one another.
7. You receive consistent messages from different levels and/or functions within the organization.
8. Your organization has been successful in building commitment for security, business continuity, disaster recovery, and emergency management projects and programs.

9. A vehicle (e.g., work group) exists to facilitate ongoing communication among related functions within your organization (BC/DR/emergency management/security/safety).
10. ALL employees at all levels are aware of the BC, DR, security, and related programs within the organization

SCORING:

Give yourself:

10 points for each TRUE response
5 points for each PARTIALLY TRUE response
0 points for each NOT TRUE response
Maximum possible points = 100

Score	Assessment
90–100	Outstanding. Congratulations to you and your organization. Continue to avoid silo formation.
70–80	Very good. Look for additional improvement opportunities, and avoid further entrenching existing silos.
50–60	Mediocre. Note which questions were answered "no," and concentrate efforts there.
30–40	Improvement needed. Further assess the situation and develop a plan for positive change.
0–20	HELP!! A great deal of work to be done. Get started now.

Source: Betty A. Kildow

EVALUATING AN EMERGENCY PREPAREDNESS AND RESPONSE PROGRAM

The following survey allows management to assess its emergency preparedness and response program.

Responding to the following twenty-five questions will provide a basic assessment of your organization's emergency preparedness and response program. Simply answer each question with a yes or no. There is no partial credit, and not knowing an answer equates to a negative response.

1. Is your building equipped with life safety systems, such as emergency lighting, a fire suppression system, a fire alarm system, and fire extinguishers?
2. Do you have an up-to-date emergency preparedness and response plan, one that has been reviewed and updated within the past six months?

3. Does your organization provide each employee with a printed copy of its emergency procedures and post the procedures in all public areas of the building(s), such as reception area, conference rooms, and break rooms?
4. Have you conducted a threat assessment analysis to determine the hazards and threats (natural disasters, technological disasters, human-caused disasters) that may impact your organization?
5. Have you instituted a comprehensive mitigation plan to eliminate or lessen the impact of identified threats?
6. Do you have an evacuation plan? If so, do you review and, as necessary, update this plan not less than annually?
7. Are all employees within your organization familiar with your emergency response plan, and do they know what to do for each type of emergency situation?
8. Does your organization provide new employee orientation to emergency procedures and not less than annual emergency response refresher training (evacuation, bomb threats, medical emergencies) for all employees?
9. In the event of an evacuation, have you determined shutdown procedures to be followed, guidelines for when to follow/not follow the procedures, and designated employees to do so?
10. Do you have employees in each location who have current certification to provide first aid and CPR?
11. Have you designated and trained employee emergency response teams (ERTs) to respond to emergency situations?
12. Do your employees know locations near your building(s) where emergency medical care will be provided after disasters?
13. Have criteria for evacuation and sheltering-in-place been established?
14. Do all employees know how they will be notified if it is necessary to evacuate or shelter-in-place?
15. Is there a system for accounting for employees following an evacuation that includes having assembly areas outside the building?
16. Does the company maintain daily records of visitors, in addition to staff, in the building?
17. Have you conducted a full-scale drill of your evacuation plan within the past six months?
18. Have you set up procedures and lines of communication to provide information to employees after a disaster occurs?
19. Have you established a damage assessment process and identified who will conduct the assessment once public safety officials declare it is safe to reenter the building?
20. Do you have a program in place to help employees address the emotional response to a disaster that impacts your organization and/or its employees?
21. Does your organization have an enforced policy that all visitors to your building(s) be escorted at all times?

GUIDANCE APPENDIX

22. Do you have a process that ensures that in an emergency or an evacuation drill special assistance will be provided to employees and visitors who have mobility, sight, hearing, or other special needs and have indicated that they require assistance?
23. Have employees been trained to recognize suspicious mail and packages, and do they know what steps to take if one is received?
24. Does your organization encourage employees to prepare their homes and families for disasters?
25. Can you personally name the members of the emergency response team for your area at work?

Give yourself five points for each yes; zero points for each no. How did you do? Although this assessment is not extensive, it provides a good indication of how well your company is doing to prepare to face the next disaster.

125 points	Congratulations to you and your organization—keep up the good work.
100–120 points	Very good. Take another look at the questions to which you responded "no" for areas for enhancement to make your emergency preparedness and response program even better.
55–95 points	Although your organization has made an effort to prepare for disasters, there's room for substantial improvement. Develop a plan to lessen the gap between where you are and where you need to be.
0–50 points	A great deal of work needs to be done—quickly. Get the right people involved, consider getting some outside help, and get started.

Source: Betty A. Kildow

ASSESSING AN ORGANIZATION'S REPUTATION

The following survey can be used to determine how stakeholders perceive your organization's reputation.

Organizational Reputation Scale

The following ten-item scale provides a quick assessment of an organization's reputation. The scale centers on trust, a key element in all evaluations of reputation. The ten items are added together to form the overall reputation score. After the items is a set of directions for scoring the items.

INSTRUCTIONS: Think about YOUR ORGANIZATION'S NAME HERE. The items below concern your impression of the organization. Circle one number for each of the questions. The responses range from 1 = STRONGLY DISAGREE to 5 = STRONGLY AGREE.

1. The organization is basically honest. 1 2 3 4 5
                                                                STRONGLY   STRONGLY
                                                                 DISAGREE   AGREE

2. The organization is concerned with the well-being of its publics. 1 2 3 4 5
      STRONGLY   STRONGLY
      DISAGREE   AGREE

3. I do trust the organization to tell the truth about the incident. 1 2 3 4 5
      STRONGLY   STRONGLY
      DISAGREE   AGREE

4. I would prefer to have NOTHING to do with this organization. 1 2 3 4 5
      STRONGLY   STRONGLY
      DISAGREE   AGREE

5. Under most circumstances, I WOULD NOT be likely to believe what the organization says. 1 2 3 4 5
      STRONGLY   STRONGLY
      DISAGREE   AGREE

6. The organization is basically DISHONEST. 1 2 3 4 5
      STRONGLY   STRONGLY
      DISAGREE   AGREE

7. I do NOT trust the organization to tell the truth about the incident. 1 2 3 4 5
      STRONGLY   STRONGLY
      DISAGREE   AGREE

8. Under most circumstances, I would be likely to believe what the organization says. 1 2 3 4 5
      STRONGLY   STRONGLY
      DISAGREE   AGREE

9. I would buy a product or service from this organization. 1 2 3 4 5
      STRONGLY   STRONGLY
      DISAGREE   AGREE

10. The organization is NOT concerned with the well-being of its publics. 1 2 3 4 5
      STRONGLY   STRONGLY
      DISAGREE   AGREE

Scoring the Survey

Five items must be reverse coded before creating the overall reputation score: items 4, 5, 6, 7, and 10. When you reverse code, a 1 becomes a 5, a 2 becomes a 4, a 3 stays the same, a 4 becomes a 2, and a 5 becomes a 1.

Once items 4, 5, 6, 7, and 10 are reverse coded, add the ten items together for a final score.

Interpreting the Scores

45 and above	Very positive reputation
38 to 44	Positive reputation
28 to 37	Average/neutral reputation (no strong feelings one way or the other)
23 to 27	Negative reputation
22 and below	Very negative reputation

ASSESSING THE CLIMATE FOR WORKPLACE AGGRESSION IN AN ORGANIZATION

Workplace Aggression Tolerance Questionnaire (WATQ)

The Workplace Aggression Tolerance Questionnaire (WATQ) is designed to assess attitudes about a wide array of workplace aggression behaviors and can be used for benchmarking and risk assessment in organizations. For benchmarking, the WATQ can serve as one form of evaluation for workplace violence training. Prior to training, people can be assessed for their tolerance of aggressive workplace behaviors. Sometime after the training, people can be assessed again to determine whether there has been any change in their tolerance of aggressive workplace behaviors. The evaluation centers on the question, "Are people in the organization perceiving workplace aggression as less appropriate after the interventions have been implemented?"

Although organizations do have workplace aggression policies, only a small percent have any formal risk assessment of workplace aggression. The WATQ provides basic workplace aggression risk assessment by identifying whether particular behaviors and/or departments in an organization have a high tolerance. A high tolerance suggests people would be willing to engage in that form of workplace aggression.

Directions: Please read the following situation and then respond to the statements that follow the story.

Imagine that someone you work with has just completed a performance appraisal/review with the manager. The person believes that the manager has been unfairly critical of his or her job performance—there are some negative comments on the appraisal that the person knows are not accurate.

The person explains to the manager that the negative comments are inaccurate. The manager refuses to make any changes. So, this unfairly negative evaluation goes into the person's record and is seen by others in management. The person knows that the comments could prevent a pay raise and/or promotion.

Below are actions the person might take in response to this meeting. For each action, indicate how appropriate or inappropriate the action would be in the workplace. Mark all your answers by circling the number that corresponds to your evaluation.

The responses range from 1 = VERY INAPPROPRIATE to 5 = VERY APPROPRIATE.

	VERY INAPPROPRIATE				VERY APPROPRIATE
1. Fail to return the manager's phone calls.	1	2	3	4	5
2. Say bad things about the manager.	1	2	3	4	5
3. Purposefully work very slowly.	1	2	3	4	5
4. Hit or kick the manager.	1	2	3	4	5
5. Refuse to talk to the manager.	1	2	3	4	5
6. Yell at the manager.	1	2	3	4	5
7. Leave whenever the manager enters the area.	1	2	3	4	5
8. Give the manager dirty looks.	1	2	3	4	5
9. Refuse requests from the manager.	1	2	3	4	5
10. Interrupt the manager when s/he speaks.	1	2	3	4	5
11. Give the manager obscene gestures.	1	2	3	4	5
12. Fail to send the manager information s/he needs.	1	2	3	4	5
13. Spread nasty rumors about the manager.	1	2	3	4	5
14. Show up late to meetings involving the manager.	1	2	3	4	5
15. Deface company property (graffiti, scratches, etc.).	1	2	3	4	5

GUIDANCE APPENDIX

	VERY INAPPROPRIATE	VERY APPROPRIATE

16. Fail to warn the manager about a potential problem. 1 2 3 4 5
17. Say bad things about the manager's opinions. 1 2 3 4 5
18. Delay work to make the manager look bad. 1 2 3 4 5
19. Waste needed resources at work. 1 2 3 4 5
20. Fail to deny false rumors about the manager. 1 2 3 4 5
21. Hide needed resources at work. 1 2 3 4 5
22. Sabotage the workplace (break equipment, disrupt heating, etc.). 1 2 3 4 5
23. Steal from the workplace. 1 2 3 4 5
24. Interrupt when the manager tries to speak. 1 2 3 4 5
25. Insult the manager. 1 2 3 4 5
26. Swear at the manager using obscene language. 1 2 3 4 5
27. Mistreat the manager's friends. 1 2 3 4 5
28. Verbally threaten the manager. 1 2 3 4 5

Types of Aggression Measured by Each Item

Physical-Active-Direct	4 (Items 4, 8, 10, and 11)
Physical-Active-Indirect	5 (Items 15, 19, 21, 22, and 23)
Physical-Passive-Direct	3 (Items 3, 7, and 24)
Physical-Passive-Indirect	2 (Items 14 and 18)
Verbal-Active-Direct	4 (Items 6, 25, 26, and 28)
Verbal-Active-Indirect	4 (Items 2, 13, 17, and 27)
Verbal-Passive-Direct	3 (Items 1, 5, and 9)
Verbal-Passive-Indirect	3 (Items 12, 16, and 20)

PANDEMIC PREPAREDNESS EVALUATION

The following survey can be used to determine how well an organization's pandemic preparedness matches with U.S. governmental guidelines.

Mad Cow, Y2K and SARS have had a profound effect on how we view potential threats. Those of us in the disaster recovery industry have particular insights into the differences of real and perceived threats. The CDC has recommended preparedness guidelines for large corporations.

The following section has adopted categories from the CDC preparedness checklist. Information you provide will help supply longitudinal data for realistic preparation. Please CHECK ONE box on the right for EACH of the questions.

	Not Started	In Progress	Completed
BUSINESS			
Identify a pandemic coordinator and/or team with defined roles and responsibilities for preparedness and response planning. The planning process should include input from labor representatives.	☐	☐	☐
Identify essential employees and other critical inputs (e.g., raw materials, suppliers, subcontractors, services/ products, and logistics) required to maintain business operations by location and function during a pandemic.	☐	☐	☐
Train and prepare ancillary workforce (e.g., contractors, employees in other job titles/descriptions, retirees).	☐	☐	☐
Develop and plan for scenarios likely to result in an increase or decrease in demand for your products and/or services during a pandemic (e.g., effect of restriction on mass gatherings, need for hygiene supplies).	☐	☐	☐
Determine potential impact of a pandemic on company business financials using scenarios that affect different product lines and/or production sites.	☐	☐	☐
Determine potential impact of a pandemic pandemic on business-related domestic and domestic and international travel (e.g., quarantines, border closures).	☐	☐	☐

	Not Started	In Progress	Completed
Find up-to-date, reliable pandemic information from community public health, emergency management, and other sources, and make sustainable links.	☐	☐	☐
Establish an emergency communications plan and revise periodically. This plan includes identification of key contacts (with backups), chain of communications (including suppliers and customers), and processes for tracking and communicating business and employee status.	☐	☐	☐
Implement an exercise/drill to test your plan, and revise periodically.	☐	☐	☐

EMPLOYEES AND CUSTOMERS

	Not Started	In Progress	Completed
Forecast and allow for employee absences during a pandemic due to factors such as personal illness, family member illness, community containment measures and quarantines, school and/or business closures, and public transportation closures.	☐	☐	☐
Implement guidelines to modify the frequency and type of face-to-face contact (e.g., handshaking. seating in meetings, office layout, shared workstations) among employees and between employees and customers (refer to CDC recommendations).	☐	☐	☐
Encourage and track annual influenza vaccination for employees.	☐	☐	☐
Evaluate employee access to and availability of health care services during a pandemic, and improve services as needed.	☐	☐	☐
Evaluate employee access to and availability of mental health and social services during a pandemic, including corporate, community, and faith-based resources, and improve services as needed.	☐	☐	☐

	Not Started	In Progress	Completed
Identify employees and key customers with special needs, and incorporate the requirements of such persons into your preparedness plan.	☐	☐	☐

POLICIES TO BE IMPLEMENTED DURING A PANDEMIC

	Not Started	In Progress	Completed
Establish policies for employee compensation and sick leave absences unique to a pandemic (e.g., nonpunitive, liberal leave), including policies on when a previously ill person is no longer infectious and can return to work after illness.	☐	☐	☐
Establish policies for flexible worksite (e.g., telecommuting) and flexible work hours (e.g., staggered shifts).	☐	☐	☐
Establish policies for preventing influenza spread at the work site (e.g., promoting respiratory hygiene/coughing etiquette, prompt exclusion of people with influenza symptoms).	☐	☐	☐
Establish policies for employees who have been exposed to pandemic influenza, are suspected to be ill, or become ill at the work site (e.g., infection control response, immediate mandatory sick leave).	☐	☐	☐
Establish policies for restricting travel to affected geographic areas (consider both domestic and international sites), evacuating employees working in or near all affected areas when an outbreak begins, and guidance			

GUIDANCE APPENDIX 389

	Not Started	In Progress	Completed
for employees returning from affected areas (refer to CDC travel recommendations).	☐	☐	☐
Set up authorities, triggers, and procedures for activating and terminating the company's response plan, altering business operations (e.g., shutting down operations in affected areas), and transferring business knowledge to key employees.	☐	☐	☐

PROTECTION FOR EMPLOYEES AND CUSTOMERS DURING A PANDEMIC:

	Not Started	In Progress	Completed
Provide sufficient and accessible infection control supplies (e.g., hand hygiene products, tissues and receptacles for their disposal) in all business locations.	☐	☐	☐
Enhance communications and information technology infrastructures as needed to support employee telecommuting and remote customer access.	☐	☐	☐
Ensure availability of medical consultation and advice for emergency response.	☐	☐	☐

COMMUNICATION AND EDUCATION FOR EMPLOYEES

	Not Started	In Progress	Completed
Develop and disseminate programs and materials covering pandemic fundamentals (e.g., signs and symptoms of influenza, modes of transmission), personal and family protection and response strategies (e.g., hand hygiene, coughing/sneezing etiquette, contingency plans).	☐	☐	☐
Anticipate employee fear and anxiety, rumors, and misinformation, and plan communications accordingly.	☐	☐	☐
Ensure that communications are culturally and linguistically appropriate.	☐	☐	☐
Disseminate information to employees about your pandemic preparedness and response plan.	☐	☐	☐

	Not Started	In Progress	Completed
Provide information for the at-home care of ill employees and family members.	☐	☐	☐
Develop platforms (e.g., hotlines, dedicated web sites) for communicating pandemic status and actions to employees, vendors, suppliers, and customers inside and outside the work site in a consistent and timely way, including redundancies in the emergency contact system.	☐	☐	☐
Identify community sources for timely and accurate pandemic information (domestic and international) and resources for obtaining countermeasures (e.g., vaccines and antivirals).	☐	☐	☐

EXTERNAL ORGANIZATIONS AND COMMUNITY HELP

	Not Started	In Progress	Completed
Collaborate with insurers, health plans, and major local health care facilities to share YOUR pandemic plans and understand their capabilities and plans.	☐	☐	☐
Collaborate with federal, state, and local public health agencies and/or emergency responders to participate in THEIR planning processes, share your pandemic plans, and understand their capabilities and plans.	☐	☐	☐
Communicate with local and/or state public health agencies and/or emergency responders about the assets and/or services YOUR business could contribute to the community.	☐	☐	☐
Share best practices with other businesses in your communities, chambers of commerce, and associations to improve community response efforts.	☐	☐	☐

Scoring Responses:

0 points for Not Started
1 point for In Progress
2 points for Completed

0 to 33	Not prepared. Your organization needs to focus on pandemic preparation.
33 to 62	Strong progress. Review to see where you scores are still below a 2 to continue preparation.
66	Top score. Your organization is fully prepared according to the CDC.

INFORMATION SECURITY POLICY CONTENT AREAS

Here is a list of the basic points covered in a typical information security policy.

- **Title.** A concise title provides users with a clear understanding of the policy's contents.
- **Version.** A consistent numbering system helps users (and the document's author) determine at a glance whether a policy has been updated. If most of the policies in the manual have a version number of 1.0, then a policy with a version of 1.1 has undergone a minor revision, whereas 2.0 represents a major revision.
- **Last review date.** This is the date that your organization's subject matter experts most recently reviewed the policy, whether they suggested any changes or not.
- **Publication date.** This is the date that the policy was most recently published, in either hardcopy or electronic format.
- **Expiration date.** On this specified date the policy will expire.
- **Classification.** A categorization of the policy document within your organization's overall document classification scheme is helpful.
- **Owner.** This party within your organization is responsible for reviewing, revising, and enforcing each policy.
- **ISO reference.** This is the section of the ISO standard (or any other recognized standard) on which the policy is based.
- **Associated standards.** A list of all standards and guidelines that reference the policy is helpful.
- **Scope.** This is to whom the policy applies. Most likely, all employees, contractors, and business partners will be expected to follow all policies. Still, certain policies could apply exclusively to employees or to contractors.
- **Reporting violations.** There should be a method—or several methods—that your organization's members can use to report a suspected violation of policy.

Noncompliance/penalties. This is a clear statement that explains the repercussions of violating the policy.

Terms and definitions. These are any technical, unique, or ambiguous terms used in the policy.

Body. The topic covered determines the length of the real "meat" of the policy.

Source: Michael Seese.

REQUESTING AN EXCEPTION TO AN INFORMATION SECURITY POLICY

Persons requesting exceptions to policy should complete and submit a form that specifies the following:

- The specific policy that their request violates
- An explanation of what they are requesting, including how it violates policy
- A thorough description of their business process, including details of how the policy interferes with their operations
- A business justification
- The impact to their operations if the exception were not granted
- A list of the alternatives considered and why they were determined to be unacceptable
- A risk assessment, including:
  - The type of data being put at risk
  - Any exposures created if the exception were granted
  - The likelihood that any given vulnerability could be exploited
  - A description of what an attacker could do if he or she exploited the vulnerability created by the exception
  - The impacts to the organization if the vulnerability were exploited
  - Any controls that will be implemented to mitigate the risks

The request is then reviewed. A denial of a request should contain an explanation for the denial. If a request is granted, the expiration date for the exception should be specified.

Source: Michael Seese.

COMMON SIGNS OF SOCIAL ENGINEERING

The following activities should be warnings to workers that a social engineer may be at work:

- Refusal to give *back* adequate information or an explanation
- Expressing a sense of urgency
- Providing names with no context

- Intimidating behavior
- Outright requesting of sensitive information

Source: Michael Seese.

STEPS FOR CONSTRUCTING THE CRISIS SENSING MECHANISM

Step 1: Identify Existing Crisis Sensing Activities

Audit your organization to determine what units are already sensing the environment. You want to avoid re-creating the wheel, so use the existing sensing activities as a foundation for your crisis sensing mechanism. Be sure to review risk assessment, issues management, and stakeholder relationship activities. Ask each organizational unit what it currently does to identify problems/opportunities internally or externally—its scanning activities.

Step 2: Assess the Existing Crisis Sensing Activities

You may need to develop new crisis sensing activities if the existing ones do not form a complete system. If key risk sources are being overlooked, you need to expand the crisis sensing activities. For example, if no effort is made to scan relevant activist groups, add that as a source. Make sure all possible sources you can think of are being scanned.

Step 3: Assess Information Gathering Techniques

Review how the information is being gathered. Pay particular attention to any coding systems used. A common weakness in information collection is a coding system that is too general and misses important details contained in the information. Consider an example of a retail store that codes news stories and blog entries about the organization. A general coding system would simply count the total number of positive and negative comments about the store. Such coding provides a global assessment of the reputation—is the reputation favorable or unfavorable? No insight is given into why the media image is favorable or unfavorable. A specific coding system might include categories such as sales staff, customer service, selection, merchandise quality, value/pricing, store appearance, and parking. The retail store would have separate evaluations for the seven categories. Store managers would know which exact areas of the store's reputation were strong and which needed improvement.

Step 4: Develop Procedures for Funneling Information

A crisis manager or team can neither process information it does not receive nor attend to prodromes it never knew about. Procedures must be developed for routing the information to the crisis manager in a timely fashion. Various parts of the organization will be responsible for different pieces of information. Some organizational units involved in scanning include security, operations and manufacturing, marketing and sales, finance, human resources, legal, customer communications and satisfaction, environmental and safety engineering, public relations/public affairs, engineering, shipping and distribution, and quality assurance. The different units must route their information to the crisis manager, who evaluates it for prodromes. Taking a cue from integrated marketing communication, the organization must share vital incoming information. The crisis manager/team becomes the center of a large crisis sensing mechanism. The crisis manager/team must be treated as a functioning unit that is integrated within the flow of organizational activities and information flow.

Step 5: Establish Evaluative Criteria

Each crisis manager/team must decide how to translate these general criteria into organizational-specific criteria it can use. The crisis manager/team must decide which criteria to use, create any additional criteria that may be needed, and very clearly define the evaluative criteria. Without clear definitions, the criteria cannot be applied consistently by the crisis manager/team. For instance, what is the difference between risk impact ratings of 3, 6, and 9? The criteria must be used consistently if the crisis team is to compare and rank order the prodromes. It takes time to develop precise criteria, but the rewards are well worth the work.

Step 6: Test the Crisis Sensing Mechanism

Developing a crisis sensing mechanism does not mean it works. The system must be tested. Tests are as simple as placing selected information into the various sensing activities and seeing whether that information reaches the crisis manager and how long it takes. The crisis sensing mechanism is a complex communication/information processing system that requires regular checking and refinement in order to maintain and to improve its efficiency.

HOW TO SHELTER-IN-PLACE

The basic steps to sheltering-in-place for a company include the following:

1. Close the business and post signs warning those outside the building.
2. Inform all those in the building, including visitors and customers, of the situation and the reason to shelter-in-place.
3. Shut and lock all windows and doors.
4. Have people assemble in designated rooms and bring the disaster supplies to these locations. The disaster supplies include nonperishable food, water, battery-powered radios, first aid supplies, batteries, flashlights, phones/phone access, garbage bags, duct tape, and plastic. The National Institute for Occupational Safety and Health (NIOSH) recommends a

GUIDANCE APPENDIX

variety of respirators that can be used during an evacuation to protect people from smoke inhalation.
5. Have people call their emergency contacts to inform them of what is happening.
6. Turn off all heating, ventilation, and/or air conditioning. An employee who has been trained in the proper way to shut down these mechanical systems should do this.
7. Seal the windows and vents with sheets of plastic and duct tape. Ideally the pieces of plastic are precut to fit the openings and employees are preassigned and trained in how to seal the windows and vents.
8. Seal around doors with plastic and duct tape.
9. Turn on the radio or television and listen for further instructions.
10. Once the all clear is given, remove plastic, restart the mechanical system, and go outside of the building until the old, possibly contaminated air in the system has been replaced with fresh, clean air.

Source: American Red Cross, "Shelter-in-Place in an Emergency," online at http://www.redcross.org/services/prepare/0,1082,0_258_,00.html (accessed 5 May 2006).

COMPONENTS OF AN ETHICS AND COMPLIANCE PROGRAM

The U.S. Department of Commerce has identified nine components that are needed for an effective ethics and compliance program:

1. A set of standards and procedures that guides employee conduct and helps stakeholders understand what to expect from a company's employees.
2. A system that holds employees accountable for living up to the program's requirements
3. Clear communication of the program and policies to employees
4. Active monitoring of employee conduct
5. Encouraging employees to seek advice when they have ethics questions
6. Due diligence in hiring employees
7. Encouraging employees to follow the policies and guidelines
8. Management taking appropriate actions when the policies and guidelines are violated
9. Regularly evaluating the program's effectiveness

COMPONENTS OF ANTICORRUPTION POLICIES

Antcorruption policies:

- Must forbid any bribes
- Must specify that intermediaries are not to use any of the money for bribes

- May allow facilitation payments under certain conditions, including payments that are customary and of low value, are approved by senior management, do not violate any laws, are reported immediately, and are recorded in company records
- Must set limits for gifts and entertainment
- Must specify no kickbacks and use accounting and monitoring procedures to ensure policy is not violated
- Must ensure that conflict of interest is to be avoided at all times

WHAT CREATES OUTRAGE IN RISK COMMUNICATION

Outrage can be created or reduced by at least twelve different factors:

1. *Voluntary versus coercive.* Risk bearers are less likely to experience outrage when they expose themselves to a risk rather than having the risk thrust upon them.
2. *Natural versus industrial.* Risk bearers have less of an emotional reaction to natural risks, such as a tornado, than to risks created by other people, such as hazardous waste.
3. *Familiar versus unfamiliar.* Risk bearers perceive familiar things as less risky than unfamiliar things.
4. *Memorable versus not memorable.* Risk bearers rate the risk higher when it is linked to some memorable event than when it is a little known event. Bhopal is memorable, increasing the perception of risk for similar pesticide production facilities.
5. *Dreaded versus not dreaded.* Risk bearers rate risk higher when that risk is linked to a dreaded outcome such as cancer.
6. *Chronics versus catastrophic.* Risk bearers rate risks they face every day, such as driving, less seriously than catastrophic events, such as airplane crashes.
7. *Knowable versus unknowable.* Risk bearers fear the unknown more than the known. A new risk should generate more outrage than a well-known risk.
8. *Control versus not in control.* Risk bearers feel more secure when they have the ability to control or regulate the risk. Again, the idea of driving versus flying in a plane fits well. People can control the car when they are driving but not the plane when a pilot is doing the flying.
9. *Fair versus unfair.* Risk bearers experience greater outrage from a risk if they perceive their burden from the risk carries a greater price than that of other people.
10. *Morally irrelevant versus morally relevant.* Risk bearers experience greater outrage if the risk is linked to immoral actions such as cutting corners to make profits than if the risk is related to some moral good such as curing a disease.

11. *Trustworthy versus untrustworthy.* Risk bearers experience less outrage if they trust the experts involved in the risk communication effort than if they lack trust in those experts.
12. *Responsive versus unresponsive.* Risk bearers feel greater outrage when the company responsible for the risk is unresponsive to their concerns.

Source: Peter M. Sandman, "Risk Communication: Facing Public Outrage," online at http://www.du.edu/~scbeckma/EPM4700/outrage.htm (accessed 27 Feb. 2007).

SPOKESPERSON TASKS FOR CRISIS MANAGEMENT

A spokesperson has four main tasks: (1) appear pleasant on camera, (2) answer questions effectively, (3) present crisis information clearly, and (4) handle difficult questions.

1. To appear trustworthy and in control, a spokesperson must maintain consistent eye contact (eye contact at least 60 percent of the time), use hand gestures to emphasize points, avoid a monotone by varying vocal qualities, have an expressive face, and avoid too many vocal fillers such as "uhs" or "ums."
2. An effective answer to a question is one that answers the question that is asked.
3. A spokesperson's answers must be clear and easy for the reporters to understand. This means avoiding jargon and technical terms.
4. A spokesperson needs to recognize the tough question and respond strategically. Any practice sessions should include some tricky questions.

BASIC GUIDES FOR USING PASSWORDS

1. Do not use words found in the dictionary, even those from another language.
2. Different systems should have different passwords, not just one.
3. Use upper- and lower-case letters.
4. Use a combination of numbers, special characters, and letters.
5. Do not use passwords derived from personal information such as birthdays.
6. Use a mnemonic to help you remember a password.

Source: Mindi McDowell, Jason Rafail, and Shawn Hernan, "Choosing and Protecting Passwords," 2004, online at http://www.us-cert.gov/cas/tips/ST04-002.html (accessed 5 March 2007).

SPYWARE SYMPTOMS

1. Endless pop-ups
2. Redirection to web sites other than the ones you selected

3. Appearance of new toolbars on your web browser
4. Appearance of new icons along the bottom of your computer screen
5. The home page of your browser changes
6. Different search engine appears when you click your search icon
7. Certain keys do not work in your browser
8. Appearance of random Window error messages
9. Noticeable drop in your computer's processing speed

Source: Mindi McDowell and Matt Lytle, "Recognizing and Avoiding Spyware," 2004, online at http://www.us-cert.gov/cas/tips/ST04-016.html (accessed 5 March 2007).

INFORMATION SECURITY BASIC GUIDANCE

The U.S. Department of Homeland Security has issued a set of guidelines for businesses. The recommendations cover employees and management.

Employees

1. Require the use of strong passwords.
2. Require that passwords be changed every forty-five to ninety days.
3. Never give your user name or password to anyone.
4. Never open e-mail attachments from people you do not know.
5. Get permission from the company's IT department before installing any of your own software or hardware.
6. Keep electronic and physical backups of the most important work.
7. Report suspicious activities or problems to the IT department.

Management/IT Department

1. Monitor, log, and evaluate any attempts to enter your network or system.
2. Create and reinforce clear policies for how employees are to use the information technologies.
3. Download patches regularly.
4. Create a layered defense that includes technical, organizational, and operational controls.
5. Use technical defenses such as firewalls, Internet content filters, and intrusion detection systems.
6. Update antivirus software regularly.
7. Change the manufacturer's default passwords.

SAMPLE CRISIS MANAGEMENT PLAN ELEMENTS

Generic Components of a Crisis Management Plan

1. Cover page
2. Introduction

3. Acknowledgments
4. Rehearsal dates
5. Crisis management team list
6. CMT contact sheet
7. Secondary contact sheet
8. Crisis risk assessment
9. CMT strategy worksheet
10. Stakeholder contact worksheet
11. Business continuity plan reference
12. Crisis control center
13. Postcrisis evaluation tools

Incident Report

Date and Time Incident Was Reported: Initial Report____

Individual Reporting the Incident: Follow-up ____

How to Contact the Reporting Individual:

Description of the Incident:

Exact Location of the Incident:

List the Personnel and Units Responding to the Incident:

Describe What Is Being Done to Address the Incident and by Whom:

List Any Follow-up Action That Is Needed:

Detail the Damage Inflicted by the Crisis:

Date and Time the Crisis Team Was Notified:

What Other Units (i.e., fire department security, EMTs, etc.) Were Contacted and When?

Crisis Management Team Strategy Worksheet

Stakeholder(s) Targeted by the Message:

- Consider the status of the current relationship with each stakeholder.
- Review the primary organizational performance expectations of each stakeholder.
- LIST STAKEHOLDER(S) HERE:

Goal of the Message:

Attach a copy of the actual message to this sheet.

Stakeholder Contact Worksheet

Date: Time:

Organizational Member Handling the Inquiry:

Channel Used to Contact the Organization:

Stakeholder's Classification (i.e., media, stockholder, community leader, etc.):

   Inquiring Person's Name and Title:

   Inquiring Person's Organizational Affiliation:

   How to Reach the Inquiring Person:

Question/Inquiry:

Response:

Any Follow-up Needed: If So, by Whom:

When:

INITIATING EMERGENCY NOTIFICATION COMMUNICATION

Notification during disruptive events such as crises, disasters, and emergencies is critical to safety. This entry reviews key aspects of an effective notification system: initiating emergency notification, reaching the right people with the right message at the right time, delivering and receiving messages, communication efficiency, and ease of use.

Initiating Emergency Notification

The first task facing the disaster manager is to initiate or accelerate the information flow. This poses a formidable challenge if the communication center or emergency operations center has been damaged or communication lines are down. Automated notification simplifies the task by reducing the number of messages that must be initiated; it allows the coordinator to issue a single message to an entire list of people—whether that group consists of employees, customers, parents of schoolchildren, first responders, or reporters. The chances of initiating and delivering messages successfully are much greater if the initiator has to send only five messages instead of five thousand. There should be multiple ways to initiate a message. At the very least, disaster coordinators need to be able to access and use their notification system through the Internet and by telephone without being at a particular computer or telephone to initiate a message. The system should also be able to deliver messages from multiple initiators. Managers need to delegate authority to several associates so that emergency notification does not fail should one person be incapacitated or unavailable.

Reaching the Right People with the Right Message at the Right Time

Nearly as important as reaching all the critical audiences is sending messages only to those who need to receive them. Sending extraneous messages in times of emergency can have serious unintended consequences, including chaos and crowding at the disaster site, an influx of unwanted phone calls, and even mass panic. To channel messages correctly, your messaging tools need to have the capacity for unlimited groups and subgroups of target audiences. Crisis communicators may have to poll employees from the seventh floor to make sure they were all safely evacuated from a fire; instruct employees in the network services division to report to a backup site the next morning; or recall certain ambulance crews but not others. Relationships between list members must be identifiable to allow crisis communicators to make selections based on these relationships—for example, contacting "all the senior managers and executives in the Chicago office."

Delivering and Receiving Communication

Communication flows constantly in two directions during a crisis: incoming and outgoing messages, information, and meta-messages. Usually, just sending out messages is not enough; the crisis manager must learn who has been contacted successfully and, sometimes, what the responses are. An automated notification system can be a great mechanism for receiving and reporting such responses. To facilitate two-way communication, the notification system should be able to receive an active response, such as a keypad entry, to confirm successful delivery of a message. The notification system should also be capable of surveying or polling recipients. For example, if first responders are being notified, the coordinator's message might ask them to press 1 if they are already at the disaster site, press 2 if they are on their way there, and press 3 if they are unavailable to respond to the emergency. Reports of all message delivery attempts, confirmations, and polling results should be easily available by Internet and fax, and summary reports need to communicate the overall picture quickly. Detailed reports show where individual follow-up is needed.

Communication Efficiency

The many automated notification services on the market today vary widely in terms of capacity, data security, and cost. With those major considerations always in mind, the greatest communications utility for a business obviously comes from an automated system having features and functionality that best match its communication needs in a crisis. Advanced features, such as conference calling and geographic targeting (which allows messages to be automatically delivered to all residents in a specific geographic area), are important to consider, because they maximize communication options for a crisis manager. Feature-rich systems are more likely to overcome common communication obstacles, such as phone line jams or loss of Internet connectivity. This collection has mentioned the value of having preplanned scenarios and premade statements to ensure expediency and message quality in the middle of a disaster. Many automated systems offer message libraries, in which created messages can be stored, as well as the functionality to create crisis scenarios connecting specific prepared messages with the exact group(s) they will be sent to when an incident occurs.

Ease of Use

An automated notification system should be easy to use without extensive training. Just as crisis communicators should not be spending hours making phone calls, they should not be trying to remember complicated command sequences or searching for user manuals. Stress and anxiety during a crisis make communication difficult, and an automated notification system needs to reduce the stress, not add to it. Ideally, all functional areas of a crisis management team—operations, security, legal, communications (public relations), management, and finance—will be able to utilize an automated system for their own unique communication requirements. Crisis management teams have other critical tasks besides sending and receiving messages. But, as we have seen, one-on-one communications are so stressful and time-consuming under the best of circumstances that they can delay a coordinated response and even lead to loss of property or life. A quality automated notification/mass notification system greatly reduces time spent on communications and frees emergency personnel to deal with crisis mitigation, response, and recovery work. Being able to initiate a few messages and receive hundreds or thousands of responses formatted in a readable report just a few minutes later is an enormous time-saver. An automated notification system is an ideal way to fill the information void quickly, while carefully delivering the right message to the right audience. The following table summarizes this information.

After a message has been initiated, it must be delivered to everyone on the delivery list. There are two important issues: whether the messages will arrive and when they will arrive. To maximize the likelihood that messages will be delivered in a timely manner, your emergency notification system should be able to send messages to all types of contact devices—phone, cell, fax, computer (e-mail and instant messaging), pager, PDA (including BlackBerry)—and in as many formats as possible (voice, text, short message service [SMS]). The notification system should permit multiple contact paths for each person on the list, and allow

Communication Challenge	Automated System
Heavy demand for information (volume)	• Able to send hundreds, or even thousands, of messages in seconds
Communication systems or contact paths are unavailable	• Utilizes multiple communication networks and paths (not all channels will be damaged)
	• Built with redundant, geo-dispersed systems that can survive regional failures
Severe time constraints; little time for analysis or investigation	• Leverages speed and delivery reliability to both disseminate information quickly and give decision makers more time to make choices
Crisis manager's normal location is inaccessible	• Notification system and contact data are accessible from anywhere
	• Administrator can delegate authority to an alternate whose office is intact
Personnel are scattered	• Multiple communication pathways maximize chances of finding and reaching your audience
One-on-one communication takes too long	• Thousands of calls can be placed in a few seconds/minutes; some systems offer conference calling and geographic targeting
Collecting information from audience is difficult	• Automated polling and real-time summaries of polling data
Inconsistent, inaccurate, or incomplete information is issued	• Identical recorded or text messages allow accurate and consistent information flow; speed and volume of automation allow rumors to be quashed quickly

Source: Robert Chandler.

a different order for each list member. (For example, member 1 might designate cell phone first, then e-mail, then fax; member 2 might choose work phone first, then home phone, then pager.) The notification system must be able to make unlimited attempts to contact each person on the list, until there is confirmation of receipt of the message or the message update.

RESOURCE APPENDIX

Books	406
Information and Physical Security	406
Security-Related Topics	407
Magazine and Journal Articles	407
Magazines	408
Newsletters	409
Web Sites	411
Business Continuity and Disaster Recovery	411
Corruption	413
Counterintelligence	413
Countersurveillance	414
Drug Testing	414
Ecoterrorism	415
Employee Background Checks	415
Ethics and Compliance	415
Food Security	415
General Security	415
Information Security	416
Insider Threat	419
Overseas Travel	420
Pandemics (Avian Influenza)	420
Shelter-in-Place	421
Social Engineering	421
Terrorism	421

Vendors	422
Data Backup	422
Internet and E-mail Monitoring	422
Mass Notification	423
Simulations for Training and Exercises	423
Security	423

BOOKS

Information and Physical Security

Bailes, Alyson J. K., and Frommelt, I., ed., *Business and Security: Public-Private Sector Relationships in a New Security Environment*. Stockholm, Sweden: SIPRI, 2004.

Borodzicz, Edward, *Risk, Crisis and Security Management*. Indianapolis, IN: Wiley Publishing, 2005.

Broder, James F., *Risk Analysis and the Security Survey, Third Edition*. Burlington, MA: Butterworth-Heinemann, 2006.

Calder, Alan, *A Business Guide to Information Security*. London, UK: Kogan, 2005.

Cumming, Neil, *Security: A Guide to Security System Design and Equipment Selection and Installation, Second Edition*. Burlington, MA: Butterworth-Heinemann, 1994.

Dalton, Dennis, *Security Management: Business Strategies for Success*. Burlington, MA: Butterworth-Heinemann, 2006.

Ghosh, Anup K., *Security & Privacy for E-Business*. New York: Wiley Publishing, 2001.

Gouin, Brian, *Security Design Consulting: The Business of Security System Design*. Burlington, MA: Butterworth-Heinemann, 2007.

Kairab, Sudhanshu, *A Practical Guide to Security Assessments*. Boca Raton, FL: Auerbach, 2004.

Khairallah, Michael, *Physical Security Systems Handbook: The Design and Implementation of Electronic Security Systems*. Burlington, MA: Butterworth-Heinemann, 2005.

Kildow, Betty A., *Frontdesk Security and Safety: An On-the-Job Guide to Handling Emergencies, Threats, and Unexpected Situations*. New York: AMACOM, 2004.

Kovacich, Gerald L., *The Information Systems Security Officer's Guide: Establishing and Managing an Information Protection Program, Second Edition*. Burlington, MA: Butterworth-Heinemann, 2003.

McCarty, Mary Pat, and Campbell, Stuart, *Security Transformation: Digital Defense Strategies to Protect Your Company's Reputation and Market Share*. New York: McGraw-Hill, 2001.

McCrie, Robert, *Security Operations Management, Second Edition*. Burlington, MA: Butterworth-Heinemann, 2006.

Peltier, Thomas R., *Information Security Policies and Procedures: A Practitioner's Reference, Second Edition*. Boca Raton, FL: Auerbach, 2004.

Peltier, Thomas R., *Risk Management for Business and Security*. Boca Raton, FL: Auerbach, 2008.

Purpura, Philip, *The Security Handbook, Second Edition*. Burlington, MA: Butterworth-Heinemann, 2002.

Wylder, John, *Strategic Information Security*. Boca Raton, FL: Auerbach, 2003.

Security-Related Topics

Apgar, David, *Risk Intelligence: Learning to Manage What We Don't Know*. Cambridge, MA: Harvard Business School Publishing, 2006.

Coombs, W. Timothy, *Code Red in the Boardroom: Crisis Management as Organizational DNA*. Westport, CT: Praeger, 2006.

Coombs, W. Timothy, *Ongoing Crisis Communication: Planning, Managing, and Responding, Second Edition*. Thousand Oaks, CA: Sage, 2007.

Crouhy, Michel, Galai, Dan, and Mark, Robert, *The Essentials of Risk Management*. New York: McGraw-Hill, 2005.

Entis, Phyllis, *Food Safety: Old Habits and New Perspectives*. Washington, DC: ASM Press, 2007.

Fleisher, Craig S., and Blenkhorn, David L., eds., *Controversies in Competitive Intelligence: The Enduring Issues*. Westport, CT: Praeger, 2003.

Hoffman, Bruce, *Inside Terrorism*. New York: Columbia University Press, 2006.

Ledlow, Gerald R., Johnson, James A., and Jones, Walter J., *Community Preparedness and Response to Terrorism:* Vol I, *The Terrorist Threat and Community Response*. Westport, CT: Praeger, 2005.

McFall, Kathleen, *Ecoterrorism: The Next American Revolution?* Portola, CA: High Sierra Books, 2005.

McGonagle, John J., and Vella, Carolyn M., *Protecting Your Company Against Competitive Intelligence*. Westport, CT: Quorum Books, 1998.

Mitnick, Kevin, and Simon, William, *The Art of Deception: Controlling the Human Element of Security*. Indianapolis, IN: Wiley Publishing, 2002.

Odiorne, George S., *Management and the Activity Trap*. New York: Harper & Row, 1974.

Sikich, Geary W., *Integrated Business Continuity: Maintaining Resilience in Times of Uncertainty*. Tulsa, OK: PennWell Publishing, 2003.

White, Jonathan R., *Terrorism and Homeland Security, Fifth Edition*. Mason, OH: Wadsworth Publishing, 2005.

MAGAZINE AND JOURNAL ARTICLES

Chandler, Robert C., "Managing Ethical and Regulatory Compliance Contingencies: Planning and Training Guidelines," *Contingency Planning and Management 2000 Proceedings*. Flemington, NJ: Witter Publishing, 2000, pp. 6–7.

Chandler, Robert C., and Wallace, J. D., "Brief Results of the Pepperdine University Ethical Misconduct Disaster Recovery Preparedness Survey," *Disaster Recovery Journal*, 14, 3 (2001), 21–22.

Sikich, Geary W., "What Is There to Know About a Crisis," *John Liner Review*, 14, 4 (2001).

Sikich, Geary W., "Hurricane Katrina: Nature's 'Dirty Bomb' Incident? *Continuity Central*, Lead Article, 30 Sept. 2005.

Sikich, Geary W., "Business Continuity as a Strategic Initiative," *The Business Continuity Journal*, 1, 1 (2006).

Sikich, Geary W., and Stagl, John M., "Are We Missing the Point of Pandemic Planning?" *Continuity Central*, Dec. 2005.

MAGAZINES

Business Continuity Insights

Presents articles written by experts in the field on a variety of business continuity and business security topics. It is a monthly publication and the subscription is free. You can access a subscription at http://www.continuityinsights.com/Magazine/Magazine_Subscribe.html. The magazine's web site (http://www.continuityinsights.com/) offers a searchable archive of past articles.

CIO

Has a target audience of chief information officers and high-level IT executives. People can apply for a free subscription; those who do not qualify must pay for it. An online subscription form must be completed to determine eligibility for a free subscription at http://www.cio.com/subscription-services.

CPM

This publication has a subscription fee. According to its web site, "Over the past year, *CPM* has closely monitored the business continuity profession and its need to integrate with other disciplines. Earlier this year, we introduced the concept of *operations assurance*, which advocates a closer relationship with business continuity, security, emergency management, risk management, and other critical activities, all under the umbrella of good corporate governance. *CPM* 's readers told us this is an important and strategic direction for their companies. To further develop this important concept in the evolution of business and government continuity of operations, we will provide our readers with the most useful, thought-provoking information and analysis, in a more focused and timely medium." See http://www.contingencyplanning.com/magazine/index.aspx.

CSO

According to its own description, *CSO* "has been serving the information needs of top security executives since 2002. Our loyal audience regularly turns to *CSO* for up-to-date security best practices and strategic management issues through *CSO* magazine, CSOonline.com and CSO Executive Programs." People can apply online for a free subscription to *CSO* or to its newsletter at http://www.csoonline.com/read/.

Disaster Recovery Journal

Provides articles written by experts on a variety of disaster recovery, business security, and business continuity topics. Subscription is free for online delivery and has a small fee for print copies. Subscription information can be accessed at https://www.drj.com/account/index.php.

Security Dealer

Primarily for those involved in selling security systems. It does provide insight into new and developing trends in security services. See http://www.securityinfowatch.com/cover/security-dealer/1SIW.

Security Management

An award-winning magazine published by ASIS International, the preeminent international organization for security professionals. There are reduced fees for ASIS International members. The magazine's web site includes a searchable archive. The articles are written is very accessible manner. Can be found at http://www.securitymanagement.com/.

Security, Technology & Design

Primarily written for security executives. Its content includes both physical and information security and covers access control, video technology, and network-centric applications of IP-based video and ID/data management. See http://www.securityinfowatch.com/cover/security-technology-design/2SIW.

NEWSLETTERS

@Risk: The Consensus Security Alert

"The Critical Vulnerability Analysis and the Security Alert Consensus have merged to become @RISK: The Consensus Security Alert. Delivered every Monday morning, @RISK first summarizes the three to eight vulnerabilities that matter most, tells what damage they do and how to protect yourself from them, and then adds a unique feature: a summary of the actions 15 giant organizations have taken to protect their users. @RISK adds to the critical vulnerability list a complete catalog of all the new security vulnerabilities discovered during the past week. Thus in one bulletin, you get the critical ones, what others are doing to protect themselves, plus a complete list of the full spectrum of newly discovered vulnerabilities. This is also the subscription list that receives SANS Flash Alerts when they come out two or three times a year." Can be found at http://www.sans.org/newsletters/#newsbites.

CIO Newsletters

CIO has a range of newsletters available. *Advice and Opinion* provides the week's top advice and opinion postings. *CIO Careers* offers advice for careers as well as job listings. *CIO ERP* is a CIO's monthly guide for enterprise resource planning. *CIO Enterprise* provides enterprise-level technology information, news, and tools. *CIO Information Security* provides security news and information the CIO needs to know. *CIO Insider* is a guide to the latest from CIO.com covering management, technology, and your career. *CIO Leader* give updates, insights, and advice from CIO.com on your career, as well as practical tips for effective leadership and management. *CIO News Watch* reviews the week's top news stories. *CIO Open Source* is a monthly peek at what's happening in the open source realm. *CIO Research Update* gives highlights of CIO's most recent IT research. *CIO SOA* is a resource for service-oriented and enterprise architecture. *CIO Tech Poll* reports the results of CIO quarterly surveys covering IT's overall health as well as spending and trends. *CIO Whitepapers* provides information about new and upcoming white papers. *CIO Wireless* reports information on emerging wireless technologies and infrastructure, and *Webcast Roundup* is a recap of CXO Media's sponsored on-demand video series. All of the newsletters are free with registration. Can be found at http://www.cio.com/newsletters.

Continuity E-guide

E-mailed once a week to subscribers. It provides summaries of articles on a wide range of business continuity, crisis management, and emergency management topics that are collected online. The reader can click on the full articles and even see the original sources of the material. Can be found at http://www.disaster-resource.com/newsletter/subpages/signup_page.htm.

CPM-Global Assurance

Bills itself as "the one resource that offers it all: analysis, best practices, advice and the contingency planning news. This monthly e-newsletter provides you with in-depth articles authored by the industry's best, case studies, white papers, Q&As, national and international news and product and service offerings. The *CPM Industry Insider* is a great way to stay abreast on the important issues making industry headlines. This monthly e-newsletter offers a quick read of the most important news stories of the month." Can be found at http://www.contingencyplanning.com/magazine/index.aspx.

E-News

Published by *Security Management*. This online newsletter provides short articles on a variety of topics related to physical security including travel safety, background checks, and perimeter security. Can be found at http://www.securitymanagement.com/.

Enterprise Strategies

"*Enterprise Strategies* newsletters provide real-world business and technology information for managers of large, high-volume-transaction, high-availability, high-performance computer systems and infrastructures. We offer the latest industry news, analyst and user perspective, and commentary on the latest enterprise, security, and storage trends and technologies. From getting the most out of your system to preparing for security breaches, *Enterprise Strategies* newsletters provide the information and insight to cost-effectively manage your IT resources." See http://esj.com/About_Us/default.aspx.

OUCH!

"OUCH! is the first consensus monthly security awareness report for end users. It shows them what to look for and how to avoid phishing and other scams plus viruses and other malware—using the latest attacks as examples. It also provides pointers to great resources like the amazing Phishing Self-Test. 460 organizations, large and small, helped make it a useful service. More than 100 security officers check each issue for accuracy, and readability before it is distributed to the community." Can be found at http://www.sans.org/newsletters/#newsbites.

SANS NewsBites

"SANS NewsBites is a semiweekly high-level executive summary of the most important news articles that have been published on computer security during the last week. Each news item is very briefly summarized and includes a reference on the web for detailed information, if possible." Can be found at http://www.sans.org/newsletters/#newsbites.

WEB SITES

Business Continuity and Disaster Recovery

http://www.contingencyplanning.com/
This is the home page for the Contingency Planning & Management Group. "The Contingency Planning & Management Group (CPM) is your source for information and strategic advice on business continuity, emergency management and security issues." The site provides access to its two e-newsletters: *CPM Global Assurance and CPM Industry Insider.* Registration is required to access and receive the newsletters. There is a searchable archive as well. "Each of our

electronic newsletters delivers critical information straight to your inbox to help you do your job—and do it better. The CPM Group offers two conferences per year that cover business continuity and security."

http://www.continuityinsights.com/
This is the home page for Business Continuity Insights. The site provides information about business continuity and business security. Resources include white papers, news, podcasts, presentations, and the archive for *Business Continuity Insights* magazine. Business Continuity Insights presents an annual meeting.

http://www.disaster-resource.com/index.htm
This is the home page for Disaster-Resource.com. This site provides resources for those involved in business continuity, crisis management, and emergency management. "Whether you are a senior executive looking for an industry overview, an experienced manager searching for the latest trends, or a new contingency planner in need of the basics, you will find the GUIDE to be the most comprehensive source for crisis/emergency management and business continuity information. The online DISASTER RESOURCE GUIDE is set up to help you find information, vendors, organizations and many resources to help you prepare for (mitigate) or recover from any type of natural or other type of disaster. The GUIDE is to help you keep your business running, your government agency operational, no matter what!" The site allows you to search for products, services, and articles related to business continuity, crisis management, and emergency management. The articles are collected from variety of online sources. People can also subscribe to a newsletter, *Continuity e-guide*.

http://www.drii.org/DRII/
DRI International was founded in 1988 as the Disaster Recovery Institute in order to develop a base of knowledge in contingency planning and the management of risk. DRI International administers educational and certification programs for those engaged in the practice of business continuity planning and management.

http://www.drj.com/
This is the home page for *Disaster Recovery Journal*. The site provides a variety of tools for disaster recovery and business continuity including a toolbox, glossary, and planning model. The site archives news articles and its *Disaster Recovery Journal*. *DRJ* provides two conferences each year.

http://www.idra.com/index.htm
International Disaster Recovery Association is an association of users, researchers, educators, and vendors having a special interest in the voice, data, image, and sensory telecommunications aspects of contingency planning, business continuation, disaster recovery, and restoration. IDRA's mission statement is "To maintain adequate voice, data and image telecommunication services during periods of extraordinary activity and interruptions in normal operations." IDRA holds

workshops, seminars, and multiday conferences at which all topics and exhibits devoted to telecommunications disaster recovery.

Corruption

http://www.iccwbo.org/policy/anticorruption/
This web site is operated by the Anti-Corruption Commission of the International Chamber of Commerce™ (ICC). The Anti-Corruption Commission seeks to "encourage self-regulation by business in confronting issues of extortion and bribery, and to provide business input into international initiatives to fight corruption." The web site details what corruption is, how it harms people, and the ways organizations can fight corruption. There are links to important international anticorruption conventions such as the UN Convention against Corruption, the OECD Anti-Bribery Convention, and the ICC rules of conduct.

http://www.transparency.org/
This is the home page for Transparency International, "the global civil society organisation leading the fight against corruption, [which] brings people together in a powerful worldwide coalition to end the devastating impact of corruption on men, women and children around the world. TI's mission is to create change towards a world free of corruption." This is another resource to consult for information about corruption and ways to combat it. The site includes the global corruption barometer, which provides detailed information annually about corruption in various countries.

http://www.weforum.org/en/initiatives/paci/index.htm
This part of the World Economic Forum's web site is dedicated to its "Partnering Against Corruption Initiative." It provides a discussion of the program and links to other sites dedicated to reducing corruption. "The World Economic Forum is an independent international organization committed to improving the state of the world by engaging leaders in partnerships to shape global, regional and industry agendas. Incorporated as a foundation in 1971, and based in Geneva, Switzerland, the World Economic Forum is impartial and not-for-profit; it is tied to no political, partisan or national interests. The World Economic Forum is under the supervision of the Swiss Federal Government."

Counterintelligence

http://www.ncix.gov/
This web site contains information provided by the Office of the National Counterintelligence Executive (ONCIX). ONCIX is part of the Office of the Director of National Intelligence and is staffed by senior counterintelligence (CI) and other specialists from across the national intelligence and security communities. ONCIX develops, coordinates, and produces the following: annual foreign

intelligence threat assessments and other analytic CI products; annual national CI strategy for the U.S. government; priorities for CI collection, investigations, and operations; CI program budgets and evaluations that reflect strategic priorities; in-depth espionage damage assessments; and CI awareness, outreach, and training standards policies. The site provides resources on counterintelligence and corporate espionage. Security personnel can learn about new threats, especially international threats related to corporate espionage.

Countersurveillance

http://www.abraxascorp.com/default.asp
This is the web site for Abraxas Corporation, a company specializing is risk mitigation technology. Its countersurveillance work centers around TrapWire®, a software program that identifies suspicious computer activity and helps to prevent outside attacks.

http://www.adt.com/wps/portal/adt/
This is the web site for ADT, a provider of security products and services for homes and businesses. ADT has been in existence since 1874 and is the leading provider of electronic security systems in the United States. The web site provides information on how its video surveillance can be used for countersurveillance.

http://www.brinksbusinesssecurity.com/
This is the web site for Brink's Business Security. Brink's has been in business since 1859 and has a strong connection to the security industry in the United States. Brink's provides a large array of security services and equipment. Countersurveillance is one of the applications for its products and services.

http://www.questinvestigations.net/
This is the web site for Quest Consultants International. It provides a range of information and physical security services including countersurveillance. Quest was founded in 1996. The bulk of its employees are veteran ex-FBI agents.

http://www.spybusters.com/spybuster_tips.html
This web site borders on the paranoid but provides interesting information about ways people will try to steal data, including eavesdropping through phone taps and bugging rooms. Tips and equipment for counterespionage are discussed.

Drug Testing

http://www.usdoj.gov/dea/demand/dfmanual/index.html
This link is for the Guidelines for Drug-Free Workplace developed and maintained by the U.S. Drug Enforcement Administration (DEA). The guidelines are useful to any organization that has or is developing a drug testing program. The program details the laws and regulations related to drug testing and how to create an effective drug testing program.

Ecoterrorism

http://www.cdfe.org/ecoterror.htm
This web site explains ecoterrorism and is overseen by the Center for the Defense of Free Enterprise. The site describes what ecoterrorism is, why it is illegal, and some of the major groups involved in ecoterrorism.

Employee Background Checks

http://www.uschamber.com/sb/screening/default.htm?n=tb
This part of the U.S. Chamber of Commerce Small Business web site provides details on employee screening programs. It also includes information on drug screening programs and offers, for a price, a tool kit for employee screening.

Ethics and Compliance

http://www.theecoa.org//AM/Template.cfm?Section=Home
This is the home page for the Ethics & Compliance Officer Association web site. It has limited resources for nonmembers. There is a code of conduct for ethics and compliance officers along with information about education events and training. The organization seeks to create a discussion among its members about the vital issues confronting ethics and compliance officers.

Food Security

http://www.fda.gov/ora/training/orau/FoodSecurity/startpage.html#
This link connects to the Food and Drug Administration's food security awarenss program. The web site provides training free of charge. It is a useful resource for any organization involved in food production and processing. It can be used as part of a larger organizational effort to educate all employees about the need and procedures for food safety.

http://www.fsis.usda.gov/Food_Defense_&_Emergency_Response/Security_Guidelines/index.asp
This web site is part of the United States Department of Agriculture's Food Safety and Inspection Service (FSIS) and provides information on food defense and emergency response. There are resources for consumers and organizations, with the bulk of the material designed for organizations. Topics include details for developing food security plans, sample food security plans, and emergency guidance for retail food establishments.

General Security

http://www.csoonline.com/read/
This is the home page for *CSO* magazine. The site provides a searchable archive of *CSO* articles as well as podcasts and white papers.

http://www.osac.gov/ResourceLibrary/index.cfm?display=type&type=1003
This web site provides links to a variety of security-related documents developed by governmental sources. Topics include traveling abroad, crisis management, emergency preparedness, and food security.

http://www.securityfocus.com/
This web site provides articles on a variety of security-related topics. The site has a search engine. At the site people can register for the *SecurityFocus* newsletter.

http://www.securityinfowatch.com/
This web site lists a variety of other resources for those interested in security. It is a clearinghouse of security information that includes a search engine.

Information Security

http://www.asisonline.org/
According to the web site, "Founded in 1955 as the American Society for Industrial Security (ASIS), the organization officially changed its name in 2002 to ASIS International. This new name preserves our history while better reflecting the growth and expansion of the society to more than 35,000 members around the world, covering a wide array of services and specialties within the security industry." ASIS offers three certifications: Certified Protection Professional™, Professional Certified Investigator™, and Physical Security Professional™.

http://www.cerias.purdue.edu/
This is the home page for the Center for Education and Research in Information Assurance and Security (CERIAS). CERIAS provides free access to research conducted by members of these groups. CERIAS describes itself as "one of the world's leading centers for research and education in areas of information security that are crucial to the protection of critical computing and communication infrastructure. CERIAS is unique among such national centers in its multidisciplinary approach to the problems, ranging from purely technical issues (e.g., intrusion detection, network security, etc.) to ethical, legal, educational, communicational, linguistic, and economic issues, and the subtle interactions and dependencies among them."

Research by CERIAS covers eight main topics: risk management, policies, and laws; trusted social and human interactions; security awareness, education, and training; assurable software and architectures; enclave and network security; incident detection, response, and investigation; identification, authentication, and privacy; and cryptology and rights management.

http://www.cio.com/topic/1419/Security
This is the home page for CIO, a site dedicated to chief information officers. According to the web site: "Serving chief information officers and other IT leaders, CIO.com, CIO magazine, CIO Executive Programs, CIO Custom Solutions Group and the CIO Executive Council are produced by CXO Media, an award-winning

RESOURCE APPENDIX   417

business unit of International Data Group. CXO Media also produces sister publications CSO magazine and CSOonline.com, for chief security officers and other security executives. Starting in 1987 with CIO magazine, CIO's portfolio of properties has grown to provide technology and business leaders with insight and analysis on information technology trends and a keen understanding of IT's role in achieving business goals. The magazine and website have received more than 160 awards to date, including two Grand Neal Awards from the Jesse H. Neal National Business Journalism Awards and two National Magazine of the Year awards from the American Society of Publication Editors."

http://csrc.nist.gov/focus_areas.html#st
This is the web site for the Computer Security Resource Center for the National Institute of Standards and Technology. The web site contains information on four areas: cryptographic standards and applications, security testing, security research/emerging technologies, and security management and assistance.

http://www.esj.com/index.aspx
This is the home page for Enterprise Systems and provides information security materials. People can register for the *Enterprise Strategies* newsletter at the site and search its archives as well as other information security resources.

http://www.gocsi.com/
This is the home page for the Computer Security Institute. According to the web site, "Computer Security Institute serves the needs of Information Security Professionals through membership, educational events, security surveys and awareness tools. Joining CSI provides you with high quality CSI publications, discounts on CSI conferences, access to on-line archives, career development, networking opportunities and more. CSI publishes the annual CSI/FBI Computer Crime & Security Survey, which attracts widespread media attention, and holds an annual Security Survey Roadshow in various cities. CSI Awareness offers products and training to help improve awareness that includes Frontline end user awareness newsletter, World Security Challenge web-based awareness training, awareness peer groups and more. Private training from CSI is also available for organizations."

http://h30240.www3.hp.com/courses/overview.jsp?courseId=13335&courseSessionId=6367&webPageId=1000010&hhopsession.id=d945a4712fe1df09744b78eaa492&hhopsession.id=d945a4712fe1df09744b78eaa492
This link takes you to an online training course in information security called "Security Boot Camp." The training, offered free of charge by Hewlett-Packard (HP), provides useful information to people new to information security. The materials offer useful ideas for when an organization needs to communicate the information security plan to everyone in the organization. The online HP courses can be accessed at http://h30240.www3.hp.com/index.jsp?hhopsession.id=3ed8ba66e8c2c4936b2c8a238dfe

http://www.isecom.org/osstmm/
According to the web site: "The Open Source Security Testing Methodology Manual (OSSTMM) is a peer-reviewed methodology for performing security tests and metrics. The OSSTMM test cases are divided into five channels (sections) which collectively test: information and data controls, personnel security awareness levels, fraud and social engineering control levels, computer and telecommunications networks, wireless devices, mobile devices, physical security access controls, security processes, and physical locations such as buildings, perimeters, and military bases. The OSSTMM focuses on the technical details of exactly which items need to be tested, what to do before, during, and after a security test, and how to measure the results. New tests for international best practices, laws, regulations, and ethical concerns are regularly added and updated." This is an important site for anyone interested in OSSTMM. The web site is operated by the Institute for Security and Open Methodologies.

http://www.iwar.org.uk/ecoespionage/resources/security-guide/Contents.htm#Shortcut
This is a listing of resources provided by the Information Warfare Site (IWS), "an online resource that aims to stimulate debate about a range of subjects from information security to information operations and e-commerce. It is the aim of the site to develop a special emphasis on offensive and defensive information operations. IWS first went online in December 1999. Since its launch it has undergone a complete redesign and many key texts have been added." The materials would be of interest to information security personnel. It has an extended discussion of counterintelligence.

http://www.oissg.org/component/option,com_frontpage/Itemid,1/
This is the web site for Open Information System Security Group. According to the web site, "OISSG is a not-for-profit organization. Our vision is to spread information security awareness by hosting an environment where security enthusiasts from all over the globe share and build knowledge. To achieve our vision, we determine utmost professional need, allocate resources and develop, deliver, and promote programs that add value to the information security community." Of primary interest at the site is the Information Systems Security Assessment Framework (ISSAF). "The ISSAF is OISSG's flagship project. It is an effort to develop an end-to-end framework for security assessment. The ISSAF aims to provide a single point of reference for professionals involved in security assessment; it reflects and addresses the practical issues of security assessment. The ISSAF is an evolving framework and it will be further amended and updated."

http://www.ponemon.org/index.html
"The Ponemon Institute© is dedicated to advancing responsible information and privacy management practices in business and government. To achieve this objective, the Institute conducts independent research, educates leaders from the

RESOURCE APPENDIX 419

private and public sectors and verifies the privacy and data protection practices of organizations in a variety of industries." Access to the research is restricted. Membership in the Responsible Information Management Council is required to gain access to the full site.

http://www.sans.org/?portal=4803396804b714c6beb9083a113f5363
According to its web site, "The SANS (SysAdmin, Audit, Network, Security) Institute was established in 1989 as a cooperative research and education organization. Its programs now reach more than 165,000 security professionals around the world. A range of individuals from auditors and network administrators, to chief information security officers are sharing the lessons they learn and are jointly finding solutions to the challenges they face. At the heart of SANS are the many security practitioners in varied global organizations from corporations to universities working together to help the entire information security community."

SANS provides a large collection of research documents free of charge covering a variety of information security issues. SANS also is a vendor supplying information security training and certification.

http://www.sans.org/resources/policies/
This is the site for the SANS Security Policy Project. The site provides a good overview of security policies, standards, and guidelines, and offers a variety of policy templates that may be downloaded and customized by replacing <Company Name> with the name of your organization.

http://www.staysafeonline.info/index.html
This web site is administrated by the National Cyber Security Alliance. It provides information on cyber security for individuals and small businesses. This information is designed for non-echnical users and focuses more on individual security than business security.

http://www.us-cert.gov/
This is the home page for the United States Computer Emergency Response Team (US-CERT). The web site provides a wide array of information about cyber attacks and Internet security. The materials include those for both technical and nontechnical users.

Insider Threat

http://www.polarcove.com/whitepapers/insiderthreats.htm
This link is to a white paper from Polar Cove on insider threats. The web site provides interesting data on insider threats along with some solutions. "Polar Cove is a dynamic and innovative company providing a full range of security solutions to meet your needs for secure systems development, secure application development, network security, and information security related goals. We help our customers achieve their business objectives by developing secure

solutions that encompass your ideas, business needs, key skills, and current technology partnerships."

Overseas Travel

http://www.pueblo.gsa.gov/cic_text/travel/business-overseas/travel.html
This web site is a text version of the brochure "Personal Security Guidelines for the American Business Traveler Overseas" developed by the Bureau of Diplomatic Security, Overseas Security Advisory Council.

http://www.travel.state.gov/travel/travel_1744.html
This web site is part of the U.S. Department of State and has information from the the Office of American Citizens Services and Crisis Management (ACS). This office provides services to Americans traveling or residing abroad. The site contains valuable information abut traveling and living safely abroad with sections on "Tips for Traveling Abroad" and "Living Abroad Tips."

Pandemics (Avian Influenza)

http://www.businessroundtable.org//taskforces/taskforce/document.aspx?qs =68A5BF159FC49514481138A6DBE7A7A19BB6487B96C39B1
This is a pandemic flu preparation document prepared by the Business Roundtable, an association of leading chief executive officers in the United States. The document details its recommendations for pandemic flu preparedness.

http://www.cdc.gov/business/
This section of the Centers for Disease Control and Prevention (CDC) web site provides information about avian flu. The free resources include a business planning checklist and pandemic influenza tools for businesses. The site also contains information to help organizations plan for various types of disasters.

http://www.pandemicflu.gov/index.html
This web site is the central information hub for Avian influenza information provided by the United States government. Organizations can get facts about how they should plan, along with what federal, state, and local governments are doing to prepare. The "Workplace Planning" section provides a number of tools for organizations. Some are for specific industries, but there are general Avian influenza planning documents as well.

http://www.who.int/csr/disease/avian_influenza/en/
This section of the World Health Organization's (WHO) web site is dedicated to Avian influenza. The site offers general guidance for coping with the avian flu and links to other sites on the topic. A unique feature is this site's tracking of the global outbreaks of avian flu. Visitors to the site can view an up-to-date map of the

RESOURCE APPENDIX

world to see where new and old outbreaks have occurred. This global perspective is useful for international businesses.

Shelter-in-Place

http://www.epa.gov/ord/articles/shelter_in_place_qa.htm
This web site is operated by the Environmental Protection Agency and details the concept of shelter-in-place. It is an excellent resource for any organization that may need to shelter-in-place or to warn community members to do so.

Social Engineering

www.socialengineering101.com
This site offers the social engineer possibly every tool he or she ever could need, including a chat board for sharing recent successes and failures.

Terrorism

http://www.cbp.gov/xp/cgov/import/commercial_enforcement/ctpat/
This part of the U.S. Customs and Border Protection web site is dedicated to the Customs-Trade Partnership Against Terrorism (C-TPAT). Organizations can learn what C-TPAT is, how it can help their businesses, and requirements for certification for C-TPAT. This site is a must for any organization that wants to become part of C-TPAT.

http://www.cfr.org/issue/135/?type=issue
This web site is operated by the Council on Foreign Relations and provides information about terrorists and terrorism. Organizations that are exposed to terrorism will find useful information on such topics as counterterrorism, weapons of terrorism, terrorist organizations, and terrorism and the economy.

http://www.tkb.org/Country.jsp?countryCd=US
This links the United States page of the MIPT Terrorism Knowledge Base. "The MIPT Terrorism Knowledge Base® (TKB®) is the one-stop resource for comprehensive research and analysis on global terrorist incidents, terrorism-related court cases, and terrorist groups and leaders. TKB covers the history, affiliations, locations, and tactics of terrorist groups operating across the world, with over 35 years of terrorism incident data and hundreds of group and leader profiles and trials. TKB also features interactive maps, statistical summaries, and analytical tools that can create custom graphs and tables." The Terrorism Knowledge Base provides terrorist information for a wide range of countries along with information about terrorist groups and the ability to search the data for terrorist incidents worldwide. The site is very useful for a mult-national organization trying to determine the terrorist risks facing its various locations.

VENDORS

Data Backup

Circadian Force
Provides electronic vaulting with a focus on disaster recovery and business continuity. It serves organizations across the United States. Can be found on the web at http://www.circadianforce.com/index.htm.

EMC Corporation
Considers itself one of the world leaders in products, services, and solutions for information management and storage. It has a worldwide client list and helps organizations keep their most essential digital information protected, secure, and continuously available. Can be found on the web at http://www.emc.com/about/.

Iron Mountain Incorporated (IRM)
Provides organizations with data protection, records management, and information destruction services. The company also makes available its expertise in handling complex information challenges such as regulatory compliance and litigation. Can be found on the web at http://www.ironmountain.com/index.asp.

National Records Centers Inc. (NRC)
Services include records management, document storage, digital services, document destruction, computer media rotation, and vaulting. It serves organizations throughout North and Central America and Europe. The primary focus is off-site records management and storage services. See http:// www.nationalrecordscenters.com/default.asp.

Internet and E-mail Monitoring

Cyveillance
Provides services that allow organizations to monitor online activities, including e-mail, and to protect themselves from online risks. Cyveillance uses what it calls *proactive cyber intelligence*. It can monitor what is being said about your company as well as keep an eye on employee Internet and e-mail activity. Cyveillance offers comprehensive coverage online, including chat rooms and "hidden" networks of e-mail. Can be found on the web at http://www.cyveillance.com/.

Websense, Inc.
Advertises itself as a global leader in web filtering and desktop security software. Its Internet filtering is flexible and has integrated policy enforcement. Can be found on the web at http://www.websense.com/global/en/.

RESOURCE APPENDIX

Mass Notification

3n (National Notification Network)
A leading provider of mass notification for corporations, schools, nonprofits, and government agencies. Its primary service is InstCom™, a suite of products that allows for rapid communication in emergency or urgent situations. Can be found on the web at http://www.3nonline.com/index.php.

Simulations for Training and Exercises

Crisis Simulations International
Provides real-time interactive simulations for corporations and government agencies. These realistic scenarios offer a variety of real-life situations designed to test an organization's crisis, business continuity, and emergency management skills. Can be found on the web at http://www.crisissimulations.com/index.html.

Security

Polar Cove
Provides a full range of security solutions including network security, secure application development, secure system development, and information security-related goals. Polar Cove creates a solution that meets the needs of its clients. Can be found on the web at http://www.polarcove.com/index.htm.

Redkey International
Provides specialized security risk management services. Redkey's risk management focus helps its clients maintain business continuity. Can be found on the web at http://www.redkeyinternational.com/aboutus.html.

DOCUMENT APPENDIX

Security Self-Assessment Guide for
   Information Technology Systems   425
Emergency Management Guide for
   Business and Industry   529
Ready Business: Sample Emergency Plan   582
FEMA: Protecting Your Business From Disasters   588
Guidance on Preparing Workplaces for
   an Influenza Pandemic   591
Industry Self-Assessment Checklist for Food Security   620
Security Guidelines for American Enterprises Abroad   633
Personal Security Guidelines for the
   American Business Traveler Overseas   673

SECURITY SELF-ASSESSMENT GUIDE FOR INFORMATION TECHNOLOGY SYSTEMS

Marianne Swanson

INTRODUCTION

A self-assessment conducted on a system (major application or general support system) or multiple self-assessments conducted for a group of interconnected systems (internal or external to the agency) is one method used to measure information technology (IT) security assurance. IT security assurance is the degree of confidence one has that the managerial, technical and operational security measures work as intended to protect the system and the information it processes. Adequate security of these assets is a fundamental management responsibility. Consistent with Office of Management and Budget (OMB) policy, each agency must implement and maintain a program to adequately secure its information and system assets. Agency programs must: 1) assure that systems and applications operate effectively and provide appropriate confidentiality, integrity, and availability; and 2) protect information commensurate with the level of risk and magnitude of harm resulting from loss, misuse, unauthorized access, or modification.

Agencies must plan for security, ensure that the appropriate officials are assigned security responsibility, and authorize system processing prior to operations and periodically thereafter. These management responsibilities presume that responsible agency officials understand the risks and other factors that could negatively impact their mission goals. Moreover, these officials must understand the current status of security programs and controls in order to make informed judgments and investments that appropriately mitigate risks to an acceptable level.

An important element of ensuring an organizations' IT security health is performing routine self-assessments of the agency security program. For a self-assessment to be effective, a risk assessment should be conducted in conjunction with or prior to the self-assessment. A self-assessment does not eliminate the need for a risk assessment.

There are many methods and tools for agency officials to help determine the current status of their security programs relative to existing policy. Ideally many of these methods and tools would be implemented on an ongoing basis to systematically identify programmatic weaknesses and where necessary, establish targets for continuing improvement. This document provides a method to evaluate the security of unclassified systems or groups of systems; it guides the reader in performing an IT security self-assessment. Additionally, the document provides guidance on utilizing the results of the system self-assessment to ascertain the status of the agency-wide security program. The results are obtained in a form that can readily be used to determine which of the five levels specified in the Federal IT Security Assessment Framework the agency has achieved for each topic area covered in the questionnaire. For example, the group of systems under review may have reached level 4 (Tested and Evaluated Procedures and Controls) in the topic area of physical and environmental protection, but only level 3 (Implemented Procedures and Controls) in the area of logical access controls.

Self-Assessments

This self-assessment guide utilizes an extensive questionnaire (Appendix A) containing specific control objectives and suggested techniques against which the security of a system or group of interconnected systems can be measured. The questionnaire can be based primarily on an examination of relevant documentation and a rigorous examination and test of the controls. This guide does not establish new security requirements. The control objectives are abstracted directly from long-standing requirements found in statute, policy, and guidance on security and privacy. However the guide is not intended to be a comprehensive list of control objectives and related techniques. The guide should be used in conjunction with the more detailed guidance listed in Appendix B. In addition, specific technical controls, such as those related to individual technologies or vendors, are not specifically provided due to their volume and dynamic nature. It should also be noted that an agency might have additional laws, regulations, or policies that establish specific requirements for confidentiality, integrity, or availability. Each agency should decide if additional security controls should be added to the questionnaire and, if so, customize the questionnaire appropriately.

The goal of this document is to provide a standardized approach to assessing a system. This document strives to blend the control objectives found in the many requirement and guidance documents. To assist the reader, a reference source is listed after each control objective question listed in the questionnaire. Specific attention was made to the control activities found in the General Accounting Office's (GAO) Federal Information System Control Audit Manual (FISCAM). FISCAM is the document GAO auditors and agency inspector generals use when auditing an agency. When FISCAM is referenced in the questionnaire, the major category initials along with the control activity number are provided, e.g., *FISCAM SP-3.1*. The cross mapping of the two documents will form a road map between the control objectives and techniques the audit community assess and the control objectives and techniques IT security program managers and program officials need to assess. The mapping provides a common point of reference for individuals fulfilling differing roles in the assessment process. The mapping ensures that both parties are reviewing the same types of controls.

The questionnaire may be used to assess the status of security controls for a system, an interconnected group of systems, or agency-wide. These systems include information, individual systems (e.g., major applications, general support systems, mission critical systems), or a logically related grouping of systems that support operational programs (e.g., Air Traffic Control, Medicare, Student Aid). Assessing all security controls and all interconnected system dependencies provides a metric of the IT security conditions of an agency. By using the procedures outlined in Chapter 4, the results of the assessment can be used as input on the status of an agency's IT security program.

Federal IT Security Assessment Framework

The Federal IT Security Assessment Framework issued by the federal Chief Information Officer Council in November 2000 provides a tool that agencies can use to routinely evaluate the status of their IT security programs. The document established the groundwork for standardizing on five levels of security effectiveness and measurements that agencies could use to determine which of the five levels are met. By utilizing the Framework levels, an agency can prioritize agency efforts as well as use the document over time to evaluate progress. The NIST Self-Assessment Guide builds on the Framework by providing questions on specific areas of control, such as those pertaining to access and service continuity, and a means of categorizing evaluation results in the same manner as the Framework. See Appendix C for a copy of the Framework.

Audience

The control objectives and techniques presented are generic and can be applied to organizations in private and public sectors. This document can be used by all levels of management and by those individuals responsible for IT security at the system level and organization level. Additionally, internal and external auditors may use the questionnaire to guide their review of the IT security of systems. To perform the examination and testing required to complete the questionnaire, the assessor must be familiar with and able to apply a core knowledge set of IT security basics needed to protect information and systems. In some cases, especially in the area of examining and testing technical controls, assessors with specialized technical expertise will be needed to ensure that the questionnaire's answers are reliable.

Structure of this Document

Chapter 1 introduces the document and explains IT security assessments and the relationship to other documents. Chapter 2 provides a method for determining the system boundaries and criticality of the data. Chapter 3 describes the questionnaire. Chapter 4 provides guidance on using the completed system questionnaire(s) as input into obtaining an assessment of an agency-wide IT security program. Appendix A contains the questionnaire. Appendix B lists the documents used in compiling the assessment control objective questions. Appendix C contains a copy of the *Federal IT Security Assessment Framework*. Appendix D lists references used in developing this document.

SYSTEM ANALYSIS

The questionnaire is a tool for completing an internal assessment of the controls in place for a major application or a general support system. The security of every system or group of interconnected system(s) must be described in a security plan. The system may consist of a major application or be part of a general support system. The definition of major application and general support system are contained in Appendix C. Before the questionnaire can be used effectively, a determination must be made as to the boundaries of the system and the sensitivity and criticality of the information stored within, processed by, or transmitted by the system(s). A completed general support system or major application security plan, which is required under OMB Circular A-130, Appendix III, should describe the boundaries of the system and the criticality level of the data. If a plan has not been prepared for the system, the completion of this self-assessment will aid in developing the system security plan. Many of the control objectives addressed in the assessment are to be described in the system security plan. The following two sections, Section 2.1 and Section 2.2, contain excerpts from NIST Special Publication 800-18, *Guide for Developing Security Plans for Information Technology Systems*, and will assist the reader in determining the physical and logical boundaries of the system and the criticality of the information.

System Boundaries

Defining the scope of the assessment requires an analysis of system boundaries and organizational responsibilities. Networked systems make the boundaries much harder to define. Many organizations have distributed client-server architectures where servers and workstations communicate through networks. Those same networks are connected to the Internet. A system, as defined in NIST Special Publication 800-18, *Guide for Developing Security Plans for Information Technology Systems*, is identified by defining boundaries around a set of processes, communications, storage, and related resources. The elements within these boundaries constitute a single system requiring a system security plan and a security evaluation whenever a major modification to the system occurs. Each element of the system must[1]:

- Be under the **same** direct management control;
- Have the **same** function or mission objective;
- Have essentially the **same** operating characteristics and security needs; and
- Reside in the **same** general operating environment.

All components of a system need not be physically connected (e.g., [1] a group of stand-alone personal computers (PCs) in an office; [2] a group of PCs placed in employees' homes under defined telecommuting program rules; [3] a group of portable PCs provided to employees who require mobile computing capability to perform their jobs; and [4] a system with multiple identical configurations that are installed in locations with the same environmental and physical controls).

An important element of the assessment will be determining the effectiveness of the boundary controls when the system is part of a network. The boundary controls must protect the defined system or group of systems from unauthorized intrusions. If such boundary controls are not effective, then the security of the systems under review will depend on the security of the other systems connected to it. In the absence of effective boundary controls, the assessor should determine and document the adequacy of controls related to each system that is connected to the system under review.

Sensitivity Assessment

Effective use of the questionnaire presumes a comprehensive understanding of the value of the systems and information being assessed. Value can be expressed in terms of the degree of sensitivity or criticality of the systems and information relative to each of the five protection categories in section 3534(a)(1)(A) of the Government Information Security Reform provisions of the National Defense Authorization Act of 2000, i.e., integrity, confidentiality, availability, authenticity, and non-repudiation. The addition of authenticity and non-repudiation as protection categories within the Reform Act was to stress the need for these assurances as the government progresses towards a paperless workplace. There are differing opinions on what constitutes protection categories, for continuity within several NIST Special Publication 800 documents; authenticity, non-repudiation, and accountability are associated with the integrity of the information.

- ***Confidentiality***—The information requires protection from unauthorized disclosure.
- ***Integrity***—The information must be protected from unauthorized, unanticipated, or unintentional modification. This includes, but is not limited to:
    - *Authenticity*—A third party must be able to verify that the content of a message has not been changed in transit.
    - *Non-repudiation*—The origin or the receipt of a specific message must be verifiable by a third party.
    - *Accountability*—A security goal that generates the requirement for actions of an entity to be traced uniquely to that entity.
- ***Availability***—The information technology resource (system or data) must be available on a timely basis to meet mission requirements or to avoid substantial losses. Availability also includes ensuring that resources are used only for intended purposes.

When determining the value, consider any laws, regulations, or policies that establish specific requirements for integrity, confidentiality, authenticity, availability, and non-repudiation of data and information in the system. Examples might include Presidential Decision Directive 63, the Privacy Act, or a specific statute or regulation concerning the information processed (e.g., tax or census information).

Consider the information processed by the system and the need for protective measures. Relate the information processed to each of the three basic protection requirements above (**confidentiality**, **integrity**, and **availability**). In

DOCUMENT APPENDIX

addition, it is helpful to categorize the system or group of systems by sensitivity level. Three examples of such categories for sensitive unclassified information are described below:

- *High*—Extremely grave injury accrues to U.S. interests if the information is compromised; could cause loss of life, imprisonment, major financial loss, or require legal action for correction
- *Medium*—Serious injury accrues to U.S. interests if the information is compromised; could cause significant financial loss or require legal action for correction
- *Low*—Injury accrues to U.S. interests if the information is compromised; would cause only minor financial loss or require only administrative action for correction

For example, a system and its information may require a high degree of integrity and availability, yet have no need for confidentiality.

Many agencies have developed their own methods of making these determinations. Regardless of the method used, the system owner/program official is responsible for determining the sensitivity of the system and information. The sensitivity should be considered as each control objective question in the questionnaire is answered. When a determination is made to either provide more rigid controls than are addressed by the questionnaire or not to implement the control either temporarily or permanently, there is a risk based decision field in the questionnaire that can be checked to indicate that a determination was made. The determination for lesser or more stringent protection should be made due to either the sensitivity of the data and operations affected or because there are compensating controls that lessen the need for this particular control technique. It should be noted in the comments section of the questionnaire that the system security plan contains supporting documentation as to why the specific control has or has not been implemented.

QUESTIONNAIRE STRUCTURE

The self-assessment questionnaire contains three sections: cover sheet, questions, and notes. The questionnaire begins with a cover sheet requiring descriptive information about the major application, general support system, or group of interconnected systems being assessed. The questionnaire provides a hierarchical approach to assessing a system by containing critical elements and subordinate questions. The critical element level should be determined based on the answers to the subordinate questions. The critical elements are derived primarily from OMB Circular A-130. The subordinate questions address the control objectives and techniques that can be implemented to meet the critical elements. Assessors will need to carefully review the levels of subordinate control objectives and techniques in order to determine what level has been reached for the related critical element. The control objectives were obtained from the list of source documents located in Appendix B. There is flexibility in implementing the control objectives and techniques. It is feasible that not all control objectives and techniques may be needed to achieve the critical element.

The questionnaire section may be customized by the organization. An organization can add questions, require more descriptive information, and even pre-mark certain questions if applicable. For example, many agencies may have personnel security procedures that apply to all systems within the agency. The level 1 and level 2 columns in the questionnaire can be pre-marked to reflect the standard personnel procedures in place. Additional columns may be added to reflect the status of the control, i.e., planned action date, non-applicable, or location of documentation. The questionnaire should not have questions removed or questions modified to reduce the effectiveness of the control.

After each question, there is a comment field and an initial field. The comment field can be used to note the reference to supporting documentation that is attached to the questionnaire or is obtainable for that question. The initial field can be used when a risk based decision is made concerning not to implement a control or if the control is not applicable for the system. At the end of each set of questions, there is an area provided for notes. This area may be used for denoting where in a system security plan specific sections should be modified. It can be used to document the justification as to why a control objective is not being implemented fully or why it is overly rigorous. The note section may be a good place to mark where follow-up is needed or additional testing, such as penetration testing or product evaluations, needs to be initiated. Additionally, the section may reference supporting documentation on how the control objectives and techniques were tested and a summary of findings.

Questionnaire Cover Sheet

This section provides instruction on completing the questionnaire cover sheet, standardizing on how the completed evaluation should be marked, how systems are titled, and labeling the criticality of the system.

Questionnaire Control
All completed questionnaires should be marked, handled, and controlled at the level of sensitivity determined by organizational policy. It should be noted that the information contained in a completed questionnaire could easily depict where the system or group of systems is most vulnerable.

System Identification
The cover page of the questionnaire begins with the name and title of the system to be evaluated. As explained in NIST Special Publication 800-18, each

major application or general support system should be assigned a unique name/identifier.

Assigning a unique identifier to each system helps to ensure that appropriate security requirements are met based on the unique requirements for the system, and that allocated resources are appropriately applied. Further, the use of unique system identifiers is integral to the IT system investment models and analyses established under the requirements of the Information Technology Management Reform Act of 1996 (also known as the Clinger-Cohen Act). The identifiers are required by OMB Circular A-11 and used in the annual OMB budget submissions of the Exhibit 53 and 300. In light of OMB policies concerning capital planning and investment control, the unique name/identifier should remain the same throughout the life of the system to allow the organization to track completion of security requirements over time. Please see OMB Circular A-11, Section 53.7 for additional information on assigning unique identifiers. If no unique name/identifier has been assigned or is not known, contact the information resource management office for assistance.

In many cases the major application or general support system will contain interconnected systems. The connected systems should be listed and once the assessment is complete, a determination should be made and noted on the cover sheet as to whether the boundary controls are effective. The boundary controls should be part of the assessment. If the boundary controls are not adequate, the connected systems should be assessed as well.

The line below the System Name and Title requires the assessor to mark the system category (General Support or Major Application). If an agency has additional system types or system categories, i.e., mission critical or non-mission critical, the cover sheet should be customized to include them.

Purpose and Assessor Information

The purpose and objectives of the assessment should be identified. For example, the assessment is intended to gain a high-level indication of system security in preparation for a more detailed review or the assessment is intended to be a thorough and reliable evaluation for purposes of developing an action plan. The name, title, and organization of the individuals who perform the assessment should be listed. The organization should customize the cover page accordingly.

The start date and completion date of the evaluation should be listed. The length of time required to complete an evaluation will vary. The time and resources needed to complete the assessment will vary depending on the size and complexity of the system, accessibility of system and user data, and how much information is readily available for the assessors to evaluate. For example, if a system has undergone extensive testing, certification, and documentation, the self-assessment is easy to use and serves as a baseline for future evaluations. If the system has undergone very limited amounts of testing and has poor documentation, completing the questionnaire will require more time.

Criticality of Information

The level of sensitivity of information as determined by the program official or system owner should be documented using the table on the questionnaire cover sheet. If an organization has designed their own method of determining system criticality or sensitivity, the table should be replaced with the organization's criticality or sensitivity categories. The premise behind formulating the level of sensitivity is that systems supporting higher risk operations would be expected to have more stringent controls than those that support lower risk operations.

Questions

The questions are separated into three major control areas: 1) management controls, 2) operational controls, and 3) technical controls. The division of control areas in this manner complements three other NIST Special Publications: NIST Special Publication 800-12, *An Introduction to Computer Security: The NIST Handbook* (Handbook), NIST Special Publication 800-14, *Generally Accepted Principles and Practices for Securing Information Technology Systems* (Principles and Practices), and NIST Special Publication 800-18, *Guide for Developing Security Plans for Information Technology Systems* (Planning Guide). All three documents should be referenced for further information. The Handbook should be used to obtain additional detail for any of the questions (control objectives) listed in the questionnaire. The Principles and Practices document should be used as a reference to describe the security controls. The Planning Guide formed the basis for the questions listed in the questionnaire. The documents can be obtained from the NIST Computer Security Resource Center web site at the URL: http://csrc.nist.gov.

The questions portion of this document easily maps to the three NIST documents described above since the chapters in all three documents are organized by the same control areas, i.e., management, operational, and technical.

Within each of the three control areas, there are a number of topics; for example, personnel security, contingency planning, and incident response are topics found under the operational control area. There are a total of 17 topics contained in the questionnaire; each topic contains critical elements and supporting security control objectives and techniques (questions) about the system. The critical elements are derived primarily from OMB Circular A-130 and are integral to an effective IT security program. The control objectives and techniques support the critical elements. If a number of the control objectives and techniques are not implemented, the critical elements have not been met.

Each control objective and technique may or may not be implemented depending on the system and the risk associated with the system. Under each control objective and technique question, one or more of the source documents is referenced. The reference points to the specific control activity in the GAO FISCAM document or to the title of any of the other documents listed in Appendix B, Source of Control Criteria.

In order to measure the progress of effectively implementing the needed security control, five levels of effectiveness are provided for each answer to the security control question:

- Level 1 – control objective documented in a security policy
- Level 2 – security controls documented as procedures
- Level 3 – procedures have been implemented
- Level 4 – procedures and security controls are tested and reviewed
- Level 5 – procedures and security controls are fully integrated into a comprehensive program.

The method for answering the questions can be based primarily on an examination of relevant documentation and a rigorous examination and test of the controls. The review, for example, should consist of testing the access control methods in place by performing a penetration test; examining system documentation such as software change requests forms, test plans, and approvals; and examining security logs and audit trails. Supporting documentation describing what has been tested and the results of the tests add value to the assessment and will make the next review of the system easier.

Once the checklist, including all references, is completed for the first time, future assessments of the system will require considerably less effort. The completed questionnaire would establish a baseline. If this year's assessment indicates that most of the controls in place are at level 2 or level 3, then that would be the starting point for the next evaluation. More time can be spent identifying ways to increase the level of effectiveness instead of having to gather all the initial information again. Use the comment section to list whether there is supporting documentation and the notes section for any lengthy explanations.

The audit techniques to test the implementation or effectiveness of each control objective and technique are beyond the scope of this document. The GAO FISCAM document provides audit techniques that can be used to test the control objectives.

When answering the questions about whether a specific control objective has been met, consider the sensitivity of the system. The questionnaire contains a field that can be checked when a risk-based decision has been made to either reduce or enhance a security control. There may be certain situations where management will grant a waiver either because compensating controls exists or because the benefits of operating without the control (at least temporarily) outweigh the risk of waiting for full control implementation. Alternatively, there may be times when management implements more stringent controls than generally applied elsewhere. When the risk-based decision field is checked, note the reason in the comment field of the questionnaire and have management review and initial the decision. Additionally, the system security plan for the system should contain supporting documentation as to why the control has or has not been implemented.

The assessor must read each control objective and technique question and determine in partnership with the system owner and those responsible for administering the system, whether the system's sensitivity level warrants the implementation of the control stated in the question. If the control is applicable, check whether there are documented policies (level 1), procedures for implementing the control (level 2), the control has been implemented (level 3), the control has been tested and if found ineffective, remedied (level 4), and whether the control is part of an agency's organizational culture (level 5). The shaded fields in the questionnaire do not require a check mark. The five levels describing the state of the control objective provide a picture of each operational control; however, how well each one of these controls is met is subjective. Criteria have been established for each of the five levels that should be applied when determining whether the control objective has fully reached one or more of the five levels. The criteria are contained in Appendix C, *Federal IT Security Assessment Framework*.

Based on the responses to the control objectives and techniques and in partnership with the system owner and those responsible for system administration, the assessor should conclude the level of the related critical element. The conclusion should consider the relative importance of each subordinate objective/technique to achieving the critical element and the rigor with which the technique is implemented, enforced, and tested.

Applicability of Control Objectives

As stated above, the critical elements are required to be implemented; the control objectives and techniques, however, tend to be more detailed and leave room for reasonable subjective decisions. If the control does not reasonably apply to the system, then a "non-applicable" or "N/A" can be entered next to the question.

The control objectives and techniques in the questionnaire are geared for a system or group of connected systems. It is possible to use the questionnaire for a program review at an organizational level for ascertaining if the organization has policy and procedures in place (level 1 or level 2). However, to ensure all systems have implemented, tested and fully integrated the controls (level 3, level 4, and level 5), the assessment questionnaire must be applied to each individual or interconnected group of systems. Chapter 4 describes how the results of the assessment can be used as input into an IT security program review.

The policy and procedures for a control objective and technique can be found at the Department level, agency level, agency component level, or application level. To effectively assess a system, ensure that the control objectives being assessed are at the applicable level. For example, if the system being reviewed has stringent authentication procedures, the authentication procedures for the system should be assessed, instead of the agency-wide minimum authentication procedures found in the agency IT security manual.

If a topic area is documented at a high level in policy, the level 1 box should be checked in the questionnaire. If there are additional low level policies for the system, describe the policies in the comment section of the questionnaire. If a specific control is described in detail in procedures, and implemented, the level 2 and level 3 boxes should be checked in the questionnaire. Testing and reviewing controls are an essential part of securing a system. For each specific control, check whether it has been tested and/or reviewed when a significant change occurred. The goal is to have all levels checked for each control. A conceptual sample of completing the questionnaire is contained in Appendix C. The conceptual sample has evolved into the questionnaire and differs slightly, i.e., there is now a comment and initial field.

UTILIZING THE COMPLETED QUESTIONNAIRE

The questionnaire can be used for two purposes. First it can be used by agency managers who know their agency's systems and security controls to quickly gain a general understanding of where security for a system, group of systems, or the entire agency needs improvement. Second, it can be used as a guide for thoroughly evaluating the status of security for a system. The results of such thorough reviews provide a much more reliable measure of security effectiveness and may be used to 1) fulfill reporting requirements; 2) prepare for audits; and 3) identify resource needs.

Questionnaire Analysis

Because this is a self-assessment, ideally the individuals assessing the system are the owners of the system or responsible for operating or administering the system. The same individuals who completed the assessment can conduct the analysis of the completed questionnaire. By being familiar with the system, the supporting documentation, and the results of the assessment, the next step that the assessor takes is an analysis, which summarizes the findings. A centralized group, such as an agency's Information System Security Program Office, can also conduct the analysis as long as the supporting documentation is sufficient. The results of the analysis should be placed in an action plan, and the system security plan should be created or updated to reflect each control objective and technique decision.

Action Plans

How the critical element is to be implemented, i.e., specific procedures written, equipment installed and tested, and personnel trained, should be documented in an action plan. The action plan must contain projected dates, an allocation of resources, and follow-up reviews to ensure that remedial actions have been effective. Routine reports should be submitted to senior management on weaknesses identified, the status of the action plans, and the resources needed.

Agency IT Security Program Reports

Over the years, agencies have been asked to report on the status of their IT security program. The reporting requests vary in how much detail is required and in the type of information that should be reported. The completed self-assessment questionnaires are a useful resource for compiling agency reports. Below are sample topics that should be considered in an agency-wide security program report:

- Security Program Management
- Management Controls
- Operational Controls
- Technical Controls
- Planned Activities

Security Program Management

An agency's IT security program report needs to address programmatic issues such as:

- an established agency-wide security management structure,
- a documented up-to-date IT security program plan or policy (*The assessment results for level 1 provides input.*)
    - an agency-developed risk management and mitigation plan,
    - an agency-wide incident response capability,
    - an established certification and accreditation policy,
    - an agency-wide anti-virus infrastructure in place and operational at all agency facilities,
    - information security training and awareness programs established and available to all agency employees,
    - roles and relationships clearly defined and established between the agency and bureau levels of information security program management,
- an understanding of the importance of protecting mission critical information assets,
- the integration of security into the capital planning process,
- methods used to ensure that security is an integral part of the enterprise architecture (*The assessment results for the Life Cycle topic area provides input.*),
- the total security cost from this year's budget request and a breakdown of security costs by each major operating division, and
- descriptions of agency-wide guidance issued in the past year.

Management Controls, Operational Controls, and Technical Controls

The results of the completed questionnaires' 17 control topic areas can be used to summarize an agency's implementation of the management, operational, and technical controls. For the report to project an accurate picture, the results must be summarized by system type, not totaled into an overall agency grade level. For example, ten systems were assessed using the questionnaire. Five of the ten systems assessed were major applications; the other five were general support systems. The summary would separate the systems into general support systems and major applications.

By further separating them into groups according to criticality, the report stresses which systems and which control objectives require more attention based on sensitivity and criticality. Not all systems require the same level of protection; the report should reflect that diversity. The use of percentages for describing compliance (i.e., 50 percent of the major applications and 25 percent of general support systems that are high in criticality have complete and current system security plans within the past three years) can be used as long as there is a distinct division provided between the types of systems being reported.

Additionally all or a sampling of the completed questionnaires can be analyzed to determine which controls if implemented would impact the most systems. For example, if viruses frequently plague systems, a stricter firewall policy that prevents attached files in E-mail may be a solution. Also, systemic problems should be culled out. If an agency sees an influx of poor password management controls in the questionnaire results, then possibly password checkers should be used, awareness material issued, and password- aging software installed.

The report should conclude with a summary of planned IT security initiatives. The summary should include goals, actions needed to meet the goals, projected resources, and anticipated dates of completion.

APPENDIX A

System Questionnaire

System Name, Title, and Unique Identifier: _____

Major Application _____ or General Support System _____

Name of Assessors: _____

_____

_____

Date of Evaluation: _____

List of Connected Systems:

Name of System	Are boundary controls effective?	Planned action if not effective
1.		
2.		
3.		

Criticality System	Category of Sensitivity High, Medium, or Low
Confidentiality	
Integrity	
Availability	

Purpose and Objective of Assessment: _____

_____

_____

Management Controls

Management controls focus on the management of the IT security system and the management of risk for a system. They are techniques and concerns that are normally addressed by management.

Risk Management

Risk is the possibility of something adverse happening. Risk management is the process of assessing risk, taking steps to reduce risk to an acceptable level, and maintaining that level of risk. The following questions are organized according to two critical elements. The levels for each of these critical elements should be determined based on the answers to the subordinate questions.

Specific Control Objectives and Techniques	L.1 Policy	L.2 Procedures	L.3 Implemented	L.4 Tested	L.5 Integrated	Risk Based Decision Made	Comments	Initials
Risk Management *OMB Circular A-130, III*								
1.1 Critical Element: Is risk periodically assessed?								
1.1.1 Is the current system configuration documented, including links to other systems? *NIST SP 800-18*								
1.1.2 Are risk assessments performed and documented on a regular basis or whenever the system, facilities, or other conditions change? *FISCAM SP-1*								
1.1.3 Has data sensitivity and integrity of the data been considered? *FISCAM SP-1*								
1.1.4 Have threat sources, both natural and manmade, been identified? *FISCAM SP-1*								

(*Continued*)

Specific Control Objectives and Techniques	L.1 Policy	L.2 Procedures	L.3 Implemented	L.4 Tested	L.5 Integrated	Risk Based Decision Made	Comments	Initials
1.1.5 Has a list of known system vulnerabilities, system flaws, or weaknesses that could be exploited by the threat sources been developed and maintained current? *NIST SP 800-30*[2]								
1.1.6 Has an analysis been conducted that determines whether the security requirements in place adequately mitigate vulnerabilities? *NIST SP 800-30*								
1.2. Critical Element: **Do program officials understand the risk to systems under their control and determine the acceptable level of risk?**								
1.2.1 Are final risk determinations and related management approvals documented and maintained on file? *FISCAM SP-1*								

Specific Control Objectives and Techniques	L.1 Policy	L.2 Procedures	L.3 Implemented	L.4 Tested	L.5 Integrated	Risk Based Decision Made	Comments	Initials
1.2.2 Has a mission/business impact analysis been conducted? *NIST SP 800-30*								
1.2.3. Have additional controls been identified to sufficiently mitigate identified risks? *NIST SP 800-30*								

Notes:

Review of Security Controls

Routine evaluations and response to identified vulnerabilities are important elements of managing the risk of a system. The following questions are organized according to two critical elements. The levels for each of these critical elements should be determined based on the answers to the subordinate questions.

Specific Control Objectives and Techniques	L.1 Policy	L.2 Procedures	L.3 Implemented	L.4 Tested	L.5 Integrated	Risk Based Decision Made	Comments	Initials
2. Review of Security Controls *OMB Circular A-130, III FISCAM SP-5 NIST SP 800-18*								
2.1. Critical Element: Have the security controls of the system and interconnected systems been reviewed?								
2.1.1 Has the system and all network boundaries been subjected to periodic reviews? *FISCAM SP-5.1*								

Specific Control Objectives and Techniques	L.1 Policy	L.2 Procedures	L.3 Implemented	L.4 Tested	L.5 Integrated	Risk Based Decision Made	Comments	Initials
2.1.2 Has an independent review been performed when a significant change occurred? *OMB Circular A-130, III; FISCAM SP-5.1; NIST SP 800-18*								
2.1.3 Are routine self-assessments conducted? *NIST SP 800-18*								
2.1.4 Are tests and examinations of key controls routinely made, i.e., network scans, analyses of router and switch settings, penetration testing? *OMB Circular A-130, 8B3; NIST SP 800-18*								
2.1.5 Are security alerts and security incidents analyzed and remedial actions taken? *FISCAM SP 3-4; NIST SP 800-18*								

(*Continued*)

Specific Control Objectives and Techniques	L.1 Policy	L.2 Procedures	L.3 Implemented	L.4 Tested	L.5 Integrated	Risk Based Decision Made	Comments	Initials
2.2. Critical Element: Does management ensure that corrective actions are effectively implemented?								
2.2.1 Is there an effective and timely process for reporting significant weakness and ensuring effective remedial action? *FISCAM SP 5-1 and 5.2* *NIST SP 800-18*								

Notes:

446

Life Cycle

Like other aspects of an IT system, security is best managed if planned for throughout the IT system life cycle. There are many models for the IT system life cycle but most should contain five basic phases: initiation, development/acquisition, implementation, operation, and disposal. The following questions are organized according to two critical elements. The levels for each of these critical elements should be determined based on the answers to the subordinate questions.

Specific Control Objectives and Techniques	L.1 Policy	L.2 Procedures	L.3 Implemented	L.4 Tested	L.5 Integrated	Risk Based Decision Made	Comments	Initials
3. Life Cycle *OMB Circular A-130, III FISCAM CC-1.1*								
3.1. Critical Element: Has a system development life cycle methodology been developed?								
Initiation Phase								
3.1.1. Is the sensitivity of the system determined? *OMB Circular A-130, III FISCAM AC-1.1 & 1.2 NIST SP 800-18*								

(*Continued*)

Specific Control Objectives and Techniques	L.1 Policy	L.2 Procedures	L.3 Implemented	L.4 Tested	L.5 Integrated	Risk Based Decision Made	Comments	Initials
3.1.2. Does the business case document the resources required for adequately securing the system? *Clinger-Cohen*								
3.1.3 Does the Investment Review Board ensure any investment request includes the security resources needed? *Clinger-Cohen*								
3.1.4 Are authorization for software modifications documented and maintained? *FISCAM CC-1.2.*								
3.1.5 Does the budget request include the security resources required for the system? *GISRA*								
Development/Acquisition Phase								
3.1.6 During the system design, are security requirements identified? *NIST SP 800-18*								

Specific Control Objectives and Techniques	L.1 Policy	L.2 Procedures	L.3 Implemented	L.4 Tested	L.5 Integrated	Risk Based Decision Made	Comments	Initials
3.1.7 Was an initial risk assessment performed to determine security requirements? *NIST SP 800-30*								
3.1.8 Is there a written agreement with program officials on the security controls employed and residual risk? *NIST SP 800-18*								
3.1.9 Are security controls consistent with and an integral part of the IT architecture of the agency? *OMB Circular A-130, 8B3*								
3.1.10 Are the appropriate security controls with associated evaluation and test procedures developed before the procurement action? *NIST SP 800-18*								

(*Continued*)

Specific Control Objectives and Techniques	L.1 Policy	L.2 Procedures	L.3 Implemented	L.4 Tested	L.5 Integrated	Risk Based Decision Made	Comments	Initials
3.1.11 Do the solicitation documents (e.g., Request for Proposals) include security requirements and evaluation/test procedures? *NIST SP 800-18*								
3.1.12 Do the requirements in the solicitation documents permit updating security controls as new threats/vulnerabilities are identified and as new technologies are implemented? *NIST SP 800-18*								

Implementation Phase

**3.2. Critical Element:
Are changes controlled as programs progress through testing to final approval?**

Specific Control Objectives and Techniques	L.1 Policy	L.2 Procedures	L.3 Implemented	L.4 Tested	L.5 Integrated	Risk Based Decision Made	Comments	Initials
3.2.1 Are design reviews and system tests run prior to placing the system in production? *FISCAM CC-2.1* *NIST SP 800-18*								
3.2.2 Are the test results documented? *FISCAM CC-2.1* *NIST SP 800-18*								
3.2.3 Is certification testing of security controls conducted and documented? *NIST SP 800-18*								
3.2.4 If security controls were added since development, has the system documentation been modified to include them? *NIST SP 800-18*								

(*Continued*)

Specific Control Objectives and Techniques	L.1 Policy	L.2 Procedures	L.3 Implemented	L.4 Tested	L.5 Integrated	Risk Based Decision Made	Comments	Initials
3.2.5 If security controls were added since development, have the security controls been tested and the system recertified? *FISCAM CC-2.1* *NIST SP 800-18*								
3.2.6 Has the application undergone a technical evaluation to ensure that it meets applicable federal laws, regulations, policies, guidelines, and standards? *NIST SP 800-18*								
3.2.7 Does the system have written authorization to operate either on an interim basis with planned corrective action or full authorization? *NIST SP 800-18*								
Operation/Maintenance Phase								
3.2.8 Has a system security plan been developed and approved? *OMB Circular A-130, III* *FISCAM SP 2-1* *NIST SP 800-18*								

Specific Control Objectives and Techniques	L.1 Policy	L.2 Procedures	L.3 Implemented	L.4 Tested	L.5 Integrated	Risk Based Decision Made	Comments	Initials
3.2.9 If the system connects to other systems, have controls been established and disseminated to the owners of the interconnected systems? *NIST SP 800-18*								
3.2.10 Is the system security plan kept current? *OMB Circular A-130, III* *FISCAM SP 2-1* *NIST SP 800-18*								
Disposal Phase								
3.2.11 Are official electronic records properly disposed/archived? *NIST SP 800-18*								

(*Continued*)

453

Specific Control Objectives and Techniques	L.1 Policy	L.2 Procedures	L.3 Implemented	L.4 Tested	L.5 Integrated	Risk Based Decision Made	Comments	Initials
3.2.12 Is information or media purged, overwritten, degaussed, or destroyed when disposed or used elsewhere? *FISCAM AC-3.4* *NIST SP 800-18*								
3.2.13 Is a record kept of who implemented the disposal actions and verified that the information or media was sanitized? *NIST SP 800-18*								

Notes:

Authorize Processing (Certification & Accreditation)

Authorize processing (Note: Some agencies refer to this process as certification and accreditation) provides a form of assurance of the security of the system. The following questions are organized according to two critical elements. The levels for each of these critical elements should be determined based on the answers to the subordinate questions.

Specific Control Objectives and Techniques	L.1 Policy	L.2 Procedures	L.3 Implemented	L.4 Tested	L.5 Integrated	Risk Based Decision Made	Comments	Initials
Authorize Processing (Certification & Accreditation) *OMB Circular A-130, III FIPS 102*								
4.1.Critical Element: Has the system been certified/ recertified and authorized to process (accredited)?								
4.1.1 Has a technical and/ or security evaluation been completed or conducted when a significant change occurred? *NIST SP 800-18*								
4.1.2 Has a risk assessment been conducted when a significant change occurred? *NIST SP 800-18*								

(*Continued*)

455

Specific Control Objectives and Techniques	L.1 Policy	L.2 Procedures	L.3 Implemented	L.4 Tested	L.5 Integrated	Risk Based Decision Made	Comments	Initials
4.1.3 Have Rules of Behavior been established and signed by users? *NIST SP 800-18*								
4.1.4 Has a contingency plan been developed and tested? *NIST SP 800-18*								
4.1.5 Has a system security plan been developed, updated, and reviewed? *NIST SP 800-18*								
4.1.6 Are in-place controls operating as intended? *NIST SP 800-18*								
4.1.7 Are the planned and in-place controls consistent with the identified risks and the system and data sensitivity? *NIST SP 800-18*								

Specific Control Objectives and Techniques	L.1 Policy	L.2 Procedures	L.3 Implemented	L.4 Tested	L.5 Integrated	Risk Based Decision Made	Comments	Initials
4.1.8 Has management authorized interconnections to all systems (including systems owned and operated by another program, agency, organization or contractor)? *NIST 800-18*								
4.2. Critical Element: **Is the system operating on an interim authority to process in accordance with specified agency procedures?**								
4.2.1 Has management initiated prompt action to correct deficiencies? *NIST SP 800-18*								

Notes:

System Security Plan

System security plans provide an overview of the security requirements of the system and describe the controls in place or planned for meeting those requirements. The plan delineates responsibilities and expected behavior of all individuals who access the system. The following questions are organized according to two critical elements. The levels for each of these critical elements should be determined based on the answers to the subordinate questions.

Specific Control Objectives and Techniques	L.1 Policy	L.2 Procedures	L.3 Implemented	L.4 Tested	L.5 Integrated	Risk Based Decision Made	Comments	Initials
5. System security plan *OMB Circular A-130, III* *NIST SP 800-18* *FISCAM SP-2.1*								
5.1. Critical Element: **Is a system security plan documented for the system and all interconnected systems if the boundary controls are ineffective?**								
5.1.1 Is the system security plan approved by key affected parties and management? *FISCAM SP-2.1* *NIST SP 800-18*								

Specific Control Objectives and Techniques	L.1 Policy	L.2 Procedures	L.3 Implemented	L.4 Tested	L.5 Integrated	Risk Based Decision Made	Comments	Initials
5.1.2 Does the plan contain the topics prescribed in NIST Special Publication 800-18? *NIST SP 800-18*								
5.1.3 Is a summary of the plan incorporated into the strategic IRM plan? *OMB Circular A-130, III* *NIST SP 800-18*								
5.2. Critical Element: **Is the plan kept current?**								
5.2.1 Is the plan reviewed periodically and adjusted to reflect current conditions and risks? *FISCAM SP-2.1* *NIST SP 800-18*								

Notes:

OPERATIONAL CONTROLS

The operational controls address security methods focusing on mechanisms primarily implemented and executed by people (as opposed to systems). These controls are put in place to improve the security of a particular system (or group of systems). They often require technical or specialized expertise and often rely upon management activities as well as technical controls.

Personnel Security

Many important issues in computer security involve human users, designers, implementers, and managers. A broad range of security issues relates to how these individuals interact with computers and the access and authorities they need to do their jobs. The following questions are organized according to two critical elements. The levels for each of these critical elements should be determined based on the answers to the subordinate questions.

Specific Control Objectives and Techniques	L.1 Policy	L.2 Procedures	L.3 Implemented	L.4 Tested	L.5 Integrated	Risk Based Decision Made	Comments	Initials
Personnel Security *OMB Circular A-130, III*								
6.1.Critical Element: Are duties separated to ensure least privilege and individual accountability?								

Specific Control Objectives and Techniques	L.1 Policy	L.2 Procedures	L.3 Implemented	L.4 Tested	L.5 Integrated	Risk Based Decision Made	Comments	Initials
6.1.1 Are all positions reviewed for sensitivity level? *FISCAM SD-1.2 NIST SP 800-18*								
6.1.2 Are there documented job descriptions that accurately reflect assigned duties and responsibilities and that segregate duties? *FISCAM SD-1.2*								
6.1.3 Are sensitive functions divided among different individuals? *OMB Circular A-130, III FISCAM SD-1 NIST SP 800-18*								
6.1.4 Are distinct systems support functions performed by different individuals? *FISCAM SD-1.1*								
6.1.5 Are mechanisms in place for holding users responsible for their actions? *OMB Circular A-130, III FISCAM SD-2 & 3.2*								

(*Continued*)

461

Specific Control Objectives and Techniques	L.1 Policy	L.2 Procedures	L.3 Implemented	L.4 Tested	L.5 Integrated	Risk Based Decision Made	Comments	Initials
6.1.6 Are regularly scheduled vacations and periodic job/shift rotations required? *FISCAM SD-1.1* *FISCAM SP-4.1*								
6.1.7 Are hiring, transfer, and termination procedures established? *FISCAM SP-4.1* *NIST SP 800-18*								
6.1.8 Is there a process for requesting, establishing, issuing, and closing user accounts? *FISCAM SP-4.1* *NIST SP 800-18*								
6.2. Critical Element: **Is appropriate background screening for assigned positions completed prior to granting access?**								

Specific Control Objectives and Techniques	L.1 Policy	L.2 Procedures	L.3 Implemented	L.4 Tested	L.5 Integrated	Risk Based Decision Made	Comments	Initials
6.2.1 Are individuals who are authorized to bypass significant technical and operational controls screened prior to access and periodically thereafter? *OMB Circular A-130, III* *FISCAM SP-4.1*								
6.2.2 Are confidentiality or security agreements required for employees assigned to work with sensitive information? *FISCAM SP-4.1*								
6.2.3 When controls cannot adequately protect the information, are individuals screened prior to access? *OMB Circular A-130, III*								
6.2.4 Are there conditions for allowing system access prior to completion of screening? *FISCAM AC-2.2* *NIST SP 800-18*								

Notes:

Physical and Environmental Protection

Physical security and environmental security are the measures taken to protect systems, buildings, and related supporting infrastructures against threats associated with their physical environment. The following questions are organized according to three critical elements. The levels for each of these critical elements should be determined based on the answers to the subordinate questions.

Specific Control Objectives and Techniques	L.1 Policy	L.2 Procedures	L.3 Implemented	L.4 Tested	L.5 Integrated	Risk Based Decision Made	Comments	Initials
Physical and Environmental Protection								
Physical Access Control								
7.1. Critical Element: Have adequate physical security controls been implemented that are commensurate with the risks of physical damage or access?								
7.1.1 Is access to facilities controlled through the use of guards, identification badges, or entry devices such as key cards or biometrics? *FISCAM AC-3 NIST SP 800-18*								

Specific Control Objectives and Techniques	L.1 Policy	L.2 Procedures	L.3 Implemented	L.4 Tested	L.5 Integrated	Risk Based Decision Made	Comments	Initials
7.1.2 Does management regularly review the list of persons with physical access to sensitive facilities? *FISCAM AC-3.1*								
7.1.3 Are deposits and withdrawals of tapes and other storage media from the library authorized and logged? *FISCAM AC-3.1*								
7.1.4 Are keys or other access devices needed to enter the computer room and tape/media library? *FISCAM AC-3.1*								
7.1.5 Are unused keys or other entry devices secured? *FISCAM AC-3.1*								
7.1.6 Do emergency exit and re-entry procedures ensure that only authorized personnel are allowed to re-enter after fire drills, etc? *FISCAM AC-3.1*								

(Continued)

Specific Control Objectives and Techniques	L.1 Policy	L.2 Procedures	L.3 Implemented	L.4 Tested	L.5 Integrated	Risk Based Decision Made	Comments	Initials
7.1.7 Are visitors to sensitive areas signed in and escorted? *FISCAM AC-3.1*								
7.1.8 Are entry codes changed periodically? *FISCAM AC-3.1*								
7.1.9 Are physical accesses monitored through audit trails and apparent security violations investigated and remedial action taken? *FISCAM AC-4*								
7.1.10 Is suspicious access activity investigated and appropriate action taken? *FISCAM AC-4.3*								
7.1.11 Are visitors, contractors and maintenance personnel authenticated through the use of preplanned appointments and identification checks? *FISCAM AC-3.1*								

Specific Control Objectives and Techniques	L.1 Policy	L.2 Procedures	L.3 Implemented	L.4 Tested	L.5 Integrated	Risk Based Decision Made	Comments	Initials
Fire Safety Factors								
7.1.12 Are appropriate fire suppression and prevention devices installed and working? *FISCAM SC-2.2 NIST SP 800-18*								
7.1.13 Are fire ignition sources, such as failures of electronic devices or wiring, improper storage materials, and the possibility of arson, reviewed periodically? *NIST SP 800-18*								
Supporting Utilities								
7.1.14 Are heating and air-conditioning systems regularly maintained? *NIST SP 800-18*								
7.1.15 Is there a redundant air-cooling system? *FISCAM SC-2.2*								

(*Continued*)

Specific Control Objectives and Techniques	L.1 Policy	L.2 Procedures	L.3 Implemented	L.4 Tested	L.5 Integrated	Risk Based Decision Made	Comments	Initials
7.1.16 Are electric power distribution, heating plants, water, sewage, and other utilities periodically reviewed for risk of failure? *FISCAM SC-2.2* *NIST SP 800-18*								
7.1.17 Are building plumbing lines known and do not endanger system? *FISCAM SC-2.2* *NIST SP 800-18*								
7.1.18 Has an uninterruptible power supply or backup generator been provided? *FISCAM SC-2.2*								
7.1.19 Have controls been implemented to mitigate other disasters, such as floods, earthquakes, etc.? *FISCAM SC-2.2*								

Specific Control Objectives and Techniques	L.1 Policy	L.2 Procedures	L.3 Implemented	L.4 Tested	L.5 Integrated	Risk Based Decision Made	Comments	Initials
Interception of Data								
7.2. Critical Element: Is data protected from interception?								
7.2.1 Are computer monitors located to eliminate viewing by unauthorized persons? *NIST SP 800-18*								
7.2.2 Is physical access to data transmission lines controlled? *NIST SP 800-18*								
Mobile and Portable Systems								
7.3. Critical Element: Are mobile and portable systems protected?								
7.3.1 Are sensitive data files encrypted on all portable systems? *NIST SP 800-14*								
7.3.2 Are portable systems stored securely? *NIST SP 800-14*								

Notes:

Production, Input/Output Controls

There are many aspects to supporting IT operations. Topics range from a user help desk to procedures for storing, handling and destroying media. The following questions are organized according to two critical elements. The levels for each of these critical elements should be determined based on the answers to the subordinate questions.

Specific Control Objectives and Techniques	L.1 Policy	L.2 Procedures	L.3 Implemented	L.4 Tested	L.5 Integrated	Risk Based Decision Made	Comments	Initials
Production, Input/ Output Controls								
8.1. Critical Element: **Is there user support?**								
8.1.1 Is there a help desk or group that offers advice? *NIST SP 800-18*								
8.2. Critical Element: **Are there media controls?**								
8.2.1 Are there processes to ensure that unauthorized individuals cannot read, copy, alter, or steal printed or electronic information? *NIST SP 800-18*								

Specific Control Objectives and Techniques	L.1 Policy	L.2 Procedures	L.3 Implemented	L.4 Tested	L.5 Integrated	Risk Based Decision Made	Comments	Initials
8.2.2 Are there processes for ensuring that only authorized users pick up, receive, or deliver input and output information and media? *NIST SP 800-18*								
8.2.3 Are audit trails used for receipt of sensitive inputs/outputs? *NIST SP 800-18*								
8.2.4 Are controls in place for transporting or mailing media or printed output? *NIST SP 800-18*								
8.2.5 Is there internal/external labeling for sensitivity? *NIST SP 800-18*								
8.2.6 Is there external labeling with special handling instructions? *NIST SP 800-18*								
8.2.7 Are audit trails kept for inventory management? *NIST SP 800-18*								

(*Continued*)

Specific Control Objectives and Techniques	L.1 Policy	L.2 Procedures	L.3 Implemented	L.4 Tested	L.5 Integrated	Risk Based Decision Made	Comments	Initials
8.2.8 Is media sanitized for reuse? *FISCAM AC-3.4* *NIST SP 800-18*								
8.2.9 Is damaged media stored and/or destroyed? *NIST SP 800-18*								
8.2.10 Is hardcopy media shredded or destroyed when no longer needed? *NIST SP 800-18*								

Notes:

Contingency Planning

Contingency planning involves more than planning for a move offsite after a disaster destroys a facility. It also addresses how to keep an organization's critical functions operating in the event of disruptions, large and small. The following questions are organized according to three critical elements. The levels for each of these critical elements should be determined based on the answers to the subordinate questions.

Specific Control Objectives and Techniques	L.1 Policy	L.2 Procedures	L.3 Implemented	L.4 Tested	L.5 Integrated	Risk Based Decision Made	Comments	Initials
Contingency Planning *OMB Circular A-130, III*								
9.1. Critical Element: Have the most critical and sensitive operations and their supporting computer resources been identified?								
9.1.1 Are critical data files and operations identified and the frequency of file backup documented? *FISCAM SC-1.1 & 3.1 NSTSP 800-18*								
9-1.2 Are resources supporting critical operations identified? *FISCAM SC-1.2*								

(*Continued*)

473

Specific Control Objectives and Techniques	L.1 Policy	L.2 Procedures	L.3 Implemented	L.4 Tested	L.5 Integrated	Risk Based Decision Made	Comments	Initials
9.1.3 Have processing priorities been established and approved by management? *FISCAM SC-1.3*								
9.2. Critical Element: Has a comprehensive contingency plan been developed and documented?								
9.2.1 Is the plan approved by key affected parties? *FISCAM SC-3.1*								
9.2.2 Are responsibilities for recovery assigned? *FISCAM SC-3.1*								
9.2.3 Are there detailed instructions for restoring operations? *FISCAM SC-3.1*								
9.2.4 Is there an alternate processing site; if so, is there a contract or interagency agreement in place? *FISCAM SC-3.1* *NIST SP 800-18*								

Specific Control Objectives and Techniques	L.1 Policy	L.2 Procedures	L.3 Implemented	L.4 Tested	L.5 Integrated	Risk Based Decision Made	Comments	Initials
9.2.5 Is the location of stored backups identified? *NIST SP 800-18*								
9.2.6 Are backup files created on a prescribed basis and rotated off-site often enough to avoid disruption if current files are damaged? *FISCAM SC-2.1*								
9.2.7 Is system and application documentation maintained at the off-site location? *FISCAM SC-2.1*								
9.2.8 Are all system defaults reset after being restored from a backup? *FISCAM SC-3.1*								
9.2.9 Are the backup storage site and alternate site geographically removed from the primary site and physically protected? *FISCAM SC-2.1*								

(*Continued*)

Specific Control Objectives and Techniques	L.1 Policy	L.2 Procedures	L.3 Implemented	L.4 Tested	L.5 Integrated	Risk Based Decision Made	Comments	Initials
9.2.10 Has the contingency plan been distributed to all appropriate personnel? *FISCAM SC-3.1*								
9.3. Critical Element: **Are tested contingency/disaster recovery plans in place?**								
9.3.1 Is an up-to-date copy of the plan stored securely off-site? *FISCAM SC-3.1*								
9.3.2 Are employees trained in their roles and responsibilities? *FISCAM SC-2.3* *NIST SP 800-18*								
9.3.3 Is the plan periodically tested and readjusted as appropriate? *FISCAM SC-3.1* *NIST SP 800-18*								

Notes:

Hardware and System Software Maintenance

These are controls used to monitor the installation of, and updates to, hardware and software to ensure that the system functions as expected and that a historical record is maintained of changes. Some of these controls are also covered in the Life Cycle Section. The following questions are organized according to three critical elements. The levels for each of these critical elements should be determined based on the answers to the subordinate questions.

Specific Control Objectives and Techniques	L.1 Policy	L.2 Procedures	L.3 Implemented	L.4 Tested	L.5 Integrated	Risk Based Decision Made	Comments	Initials
Hardware and System Software Maintenance *OMB Circular A-130, III*								
10.1. Critical Element: Is access limited to system software and hardware?								
10.1.1 Are restrictions in place on who performs maintenance and repair activities? *OMB Circular A-130, III FISCAM SS-3.1 NIST SP 800-18*								
10.1.2 Is access to all program libraries restricted and controlled? *FISCAM CC-3.2 & 3.3*								

(Continued)

Specific Control Objectives and Techniques	L.1 Policy	L.2 Procedures	L.3 Implemented	L.4 Tested	L.5 Integrated	Risk Based Decision Made	Comments	Initials
10.1.3 Are there on-site and off-site maintenance procedures (e.g., escort of maintenance personnel, sanitization of devices removed from the site)? *NIST SP 800-18*								
10.1.4 Is the operating system configured to prevent circumvention of the security software and application controls? *FISCAM SS-1.2*								
10.1.5 Are up-to-date procedures in place for using and monitoring use of system utilities? *FISCAM SS-2.1*								

Specific Control Objectives and Techniques	L.1 Policy	L.2 Procedures	L.3 Implemented	L.4 Tested	L.5 Integrated	Risk Based Decision Made	Comments	Initials
10.2. Critical Element: **Are all new and revised hardware and software authorized, tested and approved before implementation?**								
10.2.1 Is an impact analysis conducted to determine the effect of proposed changes on existing security controls, including the required training needed to implement the control? *NIST SP 800-18*								
10.2.2 Are system components tested, documented, and approved (operating system, utility, applications) prior to promotion to production? *FISCAM SS-3.1, 3.2, & CC-2.1* *NIST SP 800-18*								

(*Continued*)

Specific Control Objectives and Techniques	L.1 Policy	L.2 Procedures	L.3 Implemented	L.4 Tested	L.5 Integrated	Risk Based Decision Made	Comments	Initials
10.2.3 Are software change request forms used to document requests and related approvals? *FISCAM CC-1.2* *NIST SP 800-18*								
10.2.4 Are there detailed system specifications prepared and reviewed by management? *FISCAM CC-2.1*								
10.2.5 Is the type of test data to be used specified, i.e., live or made up? *NIST SP 800-18*								
10.2.6 Are default settings of security features set to the most restrictive mode? *PSN Security Assessment Guidelines*								
10.2.7 Are there software distribution implementation orders including effective date provided to all locations? *FISCAM CC-2.3*								

Specific Control Objectives and Techniques	L.1 Policy	L.2 Procedures	L.3 Implemented	L.4 Tested	L.5 Integrated	Risk Based Decision Made	Comments	Initials
10.2.8 Is there version control? *NIST SP 800-18*								
10.2.9 Are programs labeled and inventoried? *FISCAM CC-3.1*								
10.2.10 Are the distribution and implementation of new or revised software documented and reviewed? *FISCAM SS-3.2*								
10.2.11 Are emergency change procedures documented and approved by management, either prior to the change or after the fact? *FISCAM CC-2.2*								
10.2.12 Are contingency plans and other associated documentation updated to reflect system changes? *FISCAM SC-2.1 NIST SP 800-18*								

(*Continued*)

Specific Control Objectives and Techniques	L.1 Policy	L.2 Procedures	L.3 Implemented	L.4 Tested	L.5 Integrated	Risk Based Decision Made	Comments	Initials
10.2.13 Is the use of copyrighted software or shareware and personally owned software/ equipment documented? *NIST SP 800-18*								
10.3. Are systems managed to reduce vulnerabilities?								
10.3.1 Are systems periodically reviewed to identify and, when possible, eliminate unnecessary services (e.g., FTP, HTTP, mainframe supervisor calls)? *NIST SP 800-18*								
10.3.2 Are systems periodically reviewed for known vulnerabilities and software patches promptly installed? *NIST SP 800-18*								

Notes:

Data Integrity

Data integrity controls are used to protect data from accidental or malicious alteration or destruction and to provide assurance to the user the information meets expectations about its quality and integrity. The following questions are organized according to two critical elements. The levels for each of these critical elements should be determined based on the answers to the subordinate questions.

Specific Control Objectives and Techniques	L.1 Policy	L.2 Procedures	L.3 Implemented	L.4 Tested	L.5 Integrated	Risk Based Decision Made	Comments	Initials
Data Integrity *OMB Circular A-130, 8B3*								
11.1. Critical Element: Is virus detection and elimination software installed and activated?								
11.1.1 Are virus signature files routinely updated? *NIST SP 800-18*								
11.1.2 Are virus scans automatic? *NIST SP 800-18*								
11.2. Critical Element: Are data integrity and validation controls used to provide assurance that the information has not been altered and the system functions as intended?								

(*Continued*)

Specific Control Objectives and Techniques	L.1 Policy	L.2 Procedures	L.3 Implemented	L.4 Tested	L.5 Integrated	Risk Based Decision Made	Comments	Initials
11.2.1 Are reconciliation routines used by applications, i.e., checksums, hash totals, record counts? *NIST SP 800-18*								
11.2.2 Is inappropriate or unusual activity reported, investigated, and appropriate actions taken? *FISCAM SS-2.2*								
11.2.3 Are procedures in place to determine compliance with password policies? *NIST SP 800-18*								
11.2.4 Are integrity verification programs used by applications to look for evidence of data tampering, errors, and omissions? *NIST SP 800-18*								
11.2.5 Are intrusion detection tools installed on the system? *NIST SP 800-18*								

Specific Control Objectives and Techniques	L.1 Policy	L.2 Procedures	L.3 Implemented	L.4 Tested	L.5 Integrated	Risk Based Decision Made	Comments	Initials
11.2.6 Are the intrusion detection reports routinely reviewed and suspected incidents handled accordingly? *NIST SP 800-18*								
11.2.7 Is system performance monitoring used to analyze system performance logs in real time to look for availability problems, including active attacks? *NIST SP 800-18*								
11.2.8 Is penetration testing performed on the system? *NIST SP 800-18*								
11.2.9 Is message authentication used? *NIST SP 800-18*								

Notes:

Documentation

The documentation contains descriptions of the hardware, software, policies, standards, procedures, and approvals related to the system and formalize the system's security controls. When answering whether there are procedures for each control objective, the question should be phrased "are there procedures for ensuring the documentation is obtained and maintained." The following questions are organized according to two critical elements. The levels for each of these critical elements should be determined based on the answers to the subordinate questions.

Specific Control Objectives and Techniques	L.1 Policy	L.2 Procedures	L.3 Implemented	L.4 Tested	L.5 Integrated	Risk Based Decision Made	Comments	Initials
Documentation *OMB Circular A-130, 8B3*								
12.1. Critical Element: Is there sufficient documentation that explains how software/ hardware is to be used?								
12.1.1 Is there vendor-supplied documentation of purchased software? *NIST SP 800-18*								
12.1.2 Is there vendor-supplied documentation of purchased hardware? *NIST SP 800-18*								

Specific Control Objectives and Techniques	L.1 Policy	L.2 Procedures	L.3 Implemented	L.4 Tested	L.5 Integrated	Risk Based Decision Made	Comments	Initials
12.1.3 Is there application documentation for in-house applications? *NIST SP 800-18*								
12.1.4 Are there network diagrams and documentation on setups of routers and switches? *NIST SP 800-18*								
12.1.5 Are there software and hardware testing procedures and results? *NIST SP 800-18*								
12.1.6 Are there standard operating procedures for all the topic areas covered in this document? *NIST SP 800-18*								
12.1.7 Are there user manuals? *NIST SP 800-18*								

(*Continued*)

Specific Control Objectives and Techniques	L.1 Policy	L.2 Procedures	L.3 Implemented	L.4 Tested	L.5 Integrated	Risk Based Decision Made	Comments	Initials
12.1.8 Are there emergency procedures? *NIST SP 800-18*								
12.1.9 Are there backup procedures? *NIST SP 800-18*								
12.2. Critical Element: Are there formal security and operational procedures documented?								
12.2.1 Is there a system security plan? *OMB Circular A-130, III* *FISCAM SP-2.1* *NIST SP 800-18*								
12.2.2 Is there a contingency plan? *NIST SP 800-18*								

Specific Control Objectives and Techniques	L.1 Policy	L.2 Procedures	L.3 Implemented	L.4 Tested	L.5 Integrated	Risk Based Decision Made	Comments	Initials
12.2.3 Are there written agreements regarding how data is shared between interconnected systems? *OMB A-130, III NIST SP 800-18*								
12.2.4 Are there risk assessment reports? *NIST SP 800-18*								
12.2.5 Are there certification and accreditation documents and a statement authorizing the system to process? *NIST SP 800-18*								

Notes:

Security Awareness, Training, and Education

People are a crucial factor in ensuring the security of computer systems and valuable information resources. Security awareness, training, and education enhance security by improving awareness of the need to protect system resources. Additionally, training develops skills and knowledge so computer users can perform their jobs more securely and build in-depth knowledge. The following questions are organized according to two critical elements. The levels for each of these critical elements should be determined based on the answers to the subordinate questions.

Specific Control Objectives and Techniques	L.1 Policy	L.2 Procedures	L.3 Implemented	L.4 Tested	L.5 Integrated	Risk Based Decision Made	Comments	Initials
Security Awareness, Training, and Education *OMB Circular A-130, III*								
13.1. Critical Element: **Have employees received adequate training to fulfill their security responsibilities?**								
13.1.1 Have employees received a copy of the Rules of Behavior? *NIST SP 800-18*								
13.1.2 Are employee training and professional development documented and monitored? *FISCAM SP-4.2*								

Specific Control Objectives and Techniques	L.1 Policy	L.2 Procedures	L.3 Implemented	L.4 Tested	L.5 Integrated	Risk Based Decision Made	Comments	Initials
13.1.3 Is there mandatory annual refresher training? *OMB Circular A-130, III*								
13.1.4 Are methods employed to make employees aware of security, i.e., posters, booklets? *NIST SP 800-18*								
13.1.5 Have employees received a copy of or have easy access to agency security procedures and policies? *NIST SP 800-18*								

Notes:

Incident Response Capability

Computer security incidents are an adverse event in a computer system or network. Such incidents are becoming more common and impact far-reaching. The following questions are organized according to two critical elements. The levels for each of these critical elements should be determined based on the answers to the subordinate questions.

Specific Control Objectives and Techniques	L.1 Policy	L.2 Procedures	L.3 Implemented	L.4 Tested	L.5 Integrated	Risk Based Decision Made	Comments	Initials
Incident Response Capability *OMB Circular A-130, III* *FISCAM SP-3.4* *NIST 800-18*								
14.1. Critical Element: **Is there a capability to provide help to users when a security incident occurs in the system?**								
14.1.1 Is a formal incident response capability available? *FISCAM SP-3.4* *NIST SP 800-18*								
14.1.2 Is there a process for reporting incidents? *FISCAM SP-3.4* *NIST SP 800-18*								

Specific Control Objectives and Techniques	L.1 Policy	L.2 Procedures	L.3 Implemented	L.4 Tested	L.5 Integrated	Risk Based Decision Made	Comments	Initials
14.1.3 Are incidents monitored and tracked until resolved? *NIST SP 800-18*								
14.1.4 Are personnel trained to recognize and handle incidents? *FISCAM SP-3.4 NIST SP 800-18*								
14.1.5 Are alerts/advisories received and responded to? *NIST SP 800-18*								
14.1.6 Is there a process to modify incident handling procedures and control techniques after an incident occurs? *NIST SP 800-18*								
14.2. Critical Element: Is incident related information shared with appropriate organizations?								

(*Continued*)

Specific Control Objectives and Techniques	L.1 Policy	L.2 Procedures	L.3 Implemented	L.4 Tested	L.5 Integrated	Risk Based Decision Made	Comments	Initials
14.2.1 Is incident information and common vulnerabilities or threats shared with owners of interconnected systems? *OMB A-130, III* *NIST SP 800-18*								
14.2.2 Is incident information shared with FedCIRC[3] concerning incidents and common vulnerabilities and threats? *OMB A-130, III* *GISRA*								
14.2.3 Is incident information reported to FedCIRC, NIPC,[4] and local law enforcement when necessary? *OMB A-130, III* *GISRA*								

Notes:

TECHNICAL CONTROLS

Technical controls focus on security controls that the computer system executes. The controls can provide automated protection for unauthorized access or misuse, facilitate detection of security violations, and support security requirements for applications and data.

Identification and Authentication

Identification and authentication is a technical measure that prevents unauthorized people (or unauthorized processes) from entering IT system. Access control usually requires that the system be able to identify and differentiate among users. The following questions are organized according to two critical elements. The levels for each of these critical elements should be determined based on the answers to the subordinate questions.

Specific Control Objectives and Techniques	L.1 Policy	L.2 Procedures	L.3 Implemented	L.4 Tested	L.5 Integrated	Risk Based Decision Made	Comments	Initials
Identification and Authentication *OMB Circular A-130, III* *FISCAM AC-2* *NIST SP 800-18*								
15.1. Critical Element: **Are users individually authenticated via passwords, tokens, or other devices?**								

(*Continued*)

495

Specific Control Objectives and Techniques	L.1 Policy	L.2 Procedures	L.3 Implemented	L.4 Tested	L.5 Integrated	Risk Based Decision Made	Comments	Initials
15.1.1 Is a current list maintained and approved of authorized users and their access? *FISCAM AC-2 NIST SP 800-18*								
15.1.2 Are digital signatures used and conform to FIPS 186-2? *NIST SP 800-18*								
15.1.3 Are access scripts with embedded passwords prohibited? *NIST SP 800-18*								
15.1.4 Is emergency and temporary access authorized? *FISCAM AC-2.2*								
15.1.5 Are personnel files matched with user accounts to ensure that terminated or transferred individuals do not retain system access? *FISCAM AC-3.2*								

Specific Control Objectives and Techniques	L.1 Policy	L.2 Procedures	L.3 Implemented	L.4 Tested	L.5 Integrated	Risk Based Decision Made	Comments	Initials
15.1.6 Are passwords changed at least every ninety days or earlier if needed? *FISCAM AC-3.2* *NIST SP 800-18*								
15.1.7 Are passwords unique and difficult to guess (e.g., do passwords require alpha numeric, upper/lower case, and special characters)? *FISCAM AC-3.2* *NIST SP 800-18*								
15.1.8 Are inactive user identifications disabled after a specified period of time? *FISCAM AC-3.2* *NIST SP 800-18*								
15.1.9 Are passwords not displayed when entered? *FISCAM AC-3.2* *NIST SP 800-18*								
15.1.10 Are there procedures in place for handling lost and compromised passwords? *FISCAM AC-3.2* *NIST SP 800-18*								

(*Continued*)

Specific Control Objectives and Techniques	L.1 Policy	L.2 Procedures	L.3 Implemented	L.4 Tested	L.5 Integrated	Risk Based Decision Made	Comments	Initials
15.1.11 Are passwords distributed securely and users informed not to reveal their passwords to anyone (social engineering)? *NIST SP 800-18*								
15.1.12 Are passwords transmitted and stored using secure protocols/algorithms? *FISCAM AC-3.2* *NIST SP 800-18*								
15.1.13 Are vendor-supplied passwords replaced immediately? *FISCAM AC-3.2* *NIST SP 800-18*								
15.1.14 Is there a limit to the number of invalid access attempts that may occur for a given user? *FISCAM AC-3.2* *NIST SP 800-18*								
15.2. Critical Element: **Are access controls enforcing segregation of duties?**								

Specific Control Objectives and Techniques	L.1 Policy	L.2 Procedures	L.3 Implemented	L.4 Tested	L.5 Integrated	Risk Based Decision Made	Comments	Initials
15.2.1 Does the system correlate actions to users? *OMB A-130, III FISCAM SD-2.1*								
15.2.2 Do data owners periodically review access authorizations to determine whether they remain appropriate? *FISCAM AC-2.1*								

Notes:

Logical Access Controls

Logical access controls are the system-based mechanisms used to designate who or what is to have access to a specific system resource and the type of transactions and functions that are permitted. The following questions are organized according to three critical elements. The levels for each of these critical elements should be determined based on the answers to the subordinate questions.

Specific Control Objectives and Techniques	L.1 Policy	L.2 Procedures	L.3 Implemented	L.4 Tested	L.5 Integrated	Risk Based Decision Made	Comments	Initials
Logical Access Controls *OMB Circular A-130, III* *FISCAM AC-3.2* *NIST SP-800-18*								
16.1. Critical Element: **Do the logical access controls restrict users to authorized transactions and functions?** *FISCAM AC-3.2* *NIST SP 800-18*								
16.1.1 Can the security controls detect unauthorized access attempts?								
16.1.2 Is there access control software that prevents an individual from having all necessary authority or information access to allow fraudulent activity without collusion? *FISCAM AC-3.2* *NIST SP 800-18*								

Specific Control Objectives and Techniques	L.1 Policy	L.2 Procedures	L.3 Implemented	L.4 Tested	L.5 Integrated	Risk Based Decision Made	Comments	Initials
16.1.3 Is access to security software restricted to security administrators? *FISCAM AC-3.2*								
16.1.4 Do workstations disconnect or screen savers lock system after a specific period of inactivity? *FISCAM AC-3.2 NIST SP 800-18*								
16.1.5 Are inactive users' accounts monitored and removed when not needed? *FISCAM AC-3.2 NIST SP 800-18*								
16.1.6 Are internal security labels (naming conventions) used to control access to specific information types or files? *FISCAM AC-3.2 NIST SP 800-18*								
16.1.7 If encryption is used, does it meet federal standards? *NIST SP 800-18*								

(*Continued*)

Specific Control Objectives and Techniques	L.1 Policy	L.2 Procedures	L.3 Implemented	L.4 Tested	L.5 Integrated	Risk Based Decision Made	Comments	Initials
16.1.8 If encryption is used, are there procedures for key generation, distribution, storage, use, destruction, and archiving? *NIST SP 800-18*								
16.1.9 Is access restricted to files at the logical view or field? *FISCAM AC-3.2*								
16.1.10 Is access monitored to identify apparent security violations and are such events investigated? *FISCAM AC-4*								
16.2. Critical Element: **Are there logical controls over network access?**								
16.2.1 Has communication software been implemented to restrict access through specific terminals? *FISCAM AC-3.2*								
16.2.2 Are insecure protocols (e.g., UDP, ftp) disabled? *PSN Security Assessment Guidelines*								

Specific Control Objectives and Techniques	L.1 Policy	L.2 Procedures	L.3 Implemented	L.4 Tested	L.5 Integrated	Risk Based Decision Made	Comments	Initials
16.2.3 Have all vendor-supplied default security parameters been reinitialized to more secure settings? *PSN Security Assessment Guidelines*								
16.2.4 Are there controls that restrict remote access to the system? *NIST SP 800-18*								
16.2.5 Are network activity logs maintained and reviewed? *FISCAM AC-3.2*								
16.2.6 Does the network connection automatically disconnect at the end of a session? *FISCAM AC-3.2*								
16.2.7 Are trust relationships among hosts and external entities appropriately restricted? *PSN Security Assessment Guidelines*								

(*Continued*)

Specific Control Objectives and Techniques	L.1 Policy	L.2 Procedures	L.3 Implemented	L.4 Tested	L.5 Integrated	Risk Based Decision Made	Comments	Initials
16.2.8 Is dial-in access monitored? *FISCAM AC-3.2*								
16.2.9 Is access to telecommunications hardware or facilities restricted and monitored? *FISCAM AC-3.2*								
16.2.10 Are firewalls or secure gateways installed? *NIST SP 800-18*								
16.2.11 If firewalls are installed do they comply with firewall policy and rules? *FISCAM AC-3.2*								
16.2.12 Are guest and anonymous accounts authorized and monitored? *PSN Security Assessment Guidelines*								

Specific Control Objectives and Techniques	L.1 Policy	L.2 Procedures	L.3 Implemented	L.4 Tested	L.5 Integrated	Risk Based Decision Made	Comments	Initials
16.2.13 Is an approved standardized log-on banner displayed on the system warning unauthorized users that they have accessed a U.S. Government system and can be punished? *FISCAM AC-3.2* *NIST SP 800-18*								
16.2.14 Are sensitive data transmissions encrypted? *FISCAM AC-3.2*								
16.2.15 Is access to tables defining network options, resources, and operator profiles restricted? *FISCAM AC-3.2*								
16.3. Critical Element: If the public accesses the system, are there controls implemented to protect the integrity of the application and the confidence of the public?								
16.3.1 Is a privacy policy posted on the web site? *OMB-99-18*								

Notes:

Audit Trails

Audit trails maintain a record of system activity by system or application processes and by user activity. In conjunction with appropriate tools and procedures, audit trails can provide individual accountability, a means to reconstruct events, detect intrusions, and identify problems. The following questions are organized under one critical element. The levels for the critical element should be determined based on the answers to the subordinate questions.

Specific Control Objectives and Techniques	L.1 Policy	L.2 Procedures	L.3 Implemented	L.4 Tested	L.5 Integrated	Risk Based Decision Made	Comments	Initials
Audit Trails *OMB Circular A-130, III FISCAM AC-4.1 NIST SP 800-18*								
17.1. Critical Element: Is activity involving access to and modification of sensitive or critical files logged, monitored, and possible security violations investigated?								
17.1.1 Does the audit trail provide a trace of user actions? *NIST SP 800-18*								

Specific Control Objectives and Techniques	L.1 Policy	L.2 Procedures	L.3 Implemented	L.4 Tested	L.5 Integrated	Risk Based Decision Made	Comments	Initials
17.1.2 Can the audit trail support after-the-fact investigations of how, when, and why normal operations ceased? *NIST SP 800-18*								
17.1.3 Is access to online audit logs strictly controlled? *NIST SP 800-18*								
17.1.4 Are off-line storage of audit logs retained for a period of time, and if so, is access to audit logs strictly controlled? *NIST SP 800-18*								
17.1.5 Is there separation of duties between security personnel who administer the access control function and those who administer the audit trail? *NIST SP 800-18*								
17.1.6 Are audit trails reviewed frequently? *NIST SP 800-18*								

(*Continued*)

Specific Control Objectives and Techniques	L.1 Policy	L.2 Procedures	L.3 Implemented	L.4 Tested	L.5 Integrated	Risk Based Decision Made	Comments	Initials
17.1.7 Are automated tools used to review audit records in real time or near real time? *NIST SP 800-18*								
17.1.8 Is suspicious activity investigated and appropriate action taken? *FISCAM AC-4.3*								
17.1.9 Is keystroke monitoring used? If so, are users notified? *NIST SP 800-18*								

Notes:

DOCUMENT APPENDIX

APPENDIX B—SOURCE OF CONTROL CRITERIA

Office of Management and Budget Circular A-130, "Management of Federal Information Resources," Section 8B3 and Appendix III, "Security of Federal Automated Information Resources."	Establishes a minimum set of controls to be included in Federal IT security programs.
Computer Security Act of 1987.	This statute set the stage for protecting systems by codifying the requirement for Government-wide IT security planning and training.
Paperwork Reduction Act of 1995.	The PRA established a comprehensive information resources management framework including security and subsumed the security responsibilities of the Computer Security Act of 1987.
Clinger-Cohen Act of 1996.	This Act linked security to agency capital planning and budget processes, established agency Chief Information Officers, and re-codified the Computer Security Act of 1987.
Presidential Decision Directive 63, "Protecting America's Critical Infrastructures."	This directive specifies agency responsibilities for protecting the nation's infrastructure, assessing vulnerabilities of public and private sectors, and eliminating vulnerabilities.
OMB Memorandum 99-18, "Privacy Policies on Federal Web Sites."	This memorandum directs Departments and Agencies to post clear privacy policies on World Wide Web sites, and provides guidance for doing so.
General Accounting Office "Federal Information System Control Audit Manual" (FISCAM).	The FISCAM methodology provides guidance to auditors in evaluating internal controls over the confidentiality, integrity, and availability of data maintained in computer-based information systems.
NIST Special Publication 800-14, "Generally Accepted Principles and Practices for Security Information Technology Systems."	This publication guides organizations on the types of controls, objectives, and procedures that comprise an effective security program.

(continued)

NIST Special Publication 800-18, "Guide for Developing Security Plans for Information Technology Systems."	This publication details the specific controls that should be documented in a system security plan.
Defense Authorization Act (P.L. 106-398) including Title X, Subtitle G, "Government Information Security Reform" (GISRA)	The act primarily addresses the program management and evaluation aspects of security.
Office of the Manager, National Communications Systems, "Public Switched Network Security Assessment Guidelines."	The guide describes a risk assessment procedure, descriptions of a comprehensive security program, and a summary checklist.
Federal Information Processing Standards.	These documents contain mandates and/or guidance for improving the utilization and management of computers and IT systems in the Federal Government.

APPENDIX C—FEDERAL INFORMATION TECHNOLOGY SECURITY ASSESSMENT FRAMEWORK

Overview

Information and the systems that process it are among the most valuable assets of any organization. Adequate security of these assets is a fundamental management responsibility. Consistent with Office of Management and Budget (OMB) policy, each agency must implement and maintain a program to adequately secure its information and system assets. Agency programs must: 1) assure that systems and applications operate effectively and provide appropriate confidentiality, integrity, and availability; and 2) protect information commensurate with the level of risk and magnitude of harm resulting from loss, misuse, unauthorized access, or modification.

Agencies must plan for security, and ensure that the appropriate officials are assigned security responsibility and authorize system processing prior to operations and periodically thereafter. These management responsibilities presume that responsible agency officials understand the risks and other factors that could negatively impact their mission goals. Moreover, these officials must understand the current status of security programs and controls in order to make informed judgments and investments that appropriately mitigate risks to an acceptable level.

The Federal Information Technology (IT) Security Assessment Framework (or Framework) provides a method for agency officials to 1) determine the current status of their security programs relative to existing policy and 2) where necessary, establish a target for improvement. It does not establish new security requirements. The Framework may be used to assess the status of security controls for a given asset or collection of assets. These assets include information, individual systems (e.g., major applications, general support systems, mission critical systems), or a logically related grouping of systems that support operational programs, or operational programs (e.g., Air Traffic Control, Medicare, Student Aid). Assessing all asset security controls and all interconnected systems that the asset depends on produces a picture of both the security condition of an agency component and of the entire agency.

The Framework comprises five levels to guide agency assessment of their security programs and assist in prioritizing efforts for improvement. Coupled with the NIST-prepared self-assessment questionnaire,[5] the Framework provides a vehicle for consistent and effective measurement of the security status for a given asset. The security status is measured by determining if specific security controls are documented, implemented, tested and reviewed, and incorporated into a cyclical review/improvement program, as well as whether unacceptable risks are identified and mitigated. The NIST questionnaire provides specific questions that identify the control criteria against which agency policies, procedures, and security controls can be compared. Appendix A contains a sample of the upcoming NIST Special Publication.

The Framework is divided into five levels: Level 1 of the Framework reflects that an asset has documented security policy. At level 2, the asset also has documented procedures and controls to implement the policy. Level 3 indicates that procedures and controls have been implemented. Level 4 shows that the procedures and controls are tested and reviewed. At level 5, the asset has procedures and controls fully integrated into a comprehensive program.

Each level represents a more complete and effective security program. OMB and the Council recognize that the security needs for the tens of thousands of Federal information systems differ. Agencies should note that testing the effectiveness of the asset and all interconnected systems that the asset depends on is essential to understanding whether risk has been properly mitigated. When an individual system does not achieve level 4, agencies should determine whether that system meets the criteria found in OMB Memorandum M00-07 (February 28, 2000) "Incorporating and Funding Security in Information Systems Investments." Agencies should seek to bring all assets to level 4 and ultimately level 5.

Integral to all security programs whether for an asset or an entire agency is a risk assessment process that includes determining the level of sensitivity of information and systems. Many agencies have developed their own methods of making these determinations. For example, the Department of Health and Human Services uses a four—track scale for confidentiality, integrity, and availability. The Department of Energy uses five groupings or "clusters" to address sensitivity.

Regardless of the method used, the asset owner is responsible for determining how sensitive the asset is, what level of risk is acceptable, and which specific controls are necessary to provide adequate security to that asset. Again, each implemented security control must be periodically tested for effectiveness. The decision to implement and the results of the testing should be documented.

Framework Description

The Federal Information Technology Security Assessment Framework (Framework) identifies five levels of IT security program effectiveness (see Figure 1). The five levels measure specific management, operational, and technical control objectives. Each of the five levels contains criteria to determine if the level is adequately implemented. For example, in level 1, all written policy should contain the purpose and scope of the policy, the individual(s) responsible for implementing the policy, and the consequences and penalties for not following the policy. The policy for an individual control must be reviewed to ascertain that the criteria for level 1 are met. Assessing the effectiveness of the individual controls, not simply their existence, is key to achieving and maintaining adequate security.

The asset owner, in partnership with those responsible for administering the information assets (which include IT systems), must determine whether the measurement criteria are being met at each level. Before making such a determination, the degree of sensitivity of information and systems must be determined by considering the requirements for confidentiality, integrity, and availability of both the information and systems—the value of information and systems is one of the major factors in risk management.

A security program may be assessed at various levels within an organization. For example, a program could be defined as an agency asset, a major application, general support system, high impact program, physical plant, mission critical system, or logically related group of systems. The Framework refers to this grouping as an asset.

The Framework describes an asset self-assessment and provides levels to guide and prioritize agency efforts as well as a basis to measure progress. In addition, the National Institute of Standards and Technology (NIST) will develop a questionnaire that gives the implementation tools for the Framework. The questionnaire will contain specific control objectives that should be applied to secure a system.

Level 1	Documented Policy
Level 2	Documented Procedures
Level 3	Implemented Procedures and Controls
Level 4	Tested and Reviewed Procedures and Controls
Level 5	Fully Integrated Procedures and Controls

Figure 1 Federal IT Security Assessment Framework

The Framework approach begins with the premise that all agency assets must meet the minimum security requirements of the Office of Management and Budget Circular A-130, "Management of Federal Resources," Appendix III, "Security of Federal Automated Information Resources" (A-130). The criteria that are outlined in the Framework and provided in detail in the questionnaire are abstracted directly from long-standing requirements found in statute, policy, and guidance on security and privacy. It should be noted that an agency might have additional laws, regulations, or policies that establish specific requirements for confidentiality, integrity, or availability. Each agency should decide if additional security controls should be added to the questionnaire and, if so, customize the questionnaire appropriately. A list of the documents that the Framework and the questionnaire draw upon is provided below.

Source of Control Criteria

Office of Management and Budget Circular A-130, "Management of Federal Information Resources," Appendix III, "Security of Federal Automated Information Resources."	Establishes a minimum set of controls to be included in Federal IT security programs.
Computer Security Act of 1987.	This statute set the stage for protecting systems by codifying the requirement for Government-wide IT security planning and training.
Paperwork Reduction Act of 1995.	The PRA established a comprehensive information resources management framework including security and subsumed the security responsibilities of the Computer Security Act of 1987.
Clinger-Cohen Act of 1996.	This Act linked security to agency capital planning and budget processes, established agency Chief Information Officers, and re-codified the Computer Security Act of 1987.
Presidential Decision Directive 63, "Protecting America's Critical Infrastructures."	This directive specifies agency responsibilities for protecting the nation's infrastructure, assessing vulnerabilities of public and private sectors, and eliminating vulnerabilities.
Presidential Decision Directive 67, "Enduring Constitutional Government and Continuity of Government."	Relates to ensuring constitutional government, continuity of operations (COOP) planning, and continuity of government (COG) operations.

OMB Memorandum 99-05, Instructions on Complying with President's Memorandum of May 14, 1998, "Privacy and Personal Information in Federal Records."	This memorandum provides instructions to agencies on how to comply with the President's Memorandum of May 14, 1998 on "Privacy and Personal Information in Federal Records."
OMB Memorandum 99-18, "Privacy Policies on Federal Web Sites."	This memorandum directs Departments and Agencies to post clear privacy policies on World Wide Web sites, and provides guidance for doing so.
OMB Memorandum 00-13, "Privacy Policies and Data Collection on Federal Web Sites."	The purpose of this memorandum is a reminder that each agency is required by law and policy to establish clear privacy policies for its web activities and to comply with those policies.
General Accounting Office "Federal Information System Control Audit Manual" (FISCAM).	The FISCAM methodology provides guidance to auditors in evaluating internal controls over the confidentiality, integrity, and availability of data maintained in computer-based information systems.
NIST Special Publication 800-14, "Generally Accepted Principles and Practices for Security Information Technology Systems."	This publication guides organizations on the types of controls, objectives, and procedures that comprise an effective security program.
NIST Special Publication 800-18, "Guide for Developing Security Plans for Information Technology Systems."	This publication details the specific controls that should be documented in a system security plan.
Federal Information Processing Standards.	This document contains legislative and executive mandates for improving the utilization and management of computers and IT systems in the Federal Government

DOCUMENTED POLICY—LEVEL 1

Description

Level 1 of the Framework includes:

- Formally documented and disseminated security policy covering agency headquarters and major components (e.g., bureaus and operating divisions). The policy may be asset specific.
- Policy that references most of the basic requirements and guidance issued from the documents listed in the Source of Control Criteria.

An asset is at level 1 if there is a formally, up-to-date documented policy that establishes a continuing cycle of assessing risk, implements effective security policies including training, and uses monitoring for program effectiveness. Such a policy may include major agency components, (e.g., bureaus and operating divisions) or specific assets.

A documented security policy is necessary to ensure adequate and cost effective organizational and system security controls. A sound policy delineates the security management structure and clearly assigns security responsibilities, and lays the foundation necessary to reliably measure progress and compliance. The criteria listed below should be applied when assessing the policy developed for the controls that are listed in the NIST questionnaire.

Criteria

Level 1 criteria describe the components of a security policy.

Criteria for Level 1

a. **Purpose and scope.** An up-to-date security policy is written that covers all major facilities and operations agency-wide or for the asset. The policy is approved by key affected parties and covers security planning, risk management, review of security controls, rules of behavior, life-cycle management, processing authorization, personnel, physical and environmental aspects, computer support and operations, contingency planning, documentation, training, incident response, access controls, and audit trails. The policy clearly identifies the purpose of the program and its scope within the organization.
b. **Responsibilities.** The security program comprises a security management structure with adequate authority, and expertise. IT security manager(s) are appointed at an overall level and at appropriate subordinate levels. Security responsibilities and expected behaviors are clearly defined for asset owners and users, information resources management and data processing personnel, senior management, and security administrators.
c. **Compliance.** General compliance and specified penalties and disciplinary actions are also identified in the policy.

DOCUMENTED PROCEDURES—LEVEL 2

Description

Level 2 of the Framework includes:

- Formal, complete, well-documented procedures for implementing policies established at level one.
- The basic requirements and guidance issued from the documents listed in the Source of Control Criteria.

An asset is at level 2 when formally documented procedures are developed that focus on implementing specific security controls. Formal procedures promote the continuity of the security program. Formal procedures also provide the foundation for a clear, accurate, and complete understanding of the program implementation. An understanding of the risks and related results should guide the strength of the control and the corresponding procedures. The procedures document the implementation of and the rigor in which the control is applied. Level 2 requires procedures for a continuing cycle of assessing risk and vulnerabilities, implementing effective security policies, and monitoring effectiveness of the security controls. Approved system security plans are in place for all assets.

Well-documented and current security procedures are necessary to ensure that adequate and cost effective security controls are implemented. The criteria listed below should be applied when assessing the quality of the procedures for controls outlined in the NIST questionnaire.

Criteria

Level 2 criteria describe the components of security procedures.

Criteria for Level 2

a. **Control areas listed and organization's position stated.** Up-to-date procedures are written that covers all major facilities and operations within the asset. The procedures are approved by key responsible parties and cover security policies, security plans, risk management, review of security controls, rules of behavior, life-cycle management, processing authorization, personnel, physical and environmental aspects, computer support and operations, contingency planning, documentation, training, incident response, access controls, and audit trails. The procedures clearly identify management's position and whether there are further guidelines or exceptions.
b. **Applicability of procedures documented.** Procedures clarify where, how, when, to, whom, and about what a particular procedure applies.
c. **Assignment of IT security responsibilities and expected behavior.** Procedures clearly define security responsibilities and expected behaviors for (1) asset owners and users, (2) information resources management and data processing personnel, (3) management, and (4) security administrators.
d. **Points of contact and supplementary information provided.** Procedures contain appropriate individuals to be contacted for further information, guidance, and compliance.

IMPLEMENTED PROCEDURES AND CONTROLS—LEVEL 3

Description

Level 3 of the Framework includes:

- Security procedures and controls that are implemented.

- Procedures that are communicated and individuals who are required to follow them.

At level 3, the IT security procedures and controls are implemented in a consistent manner and reinforced through training. Ad hoc approaches that tend to be applied on an individual or case-by-case basis are discouraged. Security controls for an asset could be implemented and not have procedures documented, but the addition of formal documented procedures at level 2 represents a significant step in the effectiveness of implementing procedures and controls at level 3. While testing the on-going effectiveness is not emphasized in level 3, some testing is needed when initially implementing controls to ensure they are operating as intended. The criteria listed below should be used to determine if the specific controls listed in the NIST questionnaire are being implemented.

Criteria

Level 3 criteria describe how an organization can ensure implementation of their security procedures.

Criteria for Level 3

a. **Owners and users are made aware of security policies and procedures.** Security policies and procedures are distributed to all affected personnel, including system/application rules and expected behaviors. Requires users to periodically acknowledge their awareness and acceptance of responsibility for security.

b. **Policies and procedures are formally adopted and technical controls installed.** Automated and other tools routinely monitor security. Established policy governs review of system logs, penetration testing, and internal/external audits.

c. **Security is managed throughout the life cycle of the system.** Security is considered in each of the life-cycle phases: initiation, development/acquisition, implementation, operation, and disposal.

d. **Procedures established for authorizing processing (certification and accreditation).** Management officials must formally authorize system operations and manage risk.

e. **Documented security position descriptions.** Skill needs and security responsibilities in job descriptions are accurately identified.

f. **Employees trained on security procedures.** An effective training and awareness program tailored for varying job functions is planned, implemented, maintained, and evaluated.

TESTED AND EVALUATED PROCEDURES AND CONTROLS—LEVEL 4

Description

Level 4 of the Framework includes:

- Routinely evaluating the adequacy and effectiveness of security policies, procedures, and controls.
- Ensuring that effective corrective actions are taken to address identified weaknesses, including those identified as a result of potential or actual security incidents or through security alerts issued by FedCIRC, vendors, and other trusted sources.

Routine evaluations and response to identified vulnerabilities are important elements of risk management, which includes identifying, acknowledging, and responding, as appropriate, to changes in risk factors (e.g., computing environment, data sensitivity) and ensuring that security policies and procedures are appropriate and are operating as intended on an ongoing basis.

Routine self-assessments are an important means of identifying inappropriate or ineffective security procedures and controls, reminding employees of their security-related responsibilities, and demonstrating management's commitment to security. Self-assessments can be performed by agency staff or by contractors or others engaged by agency management. Independent audits such as those arranged by the General Accounting Office (GAO) or an agency Inspector General (IG), are an important check on agency performance, but should not be viewed as a substitute for evaluations initiated by agency management.

To be effective, routine evaluations must include tests and examinations of key controls. Reviews of documentation, walk-throughs of agency facilities, and interviews with agency personnel, while providing useful information, are not sufficient to ensure that controls, especially computer-based controls, are operating effectively. Examples of tests that should be conducted are network scans to identify known vulnerabilities, analyses of router and switch settings and firewall rules, reviews of other system software settings, and tests to see if unauthorized system access is possible (penetration testing). Tests performed should consider the risks of authorized users exceeding authorization as well as unauthorized users (e.g., external parties, hackers) gaining access. Similar to levels 1 through 3, to be meaningful, evaluations must include security controls of interconnected assets, e.g., network supporting applications being tested.

When assets are first implemented or are modified, they should be tested and certified to ensure that controls are initially operating as intended. (This would occur at Level 3.) Requirements for subsequent testing and recertification should be integrated into an agency's ongoing test and evaluation program.

DOCUMENT APPENDIX

In addition to test results, agency evaluations should consider information gleaned from records of potential and actual security incidents and from security alerts, such as those issued by software vendors. Such information can identify specific vulnerabilities and provide insights into the latest threats and resulting risks.

The criteria listed below should be applied to each control area listed in the NIST questionnaire to determine if the asset is being effectively evaluated.

5.2 Criteria

Level 4 criteria are listed below.

Criteria for Level 4

a. **Effective program for evaluating adequacy and effectiveness of security policies, procedures, and controls.** Evaluation requirements, including requirements regarding the type and frequency of testing, should be documented, approved, and effectively implemented. The frequency and rigor with which individual controls are tested should depend on the risks that will be posed if the controls are not operating effectively. At a minimum, controls should be evaluated whenever significant system changes are made or when other risk factors, such as the sensitivity of data processed, change. Even controls for inherently low-risk operations should be tested at a minimum of every 3 years.
b. **Mechanisms for identifying vulnerabilities revealed by security incidents or security alerts.** Agencies should routinely analyze security incident records, including any records of anomalous or suspicious activity that may reveal security vulnerabilities. In addition, they should review security alerts issued by FedCIRC, vendors, and others.
c. **Process for reporting significant security weaknesses and ensuring effective remedial action.** *Such a process should provide for routine reports to senior management on weaknesses identified through testing or other means, development of action plans, allocation of needed resources, and follow-up reviews to ensure that remedial actions have been effective. Expedited processes should be implemented for especially significant weaknesses that may present undue risk if not addressed immediately.*

FULLY INTEGRATED PROCEDURES AND CONTROLS—LEVEL 5

Description

Level 5 of the Framework includes:

- A comprehensive security program that is an integral part of an agency's organizational culture.
- Decision-making based on cost, risk, and mission impact.

The consideration of IT security is pervasive in the culture of a level 5 asset. A proven life-cycle methodology is implemented and enforced and an ongoing program to identify and institutionalize best practices has been implemented. There is active support from senior management. Decisions and actions that are part of the IT life cycle include:

- Improving security program
- Improving security program procedures
- Improving or refining security controls
- Adding security controls
- Integrating security within existing and evolving IT architecture
- Improving mission processes and risk management activities

Each of these decisions result from a continuous improvement and refinement program instilled within the organization. At level 5, the understanding of mission-related risks and the associated costs of reducing these risks are considered with a full range of implementation options to achieve maximum mission cost-effectiveness of security measures. Entities should apply the principle of selecting controls that offer the lowest cost implementation while offering adequate risk mitigation, versus high cost implementation and low risk mitigation. The criteria listed below should be used to assess whether a specific control contained in the NIST questionnaire has been fully implemented.

Criteria

Level 5 criteria describe components of a fully integrated security program.

Criteria for Level 5
a. There is an active enterprise-wide security program that achieves cost-effective security.
b. IT security is an integrated practice within the asset.
c. Security vulnerabilities are understood and managed.
d. Threats are continually re-evaluated, and controls adapted to changing security environment.
e. Additional or more cost-effective security alternatives are identified as the need arises.
f. Costs and benefits of security are measured as precisely as practicable.
g. Status metrics for the security program are established and met.

FUTURE OF THE FRAMEWORK

This version of the Framework primarily addresses security management issues. It describes a process for agencies to assess their compliance with long-standing basic requirements and guidance. With the Framework in place, agencies will have an approach to begin the assessment process. The NIST questionnaire

DOCUMENT APPENDIX

provides the tool to determine whether agencies are meeting these requirements and following the guidance.

The Framework is not static; it is a living document. Revisions will focus on expanding, refining, and providing more granularity for existing criteria. In addition, the establishment of a similar companion framework devoted to the evolution of agency electronic privacy polices may be considered in time.

The Framework can be viewed as both an auditing tool and a management tool. A balance between operational needs and cost effective security for acceptable risk will need to be made to achieve an adequate level of security.

Currently, the NIST self-assessment tool is under development and will be available in 2001. Appendix A provides a sample questionnaire to assist agencies until NIST officially releases the questionnaire.

APPENDIX A

Conceptual Sample of NIST Self-Assessment Questionnaire

Below is a conceptual sample of the Hypothetical Government Agency's (HGA) completion of the NIST questionnaire for their Training Database. Before the questionnaire was completed, the sensitivity of the information stored within, processed by and transmitted by this asset was assessed. The premise behind determining the level of sensitivity is that each asset owner is responsible for determining what level of risk is acceptable, and which specific security controls are necessary to provide adequate security.

The sensitivity of this asset was determined to be high for confidentiality and low for integrity and availability. The confidentiality of the system is high due to the system containing personnel information. Employee social security numbers, course lists, and grades are contained in the system. The integrity of the database is considered low because if the information were modified by unauthorized, unanticipated or unintentional means, employees, who can read their own training file, would detect the modifications. The availability of the system is considered low because hard copies of the training forms are available as a backup.

The questionnaire was completed for the database with the understanding that security controls that protect the integrity or availability of the data did not have to be rigidly applied. The questionnaire contains a field that can be checked when a risk-based decision has been made to either reduce or enhance a security control. There may be certain situations where management will grant a waiver either because compensating controls exist or because the benefits of operating without the control (at least temporarily) outweigh the risk of waiting for full control implementation. Alternatively, there may be times where management implements more stringent controls than generally applied elsewhere. In the example provided the specific control objectives for personnel security and for authentication were assessed. The questionnaire is an excerpt and by no means contains all the questions that would be asked in the area of personnel security and authentication. For brevity, only a few questions were provided in this sample.

An analysis of the levels checked determined that the agency should target improving their background screening implementation and testing. System administrators, programmers, and managers should all have background checks completed prior to accessing the system. The decision to allow access prior to screening was made and checked in the *Risk Based Decision Made* box. Because this box was checked, there should be specific controls implemented to ensure access is not abused, i.e., access is reviewed daily through audit trails, and users have minimal system authority.

Additionally, HGA should improve implementing and testing their password procedures because of the strong need for confidentiality. Without good password management, passwords can be easily guessed and access to the system obtained. The questionnaire's list of objectives is incomplete for both personnel security controls and for authentication controls. Even though the sample is lacking many controls, the completed questionnaire clearly depicts that HGA has policies and procedures in place but there is a strong need for implementing, testing, and reviewing the procedures and controls. The sample indicates that the Training Database would be at level 2.

Category of Sensitivity	Confidentiality		Integrity		Availability	
High	X					
Medium						
Low			X		X	

Specific Control Objectives	L.1 Policy	L.2 Procedures	L.3 Implemented	L.4 Tested	L.5 Integrated	Risk Based Decision Made
Personnel Security						
Are all positions reviewed for sensitivity level?	X	X	X			
appropriate background screening for assigned positions completed prior to granting access?	X	X				X
Are there conditions for allowing system access prior to completion of screening?	X	X				

(Continued)

DOCUMENT APPENDIX

Category of Sensitivity		Confidentiality	Integrity	Availability		
High		X				
Medium						
Low			X	X		

Specific Control Objectives	L.1 Policy	L.2 Procedures	L.3 Implemented	L.4 Tested	L.5 Integrated	Risk Based Decision Models
Are sensitive functions divided among different individuals?	X	X	X			
Are mechanisms in place for holding usersresponsible for their actions?	X	X				
Are termination procedures established?	X	X				
Authentication						
Are passwords, tokens, or biometrics used?	X	X	X			
Do passwords contain alpha numeric, upper/lower case, and special characters?	X	X				
Are passwords changed at least every ninety days or earlier if needed?	X	X				
Is there guidance for handling lost and compromised passwords?	X	X				
Are passwords transmitted and stored with one-way encryption?	X	X				
Is there a limit to the number of invalid access attempts that may occur for a given user?	X	X				

REFERENCES

Automated Information Systems Security Program Handbook (Release 2.0, May 1994), Department of Health and Human Services, May 1994.

Clinger-Cohen Act of 1996 (formerly known as the Information Management Reform Act), February 10, 1996.

Computer Security Act of 1987, 40 U.S. Code 759, (Public Law 100-235), January 8, 1988.

Control Objectives for Information and Related Technology (COBIT) 3rd Edition, Information Systems Audit and Control Foundation, July 2000.

General Accounting Office, Federal Information System Control Audit Manual (FISCAM), GOA/AIMD-12.19.6, January 1999.

General Accounting Office, Information Security Risk Assessment Practices of Leading Organizations, GAO/AIMD-99-139, August 1999.

Office of Management and Budget, Security of Federal Automated Information Resources, Appendix III to OMB Circular A-130, Management of Federal Information Resources, February 8, 1996.

Office of Management and Budget, Memorandum 99-05, Instructions on Complying with President's Memorandum of May 14, 1998, Privacy and Personal Information in Federal Records, July 1, 1999.

Office of Management and Budget, Memorandum 99-18, Privacy Policies on Federal Web Sites, June 2, 1999.

Office of Management and Budget, Memorandum 00-13, Policies and Data Collection on Federal Web Sites, June 22, 2000.

Paperwork Reduction Act of 1995, 35 U.S. Code 44, January 4, 1995.

Presidential Decision Directive 63, Protecting America's Critical Infrastructures, May 22, 1998.

Presidential Decision Directive 67, Enduring Constitutional Government and Continuity of Government, October 21, 1998.

Swanson, Marianne and Barbara Guttman, NIST Special Publication 800-14, Generally Accepted Principles and Practices for Security Information Technology Systems (GSSP), Gaithersburg, MD, National Institute of Standards and Technology, September 20, 1995.

Swanson, Marianne and Federal Computer Security Program Managers' Forum Working Group, NIST Special Publication 800-18, Guide for Developing Security Plans for Information Technology Systems, Gaithersburg, MD, National Institute of Standards and Technology, December 1998.

TERMINOLOGY

Acceptable Risk is a concern that is acceptable to responsible management, due to the cost and magnitude of implementing controls.

Accreditation is synonymous with the term **authorize processing**. Accreditation is the authorization and approval granted to a major application or general support system to process in an operational environment.

It is made on the basis of a certification by designated technical personnel that the system meets pre-specified technical requirements for achieving adequate system security. See also *Authorize Processing, Certification,* and *Designated Approving Authority*.

Asset is a major application, general support system, high impact program, physical plant, mission critical system, or a logically related group of systems.

Authorize Processing occurs when management authorizes in writing a system based on an assessment of management, operational, and technical controls. By authorizing processing in a system the management official accepts the risks associated with it. See also *Accreditation, Certification,* and *Designated Approving Authority*.

Availability Protection requires backup of system and information, contingency plans, disaster recovery plans, and redundancy. Examples of systems and information requiring availability protection are time-share systems, mission-critical applications, time and attendance, financial, procurement, or life-critical.

Awareness, Training, and Education includes (1) awareness programs set the stage for training by changing organizational attitudes towards realization of the importance of security and the adverse consequences of its failure; (2) the purpose of training is to teach people the skills that will enable them to perform their jobs more effectively; and (3) education is more in-depth than training and is targeted for security professionals and those whose jobs require expertise in IT security.

Certification is synonymous with the term **authorize processing**. Certification is a major consideration prior to authorizing processing, but not the only consideration. Certification is the technical evaluation that establishes the extent to which a computer system, application, or network design and implementation meets a pre-specified set of security requirements. See also *Accreditation* and *Authorize Processing*.

General Support System is an interconnected information resource under the same direct management control that shares common functionality. It normally includes hardware, software, information, data, applications, communications, facilities, and people and provides support for a variety of users and/or applications. Individual applications support different mission-related functions. Users may be from the same or different organizations.

Individual Accountability requires individual users to be held accountable for their actions after being notified of the rules of behavior in the use of the system and the penalties associated with the violation of those rules.

Information Owner is responsible for establishing the rules for appropriate use and protection of the data/information. The information owner retains that responsibility even when the data/information are shared with other organizations.

Major Application is an application that requires special attention to security due to the risk and magnitude of the harm resulting from the loss, misuse, or unauthorized access to, or modification of, the information in the application. A breach in a major application might comprise many individual application programs and hardware, software, and telecommunications components. Major applications can be either a major software application or a combination of hardware/software where the only purpose of the system is to support a specific mission-related function.

Material Weakness or ***significant weakness*** is used to identify control weaknesses that pose a significant risk or a threat to the operations and/or assets of an audited entity. "Material weakness" is a very specific term that is defined one way for financial audits and another way for weaknesses reported under the Federal Managers Financial Integrity Act of 1982. Such weaknesses may be identified by auditors or by management.

Networks include communication capability that allows one user or system to connect to another user or system and can be part of a system or a separate system. Examples of networks include local area network or wide area networks, including public networks such as the Internet.

Operational Controls address security methods that focus on mechanisms that primarily are implemented and executed by people (as opposed to systems).

Policy a document that delineates the security management structure and clearly assigns security responsibilities and lays the foundation necessary to reliably measure progress and compliance.

Procedures are contained in a document that focuses on the security control areas and management's position.

Risk is the possibility of harm or loss to any software, information, hardware, administrative, physical, communications, or personnel resource within an automated information system or activity.

Risk Management is the ongoing process of assessing the risk to automated information resources and information, as part of a risk-based approach used to determine adequate security for a system by analyzing the threats and vulnerabilities and selecting appropriate cost-effective controls to achieve and maintain an acceptable level of risk.

Rules of Behavior are the rules that have been established and implemented concerning use of, security in, and acceptable level of risk for the system. Rules will clearly delineate responsibilities and expected behavior of all individuals with access to the system. Rules should cover such matters as work at home, dial-in access, connection to the Internet, use of copyrighted works, unofficial use of Federal government equipment, assignment and limitation of system privileges, and individual accountability.

DOCUMENT APPENDIX

Sensitive Information refers to information whose loss, misuse, or unauthorized access to or modification of could adversely affect the national interest or the conduct of Federal programs or the privacy to which individuals are entitled.

Sensitivity an information technology environment consists of the system, data, and applications that must be examined individually and in total. All systems and applications require some level of protection for confidentiality, integrity, and/or availability that is determined by an evaluation of the sensitivity of the information processed, the relationship of the system to the organizations mission, and the economic value of the system components.

System is a generic term used for briefness to mean either a major application or a general support system.

System Operational Status is either (1) Operational—system is currently in operation, (2) Under Development—system is currently under design, development, or implementation, or (3) Undergoing a Major Modification—system is currently undergoing a major conversion or transition.

Technical Controls consist of hardware and software controls used to provide automated protection to the system or applications. Technical controls operate within the technical system and applications.

Threat is an event or activity, deliberate or unintentional, with the potential for causing harm to an IT system or activity.

Vulnerability is a flaw or weakness that may allow harm to occur to an IT system or activity.

APPENDIX D—REFERENCES

Clinger-Cohen Act of 1996 (formerly known as the Information Management Reform Act), February 10, 1996.

Computer Security Act of 1987, 40 U.S. Code 759, (Public Law 100-235), January 8, 1988.

Control Objectives for Information and Related Technology (COBIT) 3rd Edition, Information Systems Audit and Control Foundation, July 2000.

Defense Authorization Act (P.L. 106-398) including Title X, Subtitle G, "Government Information Security Reform," October 28, 2000.

Department of State, Draft Best Security Practices Checklist Appendix A, January 22, 2001.

General Accounting Office, Federal Information System Control Audit Manual (FISCAM), GOA/AIMD-12.19.6, January 1999.

General Accounting Office, Information Security Risk Assessment Practices of Leading Organizations, GAO/AIMD-99-139, August 1999.

ISSO 17799, A Code of Practice for Information Security Management (British Standard 7799),

National Communications System, Public Switched Network Security Assessment Guidelines, September 2000.

Office of Management and Budget, Security of Federal Automated Information Resources, Appendix III to OMB Circular A-130, Management of Federal Information Resources, February 8, 1996.

Office of Management and Budget, Memorandum 99-05, Instructions on Complying with President's Memorandum of May 14, 1998, Privacy and Personal Information in Federal Records, July 1, 1999.

Office of Management and Budget, Memorandum 99-18, Privacy Policies on Federal Web Sites, June 2, 1999.

Office of Management and Budget, Memorandum 00-13, Policies and Data Collection on Federal Web Sites, June 22, 2000.

Paperwork Reduction Act of 1995, 35 U.S. Code 44, January 4, 1995.

Presidential Decision Directive 63, Protecting America's Critical Infrastructures, May 22, 1998.

Presidential Decision Directive 67, Enduring Constitutional Government and Continuity of Government, October 21, 1998.

Stoneburner, Gary, Draft –Rev. A NIST Special Publication 800-30, Risk Management Guide, February 16, 2001.

Swanson, Marianne and Barbara Guttman, NIST Special Publication 800-14, Generally Accepted Principles and Practices for Security Information Technology Systems (GSSP), Gaithersburg, MD, National Institute of Standards and Technology, September 20, 1995.

Swanson, Marianne and Federal Computer Security Program Managers' Forum Working Group, NIST Special Publication 800-18, Guide for Developing Security Plans for Information Technology Systems, Gaithersburg, MD, National Institute of Standards and Technology, December 1998.

NOTES

1. OMB Circular A-130, Appendix III defines general support system or "system" in similar terms.

2. Draft NIST Special Publication 800-30, "Risk Management Guidance" dated June 2001.

3. FedCIRC (Federal Computer Incident Response Capability) is the U.S. Government's focal point for handling computer security-related incidents.

4. NIPC's mission is to serve as the U.S. Government's focal point for threat assessment, warning, investigation, and response for threats or attacks against our critical infrastructures.

5. The NIST Self-assessment Questionnaire will be issued in 2001 as a NIST Special Publication.

DOCUMENT APPENDIX

EMERGENCY MANAGEMENT GUIDE FOR BUSINESS AND INDUSTRY

A Step-by-Step Approach to Emergency Planning, Response and Recovery for Companies of All Sizes

FEMA 141/October 1993

A hurricane blasts through South Florida causing more than $25 billion in damages.

A fire at a food processing plant results in 25 deaths, a company out of business, and a small town devastated.

A bombing in the World Trade Center results in six deaths, hundreds of injuries, and the evacuation of 40,000 people.

A blizzard shuts down much of the East Coast for days. More than 150 lives are lost and millions of dollars in damages incurred.

Every year emergencies take their toll on business and industry—in lives and dollars. But something can be done. Business and industry can limit injuries and damages and return more quickly to normal operations if they plan ahead.

ABOUT THIS GUIDE

This guide provides step-by-step advice on how to create and maintain a comprehensive emergency management program. It can be used by manufacturers, corporate offices, retailers, utilities or any organization where a sizable number of people work or gather.

Whether you operate from a high-rise building or an industrial complex; whether you own, rent or lease your property; whether you are a large or small company; the concepts in this guide will apply.

To begin, you need not have in-depth knowledge of emergency management. What you need is the authority to create a plan and a commitment from the chief executive officer to make emergency management part of your corporate culture.

If you already have a plan, use this guide as a resource to assess and update your plan.

The guide is organized as follows:

Section 1: 4 Steps in the Planning Process—how to form a planning team; how to conduct a vulnerability analysis; how to develop a plan; and how to implement the plan. The information can be applied to virtually any type of business or industry.

Section 2: Emergency Management Considerations—how to build such emergency management capabilities as life safety, property protection, communications and community outreach.

Section 3: Hazard-Specific Information—technical information about specific hazards your facility may face.

Section 4: Information Sources—where to turn for additional information.

WHAT IS AN EMERGENCY?

An emergency is any unplanned event that can cause deaths or significant injuries to employees, customers or the public; or that can shut down your business, disrupt operations, cause physical or environmental damage, or threaten the facility's financial standing or public image.

Obviously, numerous events can be "emergencies," including:

- Fire
- Hazardous materials incident
- Flood or flash flood
- Hurricane
- Tornado
- Winter storm
- Earthquake
- Communications failure
- Radiological accident
- Civil disturbance
- Loss of key supplier or customer
- Explosion

The term "disaster" has been left out of this document because it lends itself to a preconceived notion of a large-scale event, usually a "natural disaster." In fact, each event must be addressed within the context of the impact it has on the company and the community. What might constitute a nuisance to a large industrial facility could be a "disaster" to a small business.

WHAT IS EMERGENCY MANAGEMENT?

Emergency management is the process of preparing for, mitigating, responding to and recovering from an emergency.

Emergency management is a dynamic process. Planning, though critical, is not the only component. Training, conducting drills, testing equipment and coordinating activities with the community are other important functions.

DOCUMENT APPENDIX

MAKING THE "CASE" FOR EMERGENCY MANAGEMENT

To be successful, emergency management requires upper management support. The chief executive sets the tone by authorizing planning to take place and directing senior management to get involved.

When presenting the "case" for emergency management, avoid dwelling on the negative effects of an emergency (e.g., deaths, fines, criminal prosecution) and emphasize the positive aspects of preparedness. For example:

- It helps companies fulfill their moral responsibility to protect employees, the community and the environment.
- It facilitates compliance with regulatory requirements of Federal, State and local agencies.
- It enhances a company's ability to recover from financial losses, regulatory fines, loss of market share, damages to equipment or products or business interruption.
- It reduces exposure to civil or criminal liability in the event of an incident.
- It enhances a company's image and credibility with employees, customers, suppliers and the community.
- It may reduce your insurance premiums.

SECTION 1: 4 STEPS IN THE PLANNING PROCESS

Establish A Planning Team

There must be an individual or group in charge of developing the emergency management plan. The following is guidance for making the appointment.

Form the Team

The size of the planning team will depend on the facility's operations, requirements and resources. Usually involving a group of people is best because:

- It encourages participation and gets more people invested in the process.
- It increases the amount of time and energy participants are able to give.
- It enhances the visibility and stature of the planning process.
- It provides for a broad perspective on the issues.

Determine who can be an active member and who can serve in an advisory capacity. In most cases, one or two people will be doing the bulk of the work. At the very least, you should obtain input from all functional areas.

Remember:

- Upper management
- Line management
- Labor
- Human Resources
- Engineering and maintenance
- Safety, health and environmental affairs
- Public information officer
- Security
- Community relations
- Sales and marketing
- Legal
- Finance and purchasing

Have participants appointed in writing by upper management. Their job descriptions could also reflect this assignment.

Establish Authority

Demonstrate management's commitment and promote an atmosphere of cooperation by "authorizing" the planning group to take the steps necessary to develop a plan. The group should be led by the chief executive or the plant manager.

Establish a clear line of authority between group members and the group leader, though not so rigid as to prevent the free flow of ideas.

Issue a Mission Statement

Have the chief executive or plant manager issue a mission statement to demonstrate the company's commitment to emergency management. The statement should:

- Define the purpose of the plan and indicate that it will involve the entire organization
- Define the authority and structure of the planning group

Establish a Schedule and Budget

Establish a work schedule and planning deadlines. Timelines can be modified as priorities become more clearly defined.

Develop an initial budget for such things as research, printing, seminars, consulting services and other expenses that may be necessary during the development process.

DOCUMENT APPENDIX

Analyze Capabilities and Hazards

This step entails gathering information about current capabilities and about possible hazards and emergencies, and then conducting a vulnerability analysis to determine the facility's capabilities for handling emergencies.

Where Do You Stand Right Now?

Review Internal Plans and Policies

Documents to look for include:

- Evacuation plan
- Fire protection plan
- Safety and health program
- Environmental policies
- Security procedures
- Insurance programs
- Finance and purchasing procedures
- Plant closing policy
- Employee manuals
- Hazardous materials plan
- Process safety assessment
- Risk management plan
- Capital improvement program
- Mutual aid agreements

Meet with Outside Groups

Meet with government agencies, community organizations and utilities. Ask about potential emergencies and about plans and available resources for responding to them. Sources of information include:

- Community emergency management office
- Mayor or Community Administrator's office
- Local Emergency Planning Committee (LEPC)
- Fire Department
- Police Department
- Emergency Medical Services organizations
- American Red Cross
- National Weather Service
- Public Works Department
- Planning Commission
- Telephone companies
- Electric utilities
- Neighboring businesses

Identify Codes and Regulations

Identify applicable Federal, State and local regulations such as:

- Occupational safety and health regulations
- Environmental regulations
- Fire codes
- Seismic safety codes
- Transportation regulations
- Zoning regulations
- Corporate policies

Identify Critical Products, Services and Operations

You'll need this information to assess the impact of potential emergencies and to determine the need for backup systems. Areas to review include:

- Company products and services and the facilities and equipment needed to produce them
- Products and services provided by suppliers, especially sole source vendors
- Lifeline services such as electrical power, water, sewer, gas, telecommunications and transportation
- Operations, equipment and personnel vital to the continued functioning of the facility

Identify Internal Resources and Capabilities

Resources and capabilities that could be needed in an emergency include:

- Personnel—fire brigade, hazardous materials response team, emergency medical services, security, emergency management group, evacuation team, public information officer
- Equipment—fire protection and suppression equipment, communications equipment, first aid supplies, emergency supplies, warning systems, emergency power equipment, decontamination equipment
- Facilities—emergency operating center, media briefing area, shelter areas, first-aid stations, sanitation facilities
- Organizational capabilities—training, evacuation plan, employee support system
- Backup systems—arrangements with other facilities to provide for:
  - Payroll
  - Communications
  - Production
  - Customer services
  - Shipping and receiving
  - Information systems support

- Emergency power
- Recovery support

Identify External Resources

There are many external resources that could be needed in an emergency. In some cases, formal agreements may be necessary to define the facility's relationship with the following:

- Local emergency management office
- Fire Department
- Hazardous materials response organization
- Emergency medical services
- Hospitals
- Local and State police
- Community service organizations
- Utilities
- Contractors
- Suppliers of emergency equipment
- Insurance carriers

Do an Insurance Review

Meet with insurance carriers to review all policies. (See Section 2: Recovery and Restoration.)

Conduct a Vulnerability Analysis

The next step is to assess the vulnerability of your facility—the probability and potential impact of each emergency. Use the Vulnerability Analysis Chart in the appendix section to guide the process, which entails assigning probabilities, estimating impact and assessing resources, using a numerical system. The lower the score the better.

List Potential Emergencies

In the first column of the chart, list all emergencies that could affect your facility, including those identified by your local emergency management office. Consider both:

- Emergencies that could occur within your facility
- Emergencies that could occur in your community

Below are some other factors to consider.

- Historical—What types of emergencies have occurred in the community, at this facility and at other facilities in the area?

- Fires
- Severe weather
- Hazardous material spills
- Transportation accidents
- Earthquakes
- Hurricanes
- Tornadoes
- Terrorism
- Utility outages

- Geographic—What can happen as a result of the facility's location? Keep in mind:
  - Proximity to flood plains, seismic faults and dams
  - Proximity to companies that produce, store, use or transport hazardous materials
  - Proximity to major transportation routes and airports
  - Proximity to nuclear power plants

- Technological—What could result from a process or system failure? Possibilities include:
  - Fire, explosion, hazardous materials incident
  - Safety system failure
  - Telecommunications failure
  - Computer system failure
  - Power failure
  - Heating/cooling system failure
  - Emergency notification system failure

- Human Error—What emergencies can be caused by employee error? Are employees trained to work safely? Do they know what to do in an emergency?
  Human error is the single largest cause of workplace emergencies and can result from:
  - Poor training
  - Poor maintenance
  - Carelessness
  - Misconduct
  - Substance abuse
  - Fatigue

- Physical—What types of emergencies could result from the design or construction of the facility? Does the physical facility enhance safety? Consider:
  - The physical construction of the facility
  - Hazardous processes or byproducts
  - Facilities for storing combustibles
  - Layout of equipment

DOCUMENT APPENDIX

- Lighting
- Evacuation routes and exits
- Proximity of shelter areas

- Regulatory—What emergencies or hazards are you regulated to deal with?

Analyze each potential emergency from beginning to end. Consider what could happen as a result of:
- Prohibited access to the facility
- Loss of electric power
- Communication lines down
- Ruptured gas mains
- Water damage
- Smoke damage
- Structural damage
- Air or water contamination
- Explosion
- Building collapse
- Trapped persons
- Chemical release

Estimate Probability
In the Probability column, rate the likelihood of each emergency's occurrence. This is a subjective consideration, but useful nonetheless.

Use a simple scale of 1 to 5 with 1 as the lowest probability and 5 as the highest.

Assess the Potential Human Impact
Analyze the potential human impact of each emergency—the possibility of death or injury.

Assign a rating in the Human Impact column of the Vulnerability Analysis Chart. Use a 1 to 5 scale with 1 as the lowest impact and 5 as the highest.

Assess the Potential Property Impact
Consider the potential property for losses and damages. Again, assign a rating in the Property Impact column, 1 being the lowest impact and 5 being the highest. Consider:

- Cost to replace
- Cost to set up temporary replacement
- Cost to repair

Assess the Potential Business Impact
Consider the potential loss of market share. Assign a rating in the Business Impact column. Again, 1 is the lowest impact and 5 is the highest. Assess the impact of:

- Business interruption
- Employees unable to report to work
- Customers unable to reach facility
- Company in violation of contractual agreements
- Imposition of fines and penalties or legal costs
- Interruption of critical supplies
- Interruption of product distribution

Assess Internal and External Resources

Next assess your resources and ability to respond. Assign a score to your Internal Resources and External Resources. The lower the score the better.

To help you do this, consider each potential emergency from beginning to end and each resource that would be needed to respond. For each emergency ask these questions:

- Do we have the needed resources and capabilities to respond?
- Will external resources be able to respond to us for this emergency as quickly as we may need them, or will they have other priority areas to serve?

If the answers are yes, move on to the next assessment. If the answers are no, identify what can be done to correct the problem. For example, you may need to:

- Develop additional emergency procedures
- Conduct additional training
- Acquire additional equipment
- Establish mutual aid agreements
- Establish agreements with specialized contractors

Add the Columns

Total the scores for each emergency. The lower the score the better. While this is a subjective rating, the comparisons will help determine planning and resource priorities—the subject of the pages to follow.

Develop The Plan

You are now ready to develop an emergency management plan. This section describes how.

Plan Components

Your plan should include the following basic components.

Executive Summary

The executive summary gives management a brief overview of:

- The purpose of the plan
- The facility's emergency management policy
- Authorities and responsibilities of key personnel
- The types of emergencies that could occur
- Where response operations will be managed

Emergency Management Elements

This section of the plan briefly describes the facility's approach to the core elements of emergency management, which are:

- Direction and control
- Communications
- Life safety
- Property protection
- Community outreach
- Recovery and restoration
- Administration and logistics

These elements, which are described in detail in Section 2, are the foundation for the emergency procedures that your facility will follow to protect personnel and equipment and resume operations.

Emergency Response Procedures

The procedures spell out how the facility will respond to emergencies. Whenever possible, develop them as a series of checklists that can be quickly accessed by senior management, department heads, response personnel and employees.

Determine what actions would be necessary to:

- Assess the situation
- Protect employees, customers, visitors, equipment, vital records and other assets, particularly during the first three days
- Get the business back up and running

Specific procedures might be needed for any number of situations such as bomb threats or tornadoes, and for such functions as:

- Warning employees and customers
- Communicating with personnel and community responders
- Conducting an evacuation and accounting for all persons in the facility
- Managing response activities
- Activating and operating an emergency operations center
- Fighting fires
- Shutting down operations
- Protecting vital records
- Restoring operations

Support Documents

Documents that could be needed in an emergency include:

- Emergency call lists—lists (wallet size if possible) of all persons on and off site who would be involved in responding to an emergency, their responsibilities and their 24-hour telephone numbers
- Building and site maps that indicate:
  - Utility shutoffs
  - Water hydrants
  - Water main valves
  - Water lines
  - Gas main valves
  - Gas lines
  - Electrical cutoffs
  - Electrical substations
  - Storm drains
  - Sewer lines
  - Location of each building (include name of building, street name and number)
  - Floor plans
  - Alarm and enunciators
  - Fire extinguishers
  - Fire suppression systems
  - Exits
  - Stairways
  - Designated escape routes
  - Restricted areas
  - Hazardous materials (including cleaning supplies and chemicals)
  - High-value items
- Resource lists—lists of major resources (equipment, supplies, services) that could be needed in an emergency; mutual aid agreements with other companies and government agencies

The Development Process

The following is guidance for developing the plan.

Identify Challenges and Prioritize Activities

Determine specific goals and milestones. Make a list of tasks to be performed, by whom and when. Determine how you will address the problem areas and resource shortfalls that were identified in the vulnerability analysis.

Write the Plan

Assign each member of the planning group a section to write. Determine the most appropriate format for each section.

Establish an aggressive timeline with specific goals. Provide enough time for completion of work, but not so much as to allow assignments to linger. Establish a schedule for:

- First draft
- Review
- Second draft
- Tabletop exercise
- Final draft
- Printing
- Distribution

Establish a Training Schedule

Have one person or department responsible for developing a training schedule for your facility. For specific ideas about training, refer to Step 4.

Coordinate with Outside Organizations

Meet periodically with local government agencies and community organizations. Inform appropriate government agencies that you are creating an emergency management plan. While their official approval may not be required, they will likely have valuable insights and information to offer.

Determine State and local requirements for reporting emergencies, and incorporate them into your procedures.

Determine protocols for turning control of a response over to outside agencies. Some details that may need to be worked out are:

- Which gate or entrance will responding units use?
- Where and to whom will they report?
- How will they be identified?
- How will facility personnel communicate with outside responders?
- Who will be in charge of response activities?

Determine what kind of identification authorities will require to allow your key personnel into your facility during an emergency.

Maintain Contact with Other Corporate Offices

Communicate with other offices and divisions in your company to learn:

- Their emergency notification requirements
- The conditions where mutual assistance would be necessary
- How offices will support each other in an emergency
- Names, telephone numbers and pager numbers of key personnel

Incorporate this information into your procedures.

Review, Conduct Training and Revise

Distribute the first draft to group members for review. Revise as needed.

For a second review, conduct a tabletop exercise with management and personnel who have a key emergency management responsibility. In a conference room setting, describe an emergency scenario and have participants discuss their responsibilities and how they would react to the situation. Based on this discussion, identify areas of confusion and overlap, and modify the plan accordingly.

Seek Final Approval

Arrange a briefing for the chief executive officer and senior management and obtain written approval.

Distribute the Plan

Place the final plan in three-ring binders and number all copies and pages. Each individual who receives a copy should be required to sign for it and be responsible for posting subsequent changes.

Determine which sections of the plan would be appropriate to show to government agencies (some sections may refer to corporate secrets or include private listings of names, telephone numbers or radio frequencies).

Distribute the final plan to:

- Chief executive and senior managers
- Key members of the company's emergency response organization
- Company headquarters
- Community emergency response agencies (appropriate sections)

Have key personnel keep a copy of the plan in their homes.
Inform employees about the plan and training schedule.

Implement the Plan

Implementation means more than simply exercising the plan during an emergency. It means acting on recommendations made during the vulnerability analysis, integrating the plan into company operations, training employees and evaluating the plan.

Integrate the Plan Into Company Operations

Emergency planning must become part of the corporate culture.

Look for opportunities to build awareness; to educate and train personnel; to test procedures; to involve all levels of management, all departments and the community in the planning process; and to make emergency management part of what personnel do on a day-to-day basis.

Test how completely the plan has been integrated by asking:

- How well does senior management support the responsibilities outlined in the plan?
- Have emergency planning concepts been fully incorporated into the facility's accounting, personnel and financial procedures?
- How can the facility's processes for evaluating employees and defining job classifications better address emergency management responsibilities?
- Are there opportunities for distributing emergency preparedness information through corporate newsletters, employee manuals or employee mailings?
- What kinds of safety posters or other visible reminders would be helpful?
- Do personnel know what they should do in an emergency?
- How can all levels of the organization be involved in evaluating and updating the plan?

Conduct Training

Everyone who works at or visits the facility requires some form of training. This could include periodic employee discussion sessions to review procedures, technical training in equipment use for emergency responders, evacuation drills and full-scale exercises. Below are basic considerations for developing a training plan.

Planning Considerations

Assign responsibility for developing a training plan. Consider the training and information needs for employees, contractors, visitors, managers and those with an emergency response role identified in the plan.

Determine for a 12 month period:

- Who will be trained
- Who will do the training
- What training activities will be used
- When and where each session will take place
- How the session will be evaluated and documented

Use the Training Drills and Exercises Chart in the appendix section to schedule training activities or create one of your own.

Consider how to involve community responders in training activities.

Conduct reviews after each training activity. Involve both personnel and community responders in the evaluation process.

Training Activities

Training can take many forms:

- Orientation and Education Sessions—These are regularly scheduled discussion sessions to provide information, answer questions and identify needs and concerns.

- Tabletop Exercise—Members of the emergency management group meet in a conference room setting to discuss their responsibilities and how they would react to emergency scenarios. This is a cost-effective and efficient way to identify areas of overlap and confusion before conducting more demanding training activities.
- Walk-through Drill—The emergency management group and response teams actually perform their emergency response functions. This activity generally involves more people and is more thorough than a tabletop exercise.
- Functional Drills—These drills test specific functions such as medical response, emergency notifications, warning and communications procedures and equipment, though not necessarily at the same time. Personnel are asked to evaluate the systems and identify problem areas.
- Evacuation Drill—Personnel walk the evacuation route to a designated area where procedures for accounting for all personnel are tested. Participants are asked to make notes as they go along of what might become a hazard during an emergency, e.g., stairways cluttered with debris, smoke in the hallways. Plans are modified accordingly.
- Full-scale Exercise—A real-life emergency situation is simulated as closely as possible. This exercise involves company emergency response personnel, employsees, management and community response organizations.

Employee Training

General training for all employees should address:

- Individual roles and responsibilities
- Information about threats, hazards and protective actions
- Notification, warning and communications procedures
- Means for locating family members in an emergency
- Emergency response procedures
- Evacuation, shelter and accountability procedures
- Location and use of common emergency equipment
- Emergency shutdown procedures

The scenarios developed during the vulnerability analysis can serve as the basis for training events.

Evaluate and Modify the Plan

Conduct a formal audit of the entire plan at least once a year. Among the issues to consider are:

- How can you involve all levels of management in evaluating and updating the plan?
- Are the problem areas and resource shortfalls identified in the vulnerability analysis being sufficiently addressed?

- Does the plan reflect lessons learned from drills and actual events?
- Do members of the emergency management group and emergency response team understand their respective responsibilities? Have new members been trained?
- Does the plan reflect changes in the physical layout of the facility? Does it reflect new facility processes?
- Are photographs and other records of facility assets up to date?
- Is the facility attaining its training objectives?
- Have the hazards in the facility changed?
- Are the names, titles and telephone numbers in the plan current?
- Are steps being taken to incorporate emergency management into other facility processes?
- Have community agencies and organizations been briefed on the plan? Are they involved in evaluating the plan?

In addition to a yearly audit, evaluate and modify the plan at these times:

- After each training drill or exercise
- After each emergency
- When personnel or their responsibilities change
- When the layout or design of the facility changes
- When policies or procedures change

Remember to brief personnel on changes to the plan.

SECTION 2: EMERGENCY MANAGEMENT CONSIDERATIONS

Direction and Control

Someone must be in charge in an emergency. The system for managing resources, analyzing information and making decisions in an emergency is called direction and control.

The direction and control system described below assumes a facility of sufficient size. Your facility may require a less sophisticated system, though the principles described here will still apply.

The configuration of your system will depend on many factors. Larger industries may have their own fire team, emergency medical technicians or hazardous materials team, while smaller organizations may need to rely on mutual aid agreements. They may also be able to consolidate positions or combine responsibilities. Tenants of office buildings or industrial parks may be part of an emergency management program for the entire facility.

Emergency Management Group (EMG)

The EMG is the team responsible for the big picture. It controls all incident-related activities. The Incident Commander (IC) oversees the technical aspects of the response.

The EMG supports the IC by allocating resources and by interfacing with the community, the media, outside response organizations and regulatory agencies.

The EMG is headed by the Emergency Director (ED), who should be the facility manager. The ED is in command and control of all aspects of the emergency. Other EMG members should be senior managers who have the authority to:

- Determine the short-and long-term effects of an emergency
- Order the evacuation or shutdown of the facility
- Interface with outside organizations and the media
- Issue press releases

The relationship between the EMG and the IC is shown in Figure 1.

Incident Command System (ICS)

The ICS was developed specifically for the fire service, but its principles can be applied to all emergencies. The ICS provides for coordinated response and a clear chain of command and safe operations.

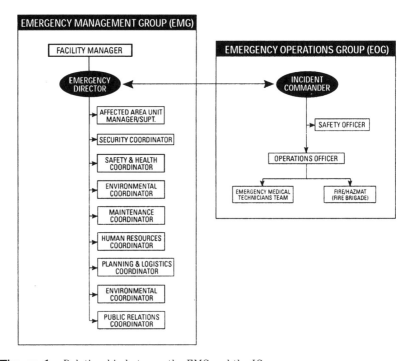

Figure 1 Relationship between the EMG and the IC.

The Incident Commander (IC) is responsible for front-line management of the incident, for tactical planning and execution, for determining whether outside assistance is needed and for relaying requests for internal resources or outside assistance through the Emergency Operations Center (EOC).

The IC can be any employee, but a member of management with the authority to make decisions is usually the best choice.

The IC must have the capability and authority to:

- Assume command
- Assess the situation
- Implement the emergency management plan
- Determine response strategies
- Activate resources
- Order an evacuation
- Oversee all incident response activities
- Declare that the incident is "over"

Emergency Operations Center (EOC)

The EOC serves as a centralized management center for emergency operations. Here, decisions are made by the EMG based upon information provided by the IC and other personnel. Regardless of size or process, every facility should designate an area where decision makers can gather during an emergency.

The EOC should be located in an area of the facility not likely to be involved in an incident, perhaps the security department, the manager's office, a conference room or the training center. An alternate EOC should be designated in the event that the primary location is not usable.

Each facility must determine its requirements for an EOC based upon the functions to be performed and the number of people involved. Ideally, the EOC is a dedicated area equipped with communications equipment, reference materials, activity logs and all the tools necessary to respond quickly and appropriately to an emergency.

Planning Considerations

To develop a direction and control system:

- Define the duties of personnel with an assigned role. Establish procedures for each position. Prepare checklists for all procedures.
- Define procedures and responsibilities for fire fighting, medical and health, and engineering.
- Determine lines of succession to ensure continuous leadership, authority and responsibility in key positions.
- Determine equipment and supply needs for each response function.
- At a minimum, assign all personnel responsibility for:
  - Recognizing and reporting an emergency
  - Warning other employees in the area

- Taking security and safety measures
- Evacuating safely
- Provide training.

Security

Isolation of the incident scene must begin when the emergency is discovered. If possible, the discoverer should attempt to secure the scene and control access, but no one should be placed in physical danger to perform these functions.

Basic security measures include:

- Closing doors or windows
- Establishing temporary barriers with furniture after people have safely evacuated
- Dropping containment materials (sorbent pads, etc.) in the path of leaking materials
- Closing file cabinets or desk drawers

Only trained personnel should be allowed to perform advanced security measures. Access to the facility, the EOC and the incident scene should be limited to persons directly involved in the response.

Coordination of Outside Response

In some cases, laws, codes, prior agreements or the very nature of the emergency require the IC to turn operations over to an outside response organization.

When this happens, the protocols established between the facility and outside response organizations are implemented. The facility's IC provides the community's IC a complete report on the situation.

The facility IC keeps track of which organizations are on-site and how the response is being coordinated. This helps increase personnel safety and accountability, and prevents duplication of effort.

Communications

Communications are essential to any business operation. A communications failure can be a disaster in itself, cutting off vital business activities.

Communications are needed to report emergencies, to warn personnel of the danger, to keep families and off-duty employees informed about what's happening at the facility to coordinate response actions and to keep in contact with customers and suppliers.

Contingency Planning

Plan for all possible contingencies from a temporary or short-term disruption to a total communications failure.

DOCUMENT APPENDIX

- Consider the everyday functions performed by your facility and the communications, both voice and data, used to support them.
- Consider the business impact if your communications were inoperable. How would this impact your emergency operations?
- Prioritize all facility communications. Determine which should be restored first in an emergency.
- Establish procedures for restoring communications systems.
- Talk to your communications vendors about their emergency response capabilities. Establish procedures for restoring services.
- Determine needs for backup communications for each business function. Options include messengers, telephones, portable microwave, amateur radios, point-to-point private lines, satellite, high-frequency radio.

Emergency Communications

Consider the functions your facility might need to perform in an emergency and the communications systems needed to support them.

Consider communications between:

- Emergency responders
- Responders and the Incident Commander (IC)
- The IC and the Emergency Operations Center (EOC)
- The IC and employees
- The EOC and outside response organizations
- The EOC and neighboring businesses
- The EOC and employees' families
- The EOC and customers
- The EOC and media

Methods of communication include:

- Messenger
- Telephone
- Two-way radio
- FAX machine
- Microwave
- Satellite
- Dial-up modems
- Local area networks
- Hand signals

Family Communications

In an emergency, personnel will need to know whether their families are okay. Taking care of one's loved ones is always a first priority.

Make plans for communicating with employees' families in an emergency. Also, encourage employees to:

- Consider how they would communicate with their families in case they are separated from one another or injured in an emergency.
- Arrange for an out-of-town contact for all family members to call in an emergency.
- Designate a place to meet family members in case they cannot get home in an emergency.

Notification

Establish procedures for employees to report an emergency. Inform employees of procedures. Train personnel assigned specific notification tasks.

Post emergency telephone numbers near each telephone, on employee bulletin boards and in other prominent locations.

Maintain an updated list of addresses and telephone and pager numbers of key emergency response personnel (from within and outside the facility).

Listen for tornado, hurricane and other severe weather warnings issued by the National Weather Service.

Determine government agencies' notification requirements in advance. Notification must be made immediately to local government agencies when an emergency has the potential to affect public health and safety.

Prepare announcements that could be made over public address systems.

Warning

Establish a system for warning personnel of an emergency. The system should:

- Be audible or within view by all people in the facility
- Have an auxiliary power supply
- Have a distinct and recognizable signal

Make plans for warning persons with disabilities. For instance, a flashing strobe light can be used to warn hearing-impaired people.

Familiarize personnel with procedures for responding when the warning system is activated.

Establish procedures for warning customers, contractors, visitors and others who may not be familiar with the facility's warning system.

Test your facility's warning system at least monthly.

Life Safety

Protecting the health and safety of everyone in the facility is the first priority during an emergency.

DOCUMENT APPENDIX

Evacuation Planning

One common means of protection is evacuation. In the case of fire, an immediate evacuation to a predetermined area away from the facility may be necessary. In a hurricane, evacuation could involve the entire community and take place over a period of days.

To develop an evacuation policy and procedure:

- Determine the conditions under which an evacuation would be necessary.
- Establish a clear chain of command. Identify personnel with the authority to order an evacuation. Designate "evacuation wardens" to assist others in an evacuation and to account for personnel.
- Establish specific evacuation procedures. Establish a system for accounting for personnel. Consider employees' transportation needs for community-wide evacuations.
- Establish procedures for assisting persons with disabilities and those who do not speak English.
- Post evacuation procedures.
- Designate personnel to continue or shut down critical operations while an evacuation is underway. They must be capable of recognizing when to abandon the operation and evacuate themselves.
- Coordinate plans with the local emergency management office.

Evacuation Routes and Exits

Designate primary and secondary evacuation routes and exits. Have them clearly marked and well lit. Post signs.

Install emergency lighting in case a power outage occurs during an evacuation. Ensure that evacuation routes and emergency exits are:

- Wide enough to accommodate the number of evacuating personnel
- Clear and unobstructed at all times
- Unlikely to expose evacuating personnel to additional hazards

Have evacuation routes evaluated by someone not in your organization.

Assembly Areas and Accountability

Obtaining an accurate account of personnel after a site evacuation requires planning and practice.

- Designate assembly areas where personnel should gather after evacuating.
- Take a head count after the evacuation. The names and last known locations of personnel not accounted for should be determined and given to the EOC. (Confusion in the assembly areas can lead to unnecessary and dangerous search and rescue operations.)

- Establish a method for accounting for non-employees such as suppliers and customers.
- Establish procedures for further evacuation in case the incident expands. This may consist of sending employees home by normal means or providing them with transportation to an off-site location.

Shelter

In some emergencies, the best means of protection is to take shelter either within the facility or away from the facility in a public building.

- Consider the conditions for taking shelter, e.g., tornado warning.
- Identify shelter space in the facility and in the community. Establish procedures for sending personnel to shelter.
- Determine needs for emergency supplies such as water, food and medical supplies.
- Designate shelter managers, if appropriate.
- Coordinate plans with local authorities.

Training and Information

Train employees in evacuation, shelter and other safety procedures. Conduct sessions at least annually or when:

- Employees are hired
- Evacuation wardens, shelter managers and others with special assignments are designated
- New equipment, materials or processes are introduced
- Procedures are updated or revised
- Exercises show that employee performance must be improved

> Provide emergency information such as checklists and evacuation maps. Post evacuation maps in strategic locations.
> Consider the information needs of customers and others who visit the facility.

Family Preparedness

Consider ways to help employees prepare their families for emergencies. This will increase their personal safety and help the facility get back up and running. Those who are prepared at home will be better able to carry out their responsibilities at work.

Property Protection

Protecting facilities, equipment and vital records is essential to restoring operations once an emergency has occurred.

DOCUMENT APPENDIX

Planning Considerations

Establish procedures for:

- Fighting fires
- Containing material spills
- Closing or barricading doors and windows
- Shutting down equipment
- Covering or securing equipment
- Moving equipment to a safe location

Identify sources of backup equipment, parts and supplies.

Designate personnel to authorize, supervise and perform a facility shutdown. Train them to recognize when to abandon the effort.

Obtain materials to carry out protection procedures and keep them on hand for use only in emergencies.

Protection Systems

Determine needs for systems to detect abnormal situations, provide warning and protect property.

Consider:

- Fire protection systems
- Lightning protection systems
- Water-level monitoring systems
- Overflow detection devices
- Automatic shutoffs
- Emergency power generation systems

Consult your property insurer about special protective systems.

Mitigation

Consider ways to reduce the effects of emergencies, such as moving or constructing facilities away from flood plains and fault zones. Also consider ways to reduce the chances of emergencies from occurring, such as changing processes or materials used to run the business.

Consider physical retrofitting measures such as:

- Upgrading facilities to withstand the shaking of an earthquake or high winds
- "Floodproofing" facilities by constructing flood walls or other flood protection devices (see Section 3 for additional information)
- Installing fire sprinkler systems
- Installing fire-resistant materials and furnishing
- Installing storm shutters for all exterior windows and doors

There are also non-structural mitigation measures to consider, including:

- Installing fire-resistant materials and furnishing
- Securing light fixtures and other items that could fall or shake loose in an emergency
- Moving heavy or breakable objects to low shelves
- Attaching cabinets and files to low walls or bolting them together
- Placing Velcro strips under typewriters, tabletop computers and television monitors
- Moving work stations away from large windows
- Installing curtains or blinds that can be drawn over windows to prevent glass from shattering onto employees
- Anchoring water heaters and bolting them to wall studs

Consult a structural engineer or architect and your community's building and zoning offices for additional information.

Facility Shutdown

Facility shutdown is generally a last resort but always a possibility. Improper or disorganized shutdown can result in confusion, injury and property damage.

Some facilities require only simple actions such as turning off equipment, locking doors and activating alarms. Others require complex shutdown procedures.

Work with department heads to establish shutdown procedures. Include information about when and how to shut off utilities.

Identify:

- The conditions that could necessitate a shutdown
- Who can order a shutdown
- Who will carry out shutdown procedures
- How a partial shutdown would affect other facility operations
- The length of time required for shutdown and restarting

Train personnel in shutdown procedures. Post procedures.

Records Preservation

Vital records may include:

- Financial and insurance information
- Engineering plans and drawings
- Product lists and specifications
- Employee, customer and supplier databases
- Formulas and trade secrets
- Personnel files

DOCUMENT APPENDIX

Preserving vital records is essential to the quick restoration of operations. Analyzing vital records involves:

1. Classifying operations into functional categories, e.g., finance, production, sales, administration
2. Determining essential functions for keeping the business up and running, such as finance, production, sales, etc.
3. Identifying the minimum information that must be readily accessible to perform essential functions, e.g., maintaining customer collections may require access to account statements
4. Identifying the records that contain the essential information and where they are located
5. Identifying the equipment and materials needed to access and use the information

Next, establish procedures for protecting and accessing vital records. Among the many approaches to consider are:

- Labeling vital records
- Backing up computer systems
- Making copies of records
- Storing tapes and disks in insulated containers
- Storing data off-site where they would not likely be damaged by an event affecting your facility
- Increasing security of computer facilities
- Arranging for evacuation of records to backup facilities
- Backing up systems handled by service bureaus
- Arranging for backup power

Community Outreach

Your facility's relationship with the community will influence your ability to protect personnel and property and return to normal operations.

This section describes ways to involve outside organizations in the emergency management plan.

Involving the Community

Maintain a dialogue with community leaders, first responders, government agencies, community organizations and utilities, including:

- Appointed and elected leaders
- Fire, police and emergency medical services personnel
- Local Emergency Planning Committee (LEPC) members

- Emergency management director
- Public Works Department
- American Red Cross
- Hospitals
- Telephone company
- Electric utility
- Neighborhood groups

Have regular meetings with community emergency personnel to review emergency plans and procedures. Talk about what you're doing to prepare for and prevent emergencies. Explain your concern for the community's welfare.

Identify ways your facility could help the community in a community-wide emergency.

Look for common interests and concerns. Identify opportunities for sharing resources and information.

Conduct confidence-building activities such as facility tours. Do a facility walk-through with community response groups.

Involve community fire, police and emergency management personnel in drills and exercises.

Meet with your neighbors to determine how you could assist each other in an emergency.

Mutual Aid Agreements

To avoid confusion and conflict in an emergency, establish mutual aid agreements with local response agencies and businesses.

These agreements should:

- Define the type of assistance
- Identify the chain of command for activating the agreement
- Define communications procedures

Include these agencies in facility training exercises whenever possible.

Community Service

In community-wide emergencies, business and industry are often needed to assist the community with:

- Personnel
- Equipment
- Shelter
- Training
- Storage
- Feeding facilities
- EOC facilities

DOCUMENT APPENDIX 557

- Food, clothing, building materials
- Funding
- Transportation

While there is no way to predict what demands will be placed on your company's resources, give some thought to how the community's needs might influence your corporate responsibilities in an emergency. Also, consider the opportunities for community service before an emergency occurs.

Public Information

When site emergencies expand beyond the facility, the community will want to know the nature of the incident, whether the public's safety or health is in danger, what is being done to resolve the problem and what was done to prevent the situation from happening.

Determine the audiences that may be affected by an emergency and identify their information needs. Include:

- The public
- The media
- Employees and retirees
- Unions
- Contractors and suppliers
- Customers
- Shareholders
- Emergency response organizations
- Regulatory agencies
- Appointed and elected officials
- Special interest groups
- Neighbors

Media Relations

In an emergency, the media are the most important link to the public. Try to develop and maintain positive relations with media outlets in your area. Determine their particular needs and interests. Explain your plan for protecting personnel and preventing emergencies.

Determine how you would communicate important public information through the media in an emergency.

- Designate a trained spokesperson and an alternate spokesperson
- Set up a media briefing area
- Establish security procedures
- Establish procedures for ensuring that information is complete, accurate and approved for public release

- Determine an appropriate and useful way of communicating technical information
- Prepare background information about the facility

When providing information to the media during an emergency:

Do's

- Give all media equal access to information.
- When appropriate, conduct press briefings and interviews. Give local and national media equal time.
- Try to observe media deadlines.
- Escort media representatives to ensure safety.
- Keep records of information released.
- Provide press releases when possible.

Don'ts

- Do not speculate about the incident.
- Do not permit unauthorized personnel to release information.
- Do not cover up facts or mislead the media.
- Do not place blame for the incident.

Recovery and Restoration

Business recovery and restoration, or business resumption, goes right to a facility's bottom line: keeping people employed and the business running.

Planning Considerations

Consider making contractual arrangements with vendors for such post-emergency services as records preservation, equipment repair, earthmoving or engineering.

Meet with your insurance carriers to discuss your property and business resumptions policies (see the next page for guidelines).

Determine critical operations and make plans for bringing those systems back on-line. The process may entail:

- Repairing or replacing equipment
- Relocating operations to an alternate location
- Contracting operations on a temporary basis

Take photographs or videotape the facility to document company assets. Update these records regularly.

Continuity of Management

You can assume that not every key person will be readily available or physically at the facility after an emergency. Ensure that recovery decisions can be made

without undue delay. Consult your legal department regarding laws and corporate bylaws governing continuity of management.
  Establish procedures for:

- Assuring the chain of command
- Maintaining lines of succession for key personnel
- Moving to alternate headquarters

  Include these considerations in all exercise scenarios.

Insurance

Most companies discover that they are not properly insured only after they have suffered a loss. Lack of appropriate insurance can be financially devastating. Discuss the following topics with your insurance advisor to determine your individual needs.

- How will my property be valued?
- Does my policy cover the cost of required upgrades to code?
- How much insurance am I required to carry to avoid becoming a co-insurer?
- What perils or causes of loss does my policy cover?
- What are my deductibles?
- What does my policy require me to do in the event of a loss?
- What types of records and documentation will my insurance company want to see? Are records in a safe place where they can be obtained after an emergency?
- To what extent am I covered for loss due to interruption of power? Is coverage provided for both on- and off-premises power interruption?
- Am I covered for lost income in the event of business interruption because of a loss? Do I have enough coverage? For how long is coverage provided? How long is my coverage for lost income if my business is closed by order of a civil authority?
- To what extent am I covered for reduced income due to customers' not all immediately coming back once the business reopens?
- How will my emergency management program affect my rates?

Employee Support

Since employees who will rely on you for support after an emergency are your most valuable asset, consider the range of services that you could provide or arrange for, including:

- Cash advances
- Salary continuation
- Flexible work hours
- Reduced work hours

- Crisis counseling
- Care packages
- Day care

Resuming Operations

Immediately after an emergency, take steps to resume operations.

- Establish a recovery team, if necessary. Establish priorities for resuming operations.
- Continue to ensure the safety of personnel on the property. Assess remaining hazards. Maintain security at the incident scene.
- Conduct an employee briefing.
- Keep detailed records. Consider audio recording all decisions. Take photographs of or video-tape the damage.
- Account for all damage-related costs. Establish special job order numbers and charge codes for purchases and repair work.
- Follow notification procedures. Notify employees' families about the status of personnel on the property. Notify off-duty personnel about work status. Notify insurance carriers and appropriate government agencies.
- Protect undamaged property. Close up building openings. Remove smoke, water and debris. Protect equipment against moisture. Restore sprinkler systems. Physically secure the property. Restore power.
- Conduct an investigation. Coordinate actions with appropriate government agencies.
- Conduct salvage operations. Segregate damaged from undamaged property. Keep damaged goods on hand until an insurance adjuster has visited the premises, but you can move material outside if it's seriously in the way and exposure to the elements won't make matters worse.
- Take an inventory of damaged goods. This is usually done with the adjuster, or the adjuster's salvor if there is any appreciable amount of goods or value. If you release goods to the salvor, obtain a signed inventory stating the quantity and type of goods being removed.
- Restore equipment and property. For major repair work, review restoration plans with the insurance adjuster and appropriate government agencies.
- Assess the value of damaged property. Assess the impact of business interruption.
- Maintain contact with customers and suppliers.

Administration and Logistics

Maintain complete and accurate records at all times to ensure a more efficient emergency response and recovery. Certain records may also be required by regulation or by your insurance carriers or prove invaluable in the case of legal action after an incident.

Administrative Actions

Administrative actions prior to an emergency include:

- Establishing a written emergency management plan
- Maintaining training records
- Maintaining all written communications
- Documenting drills and exercises and their critiques
- Involving community emergency response organizations in planning activities

Administrative actions during and after an emergency include:

- Maintaining telephone logs
- Keeping a detailed record of events
- Maintaining a record of injuries and follow-up actions
- Accounting for personnel
- Coordinating notification of family members
- Issuing press releases
- Maintaining sampling records
- Managing finances
- Coordinating personnel services
- Documenting incident investigations and recovery operations

Logistics

Before an emergency, logistics may entail:

- Acquiring equipment
- Stockpiling supplies
- Designating emergency facilities
- Establishing training facilities
- Establishing mutual aid agreements
- Preparing a resource inventory

During an emergency, logistics may entail the provision of:

- Providing utility maps to emergency responders
- Providing material safety data sheets to employees
- Moving backup equipment in place
- Repairing parts
- Arranging for medical support, food and transportation
- Arranging for shelter facilities
- Providing for backup power
- Providing for backup communications

SECTION 3: HAZARD-SPECIFIC INFORMATION

Fire

Fire is the most common of all the hazards. Every year fires cause thousands of deaths and injuries and billions of dollars in property damage.

Planning Considerations

Consider the following when developing your plan:

- Meet with the fire department to talk about the community's fire response capabilities. Talk about your operations. Identify processes and materials that could cause or fuel a fire, or contaminate the environment in a fire.
- Have your facility inspected for fire hazards. Ask about fire codes and regulations.
- Ask your insurance carrier to recommend fire prevention and protection measures. Your carrier may also offer training.
- Distribute fire safety information to employees: how to prevent fires in the workplace, how to contain a fire, how to evacuate the facility, where to report a fire.
- Instruct personnel to use the stairs—not elevators—in a fire. Instruct them to crawl on their hands and knees when escaping a hot or smoke-filled area.
- Conduct evacuation drills. Post maps of evacuation routes in prominent places. Keep evacuation routes including stairways and doorways clear of debris.
- Assign fire wardens for each area to monitor shutdown and evacuation procedures.
- Establish procedures for the safe handling and storage of flammable liquids and gases. Establish procedures to prevent the accumulation of combustible materials.
- Provide for the safe disposal of smoking materials.
- Establish a preventive maintenance schedule to keep equipment operating safely.
- Place fire extinguishers in appropriate locations.
- Train employees in use of fire extinguishers.
- Install smoke detectors. Check smoke detectors once a month, change batteries at least once a year.
- Establish a system for warning personnel of a fire. Consider installing a fire alarm with automatic notification to the fire department.
- Consider installing a sprinkler system, fire hoses and fire-resistant walls and doors.
- Ensure that key personnel are familiar with all fire safety systems.
- Identify and mark all utility shutoffs so that electrical power, gas or water can be shut off quickly by fire wardens or responding personnel.
- Determine the level of response your facility will take if a fire occurs. Among the options are:

Option 1—Immediate evacuation of all personnel on alarm.

Option 2—All personnel are trained in fire extinguisher use. Personnel in the immediate area of a fire attempt to control it. If they cannot, the fire alarm is sounded and all personnel evacuate.

Option 3—Only designated personnel are trained in fire extinguisher use.

Option 4—A fire team is trained to fight incipient-stage fires that can be controlled without protective equipment or breathing apparatus. Beyond this level fire, the team evacuates.

Option 5—A fire team is trained and equipped to fight structural fires using protective equipment and breathing apparatus.

Hazardous Materials Incidents

Hazardous materials are substances that are either flammable or combustible, explosive, toxic, noxious, corrosive, oxidizable, an irritant or radioactive.

A hazardous material spill or release can pose a risk to life, health or property. An incident can result in the evacuation of a few people, a section of a facility or an entire neighborhood.

There are a number of Federal laws that regulate hazardous materials, including: the Superfund Amendments and Reauthorization Act of 1986 (SARA), the Resource Conservation and Recovery Act of 1976 (RCRA), the Hazardous Materials Transportation Act (HMTA), the Occupational Safety and Health Act (OSHA), the Toxic Substances Control Act (TSCA) and the Clean Air Act.

Title III of SARA regulates the packaging, labeling, handling, storage and transportation of hazardous materials. The law requires facilities to furnish information about the quantities and health effects of materials used at the facility, and to promptly notify local and State officials whenever a significant release of hazardous materials occurs.

In addition to on-site hazards, you should be aware of the potential for an off-site incident affecting your operations. You should also be aware of hazardous materials used in facility processes and in the construction of the physical plant.

Detailed definitions as well as lists of hazardous materials can be obtained from the Environmental Protection Agency (EPA) and the Occupational Safety and Health Administration (OSHA).

Planning Considerations

Consider the following when developing your plan:

- Identify and label all hazardous materials stored, handled, produced and disposed of by your facility. Follow government regulations that apply to your facility. Obtain material safety data sheets (MSDS) for all hazardous materials at your location.
- Ask the local fire department for assistance in developing appropriate response procedures.

- Train employees to recognize and report hazardous material spills and releases. Train employees in proper handling and storage.
- Establish a hazardous material response plan:
  - Establish procedures to notify management and emergency response organizations of an incident.
  - Establish procedures to warn employees of an incident.
  - Establish evacuation procedures.
  - Depending on your operations, organize and train an emergency response team to confine and control hazardous material spills in accordance with applicable regulations.
- Identify other facilities in your area that use hazardous materials. Determine whether an incident could affect your facility.
- Identify highways, railroads and waterways near your facility used for the transportation of hazardous materials. Determine how a transportation accident near your facility could affect your operations.

Floods and Flash Floods

Floods are the most common and widespread of all natural disasters. Most communities in the United States can experience some degree of flooding after spring rains, heavy thunderstorms or winter snow thaws.

Most floods develop slowly over a period of days. Flash floods, however, are like walls of water that develop in a matter of minutes. Flash floods can be caused by intense storms or dam failure.

Planning Considerations
Consider the following when preparing for floods:

- Ask your local emergency management office whether your facility is located in a flood plain. Learn the history of flooding in your area. Learn the elevation of your facility in relation to steams, rivers and dams.
- Review the community's emergency plan. Learn the community's evacuation routes. Know where to find higher ground in case of a flood.
- Establish warning and evacuation procedures for the facility. Make plans for assisting employees who may need transportation.
- Inspect areas in your facility subject to flooding. Identify records and equipment that can be moved to a higher location. Make plans to move records and equipment in case of flood.
- Purchase a NOAA Weather Radio with a warning alarm tone and battery backup. Listen for flood watches and warnings.

Flood Watch—Flooding is possible. Stay tuned to NOAA radio. Be prepared to evacuate. Tune to local radio and television stations for additional information.

DOCUMENT APPENDIX

Flood Warning—Flooding is already occurring or will occur soon. Take precautions at once. Be prepared to go to higher ground. If advised, evacuate immediately.

- Ask your insurance carrier for information about flood insurance. Regular property and casualty insurance does not cover flooding.
- Consider the feasibility of floodproofing your facility. There are three basic types of methods.
  1. Permanent floodproofing measures are taken before a flood occurs and require no human intervention when flood waters rise. They include:
     - Filling windows, doors or other openings with water-resistant materials such as concrete blocks or bricks. This approach assumes the structure is strong enough to withstand flood waters.
     - Installing check valves to prevent water from entering where utility and sewer lines enter the facility.
     - Reinforcing walls to resist water pressure. Sealing walls to prevent or reduce seepage.
     - Building watertight walls around equipment or work areas within the facility that are particularly susceptible to flood damage.
     - Constructing floodwalls or levees outside the facility to keep flood waters away.
     - Elevating the facility on walls, columns or compacted fill. This approach is most applicable to new construction, though many types of buildings can be elevated.
  2. Contingent floodproofing measures are also taken before a flood but require some additional action when flooding occurs. These measures include:
     - Installing watertight barriers called flood shields to prevent the passage of water through doors, windows, ventilation shafts or other openings
     - Installing permanent water-tight doors
     - Constructing movable floodwalls
     - Installing permanent pumps to remove flood waters
  3. Emergency floodproofing measures are generally less expensive than those listed above, though they require substantial advance warning and do not satisfy the minimum requirements for watertight floodproofing as set forth by the National Flood Insurance Program (NFIP). They include:
     - Building walls with sandbags
     - Constructing a double row of walls with boards and posts to create a "crib," then filling the crib with soil
     - Constructing a single wall by stacking small beams or planks on top of each other
- Consider the need for backup systems:
  - Portable pumps to remove flood water
  - Alternate power sources such as generators or gasoline-powered pumps

- Battery-powered emergency lighting
- Participate in community flood control projects.

Hurricanes

Hurricanes are severe tropical storms with sustained winds of 74 miles per hour or greater. Hurricane winds can reach 160 miles per hour and extend inland for hundreds of miles.

Hurricanes bring torrential rains and a storm surge of ocean water that crashes into land as the storm approaches. Hurricanes also spawn tornadoes.

Hurricane advisories are issued by the National Weather Service as soon as a hurricane appears to be a threat. The hurricane season lasts from June through November.

Planning Considerations

The following are considerations when preparing for hurricanes:

- Ask your local emergency management office about community evacuation plans.
- Establish facility shutdown procedures. Establish warning and evacuation procedures. Make plans for assisting employees who may need transportation.
- Make plans for communicating with employees' families before and after a hurricane.
- Purchase a NOAA Weather Radio with a warning alarm tone and battery backup. Listen for hurricane watches and warnings.

  Hurricane Watch—A hurricane is possible within 24 to 36 hours. Stay tuned for additional advisories. Tune to local radio and television stations for additional information. An evacuation may be necessary.

  Hurricane Warning—A hurricane will hit land within 24 hours. Take precautions at once. If advised, evacuate immediately.
- Survey your facility. Make plans to protect outside equipment and structures.
- Make plans to protect windows. Permanent storm shutters offer the best protection. Covering windows with 5/8" marine plywood is a second option.
- Consider the need for backup systems:
  - Portable pumps to remove flood water
  - Alternate power sources such as generators or gasoline-powered pumps
  - Battery-powered emergency lighting
- Prepare to move records, computers and other items within your facility or to another location.

Tornadoes

Tornadoes are incredibly violent local storms that extend to the ground with whirling winds that can reach 300 mph.

DOCUMENT APPENDIX

Spawned from powerful thunderstorms, tornadoes can uproot trees and buildings and turn harmless objects into deadly missiles in a matter of seconds. Damage paths can be in excess of one mile wide and 50 miles long.

Tornadoes can occur in any state but occur more frequently in the Midwest, Southeast and Southwest. They occur with little or no warning.

Planning Considerations

The following are considerations when planning for tornadoes:

- Ask your local emergency management office about the community's tornado warning system.
- Purchase a NOAA Weather Radio with a warning alarm tone and battery backup. Listen for tornado watches and warnings.
  Tornado Watch—Tornadoes are likely. Be ready to take shelter. Stay tuned to radio and television stations for additional information.
  Tornado Warning—A tornado has been sighted in the area or is indicated by radar. Take shelter immediately.
- Establish procedures to inform personnel when tornado warnings are posted. Consider the need for spotters to be responsible for looking out for approaching storms.
- Work with a structural engineer or architect to designate shelter areas in your facility. Ask your local emergency management office or National Weather Service office for guidance.
- Consider the amount of space you will need. Adults require about six square feet of space; nursing home and hospital patients require more.
- The best protection in a tornado is usually an underground area. If an underground area is not available, consider:
  - Small interior rooms on the lowest floor and without windows
  - Hallways on the lowest floor away from doors and windows
  - Rooms constructed with reinforced concrete, brick or block with no windows and a heavy concrete floor or roof system overhead
  - Protected areas away from doors and windows

Note: Auditoriums, cafeterias and gymnasiums that are covered with a flat, wide-span roof are not considered safe.

- Make plans for evacuating personnel away from lightweight modular offices or mobile home-size buildings. These structures offer no protection from tornadoes.
- Conduct tornado drills.
- Once in the shelter, personnel should protect their heads with their arms and crouch down.

Severe Winter Storms

Severe winter storms bring heavy snow, ice, strong winds and freezing rain. Winter storms can prevent employees and customers from reaching the facility, leading to a temporary shutdown until roads are cleared. Heavy snow and ice can also cause structural damage and power outages.

Planning Considerations

Following are considerations for preparing for winter storms:

- Listen to NOAA Weather Radio and local radio and television stations for weather information:
    Winter Storm Watch—Severe winter weather is possible.
    Winter Storm Warning—Severe winter weather is expected.
    Blizzard Warning—Severe winter weather with sustained winds of at least 35 mph is expected.
    Traveler's Advisory—Severe winter conditions may make driving difficult or dangerous.
- Establish procedures for facility shutdown and early release of employees.
- Store food, water, blankets, battery-powered radios with extra batteries and other emergency supplies for employees who become stranded at the facility.
- Provide a backup power source for critical operations.
- Arrange for snow and ice removal from parking lots, walkways, loading docks, etc.

Earthquakes

Earthquakes occur most frequently west of the Rocky Mountains, although historically the most violent earthquakes have occurred in the central United States. Earthquakes occur suddenly and without warning.

Earthquakes can seriously damage buildings and their contents; disrupt gas, electric and telephone services; and trigger landslides, avalanches, flash floods, fires and huge ocean waves called tsunamis. Aftershocks can occur for weeks following an earthquake.

In many buildings, the greatest danger to people in an earthquake is when equipment and non-structural elements such as ceilings, partitions, windows and lighting fixtures shake loose.

Planning Considerations

Following are guidelines for preparing for earthquakes:

- Assess your facility's vulnerability to earthquakes. Ask local government agencies for seismic information for your area.

- Have your facility inspected by a structural engineer. Develop and prioritize strengthening measures. These may include:
  - Adding steel bracing to frames
  - Adding sheer walls to frames
  - Strengthening columns and building foundations
  - Replacing unreinforced brick filler walls
- Follow safety codes when constructing a facility or making major renovations.
- Inspect non-structural systems such as air conditioning, communications and pollution control systems. Assess the potential for damage. Prioritize measures to prevent damages.
- Inspect your facility for any item that could fall, spill, break or move during an earthquake. Take steps to reduce these hazards:
  - Move large and heavy objects to lower shelves or the floor. Hang heavy items away from where people work.
  - Secure shelves, filing cabinets, tall furniture, desktop equipment, computers, printers, copiers and light fixtures.
  - Secure fixed equipment and heavy machinery to the floor. Larger equipment can be placed on casters and attached to tethers which attach to the wall.
  - Add bracing to suspended ceilings, if necessary.
  - Install safety glass where appropriate.
  - Secure large utility and process piping.
- Keep copies of design drawings of the facility to be used in assessing the facility's safety after an earthquake.
- Review processes for handling and storing hazardous materials. Have incompatible chemicals stored separately.
- Ask your insurance carrier about earthquake insurance and mitigation techniques.
- Establish procedures to determine whether an evacuation is necessary after an earthquake.
- Designate areas in the facility away from exterior walls and windows where occupants should gather after an earthquake if an evacuation is not necessary.
- Conduct earthquake drills. Provide personnel with the following safety information:
  - In an earthquake, if indoors, stay there. Take cover under a sturdy piece of furniture or counter, or brace yourself against an inside wall. Protect your head and neck.
  - If outdoors, move into the open, away from buildings, street lights and utility wires.
  - After an earthquake, stay away from windows, skylights and items that could fall. Do not use the elevators.
  - Use stairways to leave the building if it is determined that a building evacuation is necessary.

Technological Emergencies

Technological emergencies include any interruption or loss of a utility service, power source, life support system, information system or equipment needed to keep the business in operation.

Planning Considerations

The following are suggestions for planning for technological emergencies:

- Identify all critical operations, including:
  - Utilities including electric power, gas, water, hydraulics, compressed air, municipal and internal sewer systems, wastewater treatment services
  - Security and alarm systems, elevators, lighting, life support systems, heating, ventilation and air conditioning systems, electrical distribution system
  - Manufacturing equipment, pollution control equipment
  - Communication systems, both data and voice computer networks
  - Transportation systems including air, highway, railroad and waterway
- Determine the impact of service disruption.
- Ensure that key safety and maintenance personnel are thoroughly familiar with all building systems.
- Establish procedures for restoring systems. Determine need for backup systems.
- Establish preventive maintenance schedules for all systems and equipment.

SECTION 4: INFORMATION SOURCES

Additional Readings from FEMA

The following publications can be obtained from FEMA by writing to: FEMA, Publications, P.O. Box 2012, Jessup, MD 20794-2012.

- Principal Threats Facing Communities and Local Emergency Management Coordinators (FEMA 191)—Statistics and analyses of natural disasters and man-made threats in the U.S.
- Floodproofing Non-Residential Structures (FEMA 102)—Technical information for building owners, designers and contractors on floodproofing techniques (200 pages).
- Non-Residential Floodproofing—Requirements and Certification for Buildings Located in Flood Hazard Areas in Accordance with the National Flood Insurance Program (FIA-TB-3)—Planning and engineering considerations for floodproofing new commercial buildings.
- Building Performance: Hurricane Andrew in Florida (FIA 22)—Technical guidance for enhancing the performance of buildings in hurricanes.

DOCUMENT APPENDIX

- Building Performance: Hurricane Iniki in Hawaii (FIA 23)—Technical guidance for reducing hurricane and flood damage.
- Answers to Questions About Substantially Damaged Buildings (FEMA 213)—Information about regulations and policies of the National Flood Insurance Program regarding substantially damaged buildings (25 pages).
- Design Guidelines for Flood Damage Reduction (FEMA 15)—A study on land use, watershed management, design and construction practices in floodprone areas.
- Comprehensive Earthquake Preparedness Planning Guidelines: Corporate (FEMA 71)—Earthquake planning guidance for corporate safety officers and managers.

Ready-to-Print Brochure Mechanicals for Your Employee Safety Program

You can provide your employees and customers with life-saving information from FEMA and the American Red Cross. Available at no charge is ready-to-print artwork for a series of brochures on disaster preparedness and family safety.

Select any of the brochures below, and you'll receive camera-ready materials, printing instructions and ideas for adding your own logo or sponsor message. Write to: Camera-ready Requests, Community & Family Preparedness Program, 500 C Street, SW Washington, DC 20472.

- Your Family Disaster Plan—A 4-step plan for individuals and families on how to prepare for any type of disaster.
- Emergency Preparedness Checklist—An action checklist on home safety, evacuation and disaster preparedness.
- Your Family Disaster Supplies Kit—A checklist of emergency supplies for the home and car.
- Helping Children Cope With Disaster—Practical advice on how to help children deal with the stress of disaster.

Emergency Management Offices

FEMA Headquarters
Federal Emergency Management Agency, 500 C Street, SW, Washington, DC 20472, (202)646-2500.

FEMA Regional Offices

- Region 1: Boston (617)223-9540
- Region 2: New York (212)225-7209

- Region 3: Philadelphia (215)931-5500
- Region 4: Atlanta (404)853-4200
- Region 5: Chicago (312)408-5500
- Region 6: Denton, TX (817)898-5104
- Region 7: Kansas City, MO (816)283-7061
- Region 8: Denver (303)235-1813
- Region 9: San Francisco (415)923-7100
- Region 10: Bothell, WA (206)487-4604

State Emergency Management Agencies
(FEMA region numbers are in parentheses.)

Alabama (4)
Alabama Emergency Management Agency
5898 S. County Rd. 41 Drawer 2160
Clanton, AL 35045-5160
(205)280-2201

Alaska (10)
Department of Military & Veteran Affairs
P.O. Box 5750
Camp Denali, AK 99595-5750
(907)428-7000

Arizona(9)
Arizona Division of Emergency Services
National Guard Bldg.
5636 E. McDowell Rd.
Phoenix, AZ 85008
(602)231-6245

Arkansas (6)
Office of Emergency Services
P.O. Box 758
Conway, AR 72032
(501)321-5601

California (9)
Office of Emergency Services
2800 Meadowview Rd.
Sacramento, CA 95823
(916)262-1816

Colorado (8)
Colorado Office of Emergency Management
Camp George West
Golden, CO 80401
(303)273-1622

Connecticut (1)
Connecticut Office of Emergency Management
360 Broad St.
Hartford, CT 06105
(203)566-3180

Delaware (3)
Division of Emergency Planning and Operations
P.O. Box 527
Delaware City, DE 19706
(302) 326-6000

District of Columbia (3)
Office of Emergency Preparedness
200 14th St., NW, 8th Floor
Washington, DC 20009
(202)727-3159

Florida (4)
Division of Emergency Management
2555 Shumar Oak Blvd.
Tallahassee, FL 32399-2100
(904)413-9969

Georgia (4)
Georgia Emergency Management Agency
P.O. Box 18055
Atlanta, GA 30316-0055
(404)635-7001

Hawaii (9)
State Civil Defense
3949 Diamond Head Rd.
Honolulu, HI 96816-4495
(808)733-4300

Idaho (10)
Bureau of Disaster Services
650 W. State St.
Boise, ID 83720
(208)334-2336

Illinois (5)
Illinois Emergency Management Agency
110 E. Adams St.
Springfield, IL 62706
(217)782-2700

Indiana (5)
Indiana Emergency Management Agency
State Office Bldg., Room E-208
302 W. Washington St.
Indianapolis, IN 46204
(317)232-3980

Iowa (7)
Iowa Emergency Management Division
Hoover State Office Bldg.
Level A, Room 29
Des Moines, IA 50319
(515)281-3231

Kansas (7)
Division of Emergency Preparedness
2800 S.W. Topeka Blvd
Topeka, KS 66611-1401
(913)274-1401

Kentucky (4)
Kentucky Disaster and Emergency Services
100 Minutemen Pkwy
Frankfort, KY 40601-6168
(502)564-8682

Louisiana (6)
Office of Emergency Preparedness
Department of Public Safety
LA Military Dept.
P.O. Box 44217
Capitol Station
Baton Rouge, LA 70804
(504)342-5470

Maine (1)
Maine Emergency Management Agency
72 State House Station
Augusta, ME 04333-0072
(207)287-4080

Maryland (3)
Maryland Emergency
Management and Civil Defense Agency
Two Sudbrook Ln., East
Pikesville, MD 21208
(410)486-4422

Massachusetts (1)
Massachusetts Emergency
Management Agency
P.O. Box 1496
Framingham, MA 01701-0317
(508)820-2000

Michigan (5)
Emergency Management Division
Michigan State Police
300 S. Washington Sq.
Suite 300
Lansing, MI 48913
(517)366-6198

Minnesota (5)
Division of Emergency Services
Department of Public Safety State Capitol,
B-5 St. Paul, MN 55155
(612)296-0450

Mississippi (4)
Mississippi Emergency
Management Agency
P.O. Box 4501, Fondren Station
Jackson, MS 39296
(601)352-9100

Missouri (7)
State Emergency Management Agency
P.O. Box 116
Jefferson City, MO 65102
(573)526-9101

Montana (8)
Emergency Management Specialist
Disaster and Emergency Services
P.O. Box 4789
Helena, MT 59604-4789
(406)444-6911

Nebraska
Nebraska Civil Defense Agency
National Guard Center
1300 Military Road
Lincoln, NE 68508-1090
(402)471-7410

Nevada (9)
Nevada Division of Emergency Services
2525 S. Carson St.
Carson City, NV 89710
(702)687-4240

New Hampshire (1)
Governor's Office of Emergency Management
State Office Park South
107 Pleasant St.
Concord, NH 03301-3809
(603)271-2231

New Jersey (2)
Office of Emergency Management
P.O. Box 7068
W. Trenton, NJ 08628-0068
(609)538-6050

New Mexico (6)
Emergency Planning and Coordination
Department of Public Safety
4491 Cerrillos Rd.
P.O. Box 1628
Santa Fe, NM 87504-1628
(505)827-9222

New York (2)
State Emergency Management Office
Bldg. #22, Suite 101
Albany, NY 12226-2251
(518)457-2222

North Carolina
Division of Emergency Management
116 West Jones St.
Raleigh, NC 27603-1335
(919)733-5406

North Dakota (8)
North Dakota Division of Emergency Management
P.O. Box 5511
Bismarck, ND 58502-5511
(701)328-3300

Ohio (5)
Ohio Emergency Management Agency
2825 W. Dublin Granville Rd.
Columbus, OH 43235-2206
(614)889-7150

Oklahoma (6)
Oklahoma Civil Defense
P.O. Box 53365
Oklahoma City, OK 73152-3365
(405)521-2481

Oregon (10)
Emergency Management Division
Oregon State Executive Department
595 Cottage St., NE
Salem, OR 97310
(503)378-2911

Pennsylvania (3)
Pennsylvania Emergency
Management Agency
P.O. Box 3321
Harrisburg, PA 17105-3321
(717)651-2007

Puerto Rico (2)
State Civil Defense
Commonwealth of Puerto Rico
P.O. Box 5127
San Juan, PR 00906
(809)724-0124

Rhode Island (1)
Rhode Island Emergency Management Agency
675 New London Avenue
Cranston, RI 02920
(401)946-9996

South Carolina (4)
South Carolina Emergency
Management Division
1429 Senate St., Rutledge Bldg.
Columbia, SC 29201-3782
(803)734-8020

South Dakota (8)
Division of Emergency and Disaster Services
State Capitol, 500 East Capitol
Pierre, SD 57501
(605)773-3231

Tennessee (4)
Tennessee Emergency
Management Agency
3041 Sidco Dr. P.O. 41502
Nashville, TN 37204-1502
(615)741-6528

Texas (6)
Division of Emergency
Management
P.O. Box 4087
Austin, TX 78773-0001
(512)424-2000

Utah (8)
Division of Comprehensive Emergency Management
Sate Office Bldg., Room 1110
Salt Lake City, UT 84114
(801)538-3400

Vermont (1)
Vermont Emergency Management Agency
Dept. of Public Safety
Waterbury State Complex
103 S. Main St.
Waterbury, VT 05671-2101
(802)244-8271

Virgin Islands (2)
Territorial Emergency Management Agency
A & Q Building # 2c Estate Content
St Thomas, VI 00820
(809)773-2244

Virginia (3)
Department of Emergency Services
P.O. Box 40955
Richmond, VA 23225-6491
(804)674-2497

Washington (10)
Division of Emergency Management
4220 E. Martin Way, MS-PT 11
Olympia, WA 98504-0955
(360)923-4505

West Virginia (3)
West Virginia Office of Emergency Services
State Capitol Complex
Room EB80
Charleston, WV 25305-0360
(304)558-5380

Wisconsin (5)
Division of Emergency Government
2400 Wright St. P.O. Box 7865
Madison, WI 53707
(608)242-3232

Wyoming (8)
Wyoming Emergency Management Agency
P.O. Box 1709
Cheyenne, WY 82003
(307)777-7566

Vulnerability Analysis Chart

TYPE OF EMERGENCY	Probability High 5 ↔ 1 Low	Human Impact High Impact 5 ↔ 1 Low Impact	Property Impact	Business Impact	Internal Resources Weak Resources 5 ↔ 1 Strong Resources	External Resources	Total

The lower the score the better

Training Drills and Exercises

	January	February	March	April	May	June	July	August	September	October	November	December
MANAGEMENT ORIENTATION/REVIEW												
EMPLOYEE ORIENTATION/REVIEW												
CONTRACTOR ORIENTATION/REVIEW												
COMMUNITY/MEDIA ORIENTATION/REVIEW												
MANAGEMENT TABLETOP EXERCISE												
RESPONSE TEAM TABLETOP EXERCISE												
WALK-THROUGH DRILL												
FUNCTIONAL DRILLS												
EVACUATION DRILL												
FULL-SCALE EXERCISE												

READY BUSINESS: SAMPLE EMERGENCY PLAN

SAMPLE BUSINESS CONTINUITY AND DISASTER PREPAREDNESS PLAN

❏ PLAN TO STAY IN BUSINESS

_____
Business Name

_____
Address

_____
City, State

_____
Telephone Number

The following person is our primary crisis manager and will serve as the company spokesperson in an emergency.

_____
Primary Emergency Contact

_____
Telephone Number

_____
Alternative Number

_____
E-mail

❏ EMERGENCY CONTACT INFORMATION

_____
Dial 9-1-1 in an Emergency

_____
Non-Emergency Police/Fire

_____
Insurance Provider

If this location is not accessible we will operate from location below:

_____
Business Name

_____
Address

_____
City, State

_____
Telephone Number

If the person is unable to manage the crisis, the person below will succeed in management:

_____
Secondary Emergency Contact

_____
Telephone Number

_____
Alternative Number

_____
E-mail

DOCUMENT APPENDIX

☐ BE INFORMED

The following natural and man-made disasters could impact our business.

- _____
- _____
- _____
- _____
- _____

☐ EMERGENCY PLANNING TEAM

The following people will participate in emergency planning and crisis management.

- _____
- _____
- _____
- _____
- _____

☐ WE PLAN TO COORDINATE WITH OTHERS

The following people from neighboring businesses and our building management will participate on our emergency planning team.

- _____
- _____
- _____
- _____
- _____

☐ OUR CRITICAL OPERATIONS

The following is a prioritized list of our critical operations, staff and procedures we need to recover from a disaster.

Operation	Staff in Charge	Action Plan
_____	_____	_____
_____	_____	_____
_____	_____	_____
_____	_____	_____
_____	_____	_____

❐ SUPPLIERS AND CONTRACTORS

Company Name: _____

Street Address: _____

City: _____ State: _____ Zip Code: _____

Phone: _____ Fax: _____ E-Mail: _____

Contact Name: _____ Account Number: _____

Materials/Service Provided: _____

If this company experiences a disaster, we will obtain supplies/materials from the following:

Company Name: _____

Street Address: _____

City: _____ State: _____ Zip Code: _____

Phone: _____ Fax: _____ E-Mail: _____

Contact Name: _____ Account Number: _____

Materials/Service Provided: _____

If this company experiences a disaster, we will obtain supplies/materials from the following:

Company Name: _____

Street Address: _____

City: _____ State: _____ Zip Code: _____

Phone: _____ Fax: _____ E-Mail: _____

Contact Name: _____ Account Number: _____

Materials/Service Provided: _____

DOCUMENT APPENDIX 585

❏ EVACUATION PLAN FOR _____ LOCATION
(Insert address)

- We have developed these plans in collaboration with neighboring businesses and building owners to avoid confusion or gridlock.
- We have located, copied and posted building and site maps.
- Exits are clearly marked.
- We will practice evacuation procedures _____ times a year.

If we must leave the workplace quickly:

_____
_____

1. Warning System: _____

   We will test the warning system and record results _____ times a year.

2. Assembly Site: _____

3. Assembly Site Manager & Alternate: _____

   a. Responsibilities Include:

   _____
   _____
   _____

4. Shut Down Manager & Alternate: _____

   a. Responsibilities Include:

   _____
   _____
   _____

5. _____ is responsible for issuing all clear.

☐ SHELTER-IN-PLACE PLAN FOR _____ LOCATION
(Insert address)

- We have talked to co-workers about which emergency supplies, if any, the company will provide in the shelter location and which supplies individuals might consider keeping in a portable kit personalized for individual needs.
- We will practice shelter procedures _____ times a year.

If we must take shelter quickly

_____

_____

1. Warning System: _____

    We will test the warning system and record results _____ times a year.

2. Storm Shelter Location: _____

3. "Seal the Room" Shelter Location: _____

4. Shelter Manager & Alternate:

    a. Responsibilities Include:

    _____

    _____

    _____

5. Shut Down Manager & Alternate: _____

    a. Responsibilities Include:

    _____

    _____

    _____

6. _____ is responsible for issuing all clear.

DOCUMENT APPENDIX

☐ COMMUNICATIONS

We will communicate our emergency plans with co-workers in the following way:

_____

_____

In the event of a disaster we will communicate with employees in the following way:

_____

_____

☐ CYBER SECURITY

To protect our computer hardware, we will:

_____

To protect our computer software, we will:

_____

If our computers are destroyed, we will use back-up computers at the following location:

_____

☐ RECORDS BACK-UP

_____ is responsible for backing up our critical records including payroll and accounting systems.

Back-up records including a copy of this plan, site maps, insurance policies, bank account records and computer back ups are stored onsite _____

Another set of back-up records is stored at the following off-site location:

_____

If our accounting and payroll records are destroyed, we will provide for continuity in the following ways:

_____

❐ EMPLOYEE EMERGENCY CONTACT INFORMATION

The following is a list of our co-workers and their individual emergency contact information:

_____   _____   _____

_____   _____   _____

_____   _____   _____

_____   _____   _____

❐ ANNUAL REVIEW

We will review and update this business continuity and disaster plan in _____.

FEMA: PROTECTING YOUR BUSINESS FROM DISASTERS

ARE YOU AT RISK?

If you aren't sure whether your business is at risk from disasters caused by natural hazards, check with your local building official, city engineer, or planning and zoning administrator. They can tell you whether you are in an area where hurricanes, floods, earthquakes, wildfires, or tornadoes are likely to occur. Also, they usually can tell you how to protect your business.

WHAT YOU CAN DO

Protecting your business from disasters caused by natural hazards can involve a variety of actions, from inspecting and maintaining your buildings to installing protective devices. Most of these actions, especially those that affect the structure of your buildings or their utility systems, should be carried out by qualified maintenance staff or professional contractors licensed to work in your state, country, or city. One example of disaster protection is safely storing the important documents, electronic files, raw materials, and inventory required for the operation of your business.

PROTECT BUSINESS RECORDS AND INVENTORY

Most businesses keep on-site records and files (both hardcopy and electronic) that are essential to normal operations. Some businesses also store raw materials and product inventory. The loss of essential records, files, and other materials during a disaster is commonplace and can not only add to your damage costs, but also delay your return to normal operations. The longer your business is not operating, the more likely you are to lose customers permanently to your competitors.

To reduce your vulnerability, determine which records, files, and materials are most important; consider their vulnerability to damage during different types of disasters (such as floods, hurricanes, and earthquakes); and take steps to protect them, including the following:

- raising computers above the flood level and moving them away from large windows
- moving heavy and fragile objects to low shelves
- storing vital documents (plans, legal papers, etc.) in a secure off-site location
- regularly backing up vital electronic files (such as billing and payroll records and customer lists) and storing backup copies in a secure off-site location
- securing equipment that could move or fall during an earthquake.

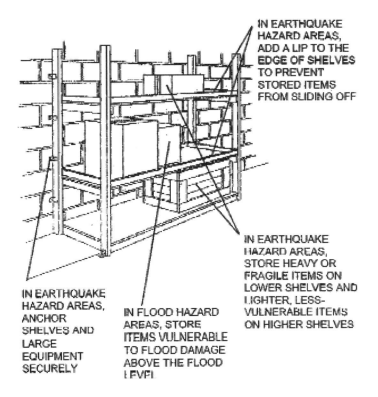

IN EARTHQUAKE HAZARD AREAS, ADD A LIP TO THE EDGE OF SHELVES TO PREVENT STORED ITEMS FROM SLIDING OFF

IN EARTHQUAKE HAZARD AREAS, STORE HEAVY OR FRAGILE ITEMS ON LOWER SHELVES AND LIGHTER, LESS-VULNERABLE ITEMS ON HIGHER SHELVES

IN EARTHQUAKE HAZARD AREAS, ANCHOR SHELVES AND LARGE EQUIPMENT SECURELY

IN FLOOD HAZARD AREAS, STORE ITEMS VULNERABLE TO FLOOD DAMAGE ABOVE THE FLOOD LEVEL

TIPS

Keep these points in mind when you protect business records and inventory:

- Make sure you are aware of the details of your flood insurance and other hazard insurance policies, specifically which items and contents are covered and under what conditions. For example, if you have a home business, you may need two flood insurance policies, a home policy and a separate business policy, depending on the percentage of the total square footage of your house that is devoted to business use. Check with your insurance agent if you have questions about any of your policies.
- When you identify equipment susceptible to damage, consider the location of the equipment. For example, equipment near a hot water tank or pipes could be damaged if the pipes burst during an earthquake, and equipment near large windows could be damaged during hurricanes.
- Assign disaster mitigation duties to your employees. For example, some employees could be responsible for securing storage bins and others for backing up computer files and delivering copies to a secure location.
- You may want to consider having other offices of your company, or a contractor, perform some administrative duties, such as maintaining payroll records or providing customer service.
- Estimate the cost of repairing or replacing each essential piece of equipment in your business. Your estimates will help you assess your vulnerability and focus your efforts.
- For both insurance and tax purposes, you should maintain written and photographic inventories of all important materials and equipment. The inventory should be stored in a safety deposit box or other secure location.

ESTIMATED COST

The cost of these measures will depend on the size and contents of your business, the nature of the potential hazards, and the effort required to ensure proper protection. In some instances, you may need to buy new equipment, such as a backup tape drive system.

OTHER SOURCES OF INFORMATION

Emergency Management Guide for Business & Industry, FEMA, 1996.
Separate Flood Insurance a Must, FEMA, 1996.
To obtain copies of FEMA documents, call FEMA Publications at 1-800-480-2520. Information is also available on the World Wide Web at http//:www.fema.gov.

GUIDANCE ON PREPARING WORKPLACES FOR AN INFLUENZA PANDEMIC

U.S. Department of Labor, Occupational Safety and Health Administration

INTRODUCTION

A pandemic is a global disease outbreak. An influenza pandemic occurs when a new influenza virus emerges for which there is little or no immunity in the human population, begins to cause serious illness and then spreads easily person-to-person worldwide. A worldwide influenza pandemic could have a major effect on the global economy, including travel, trade, tourism, food, consumption and eventually, investment and financial markets. Planning for pandemic influenza by business and industry is essential to minimize a pandemic's impact. Companies that provide critical infrastructure services, such as power and telecommunications, also have a special responsibility to plan for continued operation in a crisis and should plan accordingly. As with any catastrophe, having a contingency plan is essential.

In the event of an influenza pandemic, employers will play a key role in protecting employees' health and safety as well as in limiting the impact on the economy and society. Employers will likely experience employee absences, changes in patterns of commerce and interrupted supply and delivery schedules. Proper planning will allow employers in the public and private sectors to better protect their employees and lessen the impact of a pandemic on society and the economy. As stated in the President's *National Strategy for Pandemic Influenza*, all stakeholders must plan and be prepared.

The Occupational Safety and Health Administration (OSHA) developed this pandemic influenza planning guidance based upon traditional infection control and industrial hygiene practices. It is important to note that there is currently no pandemic; thus, this guidance is intended for planning purposes and is not specific to a particular viral strain. Additional guidance may be needed as an actual pandemic unfolds and more is known about the characteristics of the virulence of the virus, disease transmissibility, clinical manifestation, drug susceptibility, and risks to different age groups and subpopulations. Employers and employees should use this planning guidance to help identify risk levels in

> This guidance is advisory in nature and informational in content. It is not a standard or a regulation, and it neither creates new legal obligations nor alters existing obligations created by OSHA standards or the *Occupational Safety and Health Act* (OSH Act). Pursuant to the OSH Act, employers must comply with hazard-specific safety and health standards as issued and enforced either by OSHA or by an OSHA approved State Plan. In addition, Section 5(a)(1) of the OSH Act, the General Duty Clause, requires employers to provide their employers with a workplace free from recognized hazards likely to cause death or serious physical harm. Employers can be cited for violating the General Duty Clause if there is a recognized hazard and they do not take reasonable steps to prevent or abate the hazard. However, failure to implement any recommendations in this guidance is not, in itself, a violation of the General Duty Clause. Citations can only be based on standards, regulations, or the General Duty Clause.

workplace settings and appropriate control measures that include good hygiene, cough etiquette, social distancing, the use of personal protective equipment, and staying home from work when ill. Up-to-date information and guidance is available to the public through the www.pandemicflu.gov website.

THE DIFFERENCE BETWEEN SEASONAL, PANDEMIC INFLUENZA AND AVIAN INFLUENZA

Seasonal influenza refers to the periodic outbreaks of respiratory illness in the fall and winter in the United States. Outbreaks are typically limited; most people have some immunity to the circulating strain of the virus. A vaccine is prepared in advance of the seasonal influenza; it is designed to match the influenza viruses most likely to be circulating in the community. Employees living abroad and international business travelers should note that other geographic areas (for example, the Southern Hemisphere) have different influenza seasons which may require different vaccines.

Pandemic influenza refers to a worldwide outbreak of influenza among people when a new strain of the virus emerges that has the ability to infect humans and to spread from person to person. During the early phases of an influenza pandemic, people might not have any natural Immunity to the new strain; so the disease would spread rapidly among the population. A vaccine to protect people against illness from a pandemic influenza virus may not be widely available until many months after an influenza pandemic begins. It is important to emphasize that there currently is no influenza pandemic. However, pandemics have occurred throughout history and many scientists believe that it is only a matter of time before another one occurs. Pandemics can vary in severity from something that seems simply like a bad flu season to an especially severe influenza pandemic that could lead to high levels of illness, death, social disruption and economic loss. It is Impossible to predict when the next pandemic will occur or whether it will be mild or severe.

Avian influenza (AI)—also known as the bird flu—is caused by virus that infects wild birds and domestic poultry. Some forms of the avian influenza are worse than others. Avian influenza viruses are generally divided into two groups: low pathogenic avian influenza and highly pathogenic avian influenza. Low pathogenic avian influenza naturally occurs in wild birds and can spread to domestic birds. In most cases, it causes no signs of infection or only minor symptoms in birds. In general, these low path strains of the virus pose little threat to human health. Low pathogenic avian influenza virus H5 and H7 strains have the potential to mutate into highly pathogenic avian influenza and are, therefore, closely monitored. Highly pathogenic avian influenza spreads rapidly and has a high death rate in birds. Highly pathogenic avian influenza of the H5N1 strain is rapidly spreading in birds in some parts of the world.

Highly pathogenic H5N1 is one of the few avian influenza viruses to have crossed the species barrier to infect humans and it is the most deadly of those that have crossed the barrier. Most cases of H5N1 influenza infection in humans have resulted from contact with infected poultry or surfaces contaminated with secretions/excretions from infected birds.

As of February 2007, the spread of H5N1 virus from person to person has been limited to rare, sporadic cases. Nonetheless, because all influenza viruses have the ability to change, scientists are concerned that H5N1 virus one day could be able to sustain human to human transmission. Because these viruses do not commonly infect humans, there is little or no immune protection against them in the human population. If H5N1 virus were to gain the capacity to sustain transmission from person to person, a pandemic could begin.

An update on what is currently known about avian flu can be found at www.pandemicflu.gov.

HOW A SEVERE PANDEMIC INFLUENZA COULD AFFECT WORKPLACES

Unlike natural disasters or terrorist events, an influenza pandemic will be widespread, affecting multiple areas of the United States and other countries at the same time. A pandemic will also be an extended event, with multiple waves of outbreaks in the same geographic area; each outbreak could last from 6 to 8 weeks. Waves of outbreaks may occur over a year or more. Your workplace will likely experience:

- *Absenteeism*—A pandemic could affect as many as 40 percent of the workforce during periods of peak influenza illness. Employees could be absent because they are sick, must care for sick family members or for children if schools or day care centers are closed, are afraid to come to work, or the employer might not be notified that the employee has died.
- *Change in patterns of commerce*—During a pandemic, consumer demand for items related to infection control is likely to increase dramatically, while consumer interest in other goods may decline. Consumers may also change the ways in which they shop as a result of the pandemic. Consumers may try to shop at off-peak hours to reduce contact with other people, show increased interest in home delivery services, or prefer other options, such as drive-through service, to reduce person-to-person contact.
- *Interrupted supply/delivery*—Shipments of items from those geographic areas severely affected by the pandemic may be delayed or cancelled.

WHO SHOULD PLAN FOR A PANDEMIC

To reduce the impact of a pandemic on your operations, employees, customers and the general public, it is important for all businesses and organizations to begin continuity planning for a pandemic now. Lack of continuity planning can result in a cascade of failures as employers attempt to address challenges of a pandemic with insufficient resources and employees who might not be adequately trained in the jobs they will be asked to perform. Proper planning will allow employers to better protect their employees and prepare for changing patterns of commerce and potential disruptions in supplies or services. Important tools for pandemic planning for employers are located at www.pandemicflu.gov.

The U.S. government has placed a special emphasis on supporting pandemic influenza planning for public and private sector businesses deemed to be critical industries and key resources (CI/KR). Critical infrastructure are the thirteen sectors that provide the production of essential goods and services, interconnectedness and operability, public safety, and security that contribute to a strong national defense and thriving economy. Key resources are facilities, sites, and groups of organized people whose destruction could cause large-scale injury, death, or destruction of property and/or profoundly damage our national prestige and confidence. With 85 percent of the nation's critical infrastructure in the hands of the private sector, the business community plays a vital role in ensuring national pandemic preparedness and response. Additional guidance for CI/KR business is available at: www.pandemicflu.gov/plan/pdf/CIKRpandemicInfluenza Guide.pdf.

DOCUMENT APPENDIX

Critical Infrastructure and Key Resources

Key Resources

- Government Facilities
- Dams
- Commercial Facilities
- Nuclear Power Plants

Critical Infrastructure

- Food and Agriculture
- Public Health and Healthcare
- Banking and Finance
- Chemical and Hazardous Materials
- Defense Industrial Base
- Water
- Energy
- Emergency Services
- Information Technology
- Telecommunications
- Postal and Shipping
- Transportation
- National Monuments and Icons

HOW INFLUENZA CAN SPREAD BETWEEN PEOPLE

Influenza is thought to be primarily spread through large droplets (droplet transmission) that directly contact the nose, mouth or eyes. These droplets are produced when infected people cough, sneeze or talk, sending the relatively large infectious droplets and very small sprays (aerosols) into the nearby air and into contact with other people. Large droplets can only travel a limited range; therefore, people should limit close contact (within 6 feet) with others when possible. To a lesser degree, human influenza is spread by touching objects contaminated with influenza viruses and then transferring the infected material from the hands to the nose, mouth or eyes. Influenza may also be spread by very small infectious particles (aerosols) traveling in the air. The contribution of each route of exposure to influenza transmission is uncertain at this time and may vary based upon the characteristics of the influenza strain.

CLASSIFYING EMPLOYEE EXPOSURE TO PANDEMIC INFLUENZA AT WORK

Employee risks of occupational exposure to influenza during a pandemic may vary from very high to high, medium, or lower (caution) risk. The level of risk depends in part on whether or not jobs require close proximity to people potentially infected with the pandemic influenza virus, or whether they are required to have either repeated or extended contact with known or suspectedsources of pandemic influenza virus such as coworkers, the general public, outpatients, school children or other such individuals or groups.

- *Very high exposure risk* occupations are those with high potential exposure to high concentrations of known or suspected sources of pandemic influenza during specific medical or laboratory procedures.
- *High exposure risk* occupations are those with high potential for exposure to known or suspected sources of pandemic influenza virus.
- *Medium exposure risk* occupations include jobs that require frequent, close contact (within 6 feet) exposures to known or suspected sources of pandemic influenza virus such as coworkers, the general public, outpatients, school children or other such individuals or groups.
- *Lower exposure risk (caution)* occupations are those that do not require contact with people known to be infected with the pandemic virus, nor frequent close contact (within 6 feet) with the public. Even at lower risk levels, however, employers should be cautious and develop preparedness plans to minimize employee infections.

Employers of critical infrastructure and key resource employees (such as law enforcement, emergency response, or public utility employees) may consider upgrading protective measures for these employees beyond what would be suggested by their exposure risk due to the necessity of such services for the functioning of society as well as the potential difficulties in replacing them during a pandemic (for example, due to extensive training or licensing requirements).

To help employers determine appropriate work practices and precautions, OSHA has divided workplaces and work operations into four risk zones, according to the likelihood of employees' occupational exposure to pandemic influenza. We show these zones in the shape of a pyramid to represent how the risk will likely be distributed (see page 11). The vast majority of American workplaces are likely to be in the medium exposure risk or lower exposure risk (caution) groups.

HOW TO MAINTAIN OPERATIONS DURING A PANDEMIC

As an employer, you have an important role in protecting employee health and safety, and limiting the impact of an influenza pandemic. It is important to work with community planners to integrate your pandemic plan into local and state planning, particularly if your operations are part of the nation's critical infrastructure or key resources. Integration with local community planners will allow you to access resources and information promptly to maintain operations and keep your employees safe.

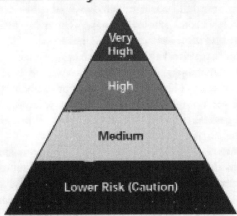

Occupational Risk Pyramid for Pandemic Influenza

Very High Exposure Risk:
- Healthcare employees (for example, doctors, nurses, dentists) performing aerosol-generating procedures on known or suspected pandemic patients (for example, cough induction procedures, bronchoscopies, some dental procedures, or invasive specimen collection).
- Healthcare or laboratory personnel collecting or handling specimens from known or suspected pandemic patients (for example, manipulating cultures from known or suspected pandemic influenza patients).

High Exposure Risk:
- Healthcare delivery and support staff exposed to known or suspected pandemic patients (for example, doctors, nurses, and other hospital staff that must enter patients' rooms).
- Medical transport of known or suspected pandemic patients in enclosed vehicles (for example, emergency medical technicians).
- Performing autopsies on known or suspected pandemic patients (for example, morgue and mortuary employees).

Medium Exposure Risk:
- Employees with high-frequency contact with the general population (such as schools, high population density work environments, and some high volume retail).

Lower Exposure Risk (Caution):
- Employees who have minimal occupational contact with the general public and other coworkers (for example, office employees).

Develop a Disaster Plan

Develop a disaster plan that includes pandemic preparedness (See www.pandemicflu.gov/plan/businesschecklist.html) and review it and conduct drills regularly.

- Be aware of and review federal, state and local health department pandemic influenza plans. Incorporate appropriate actions from these plans into workplace disaster plans.
- Prepare and plan for operations with a reduced workforce.
- Work with your suppliers to ensure that you can continue to operate and provide services.
- Develop a sick leave policy that does not penalize sick employees, thereby encouraging employees who have influenza-related symptoms (e.g., fever, headache, cough, sore throat, runny or stuffy nose, muscle aches, or upset stomach) to stay home so that they do not infect other employees. Recognize that employees with ill family members may need to stay home to care for them.
- Identify possible exposure and health risks to your employees. Are employees potentially in contact with people with influenza such as in a hospital or clinic? Are your employees expected to have a lot of contact with the general public?
- Minimize exposure to fellow employees or the public. For example, will more of your employees work from home? This may require enhancement of technology and communications equipment.
- Identify business-essential positions and people required to sustain business-necessary functions and operations. Prepare to cross-train or develop ways to function in the absence of these positions. It is recommended that employers train three or more employees to be able to sustain business-necessary functions and operations, and communicate the expectation for available employees to perform these functions if needed during a pandemic.
- Plan for downsizing services but also anticipate any scenario which may require a surge in your services.
- Recognize that, in the course of normal daily life, all employees will have non-occupational risk factors at home and in community settings that should be reduced to the extent possible. Some employees will also have individual risk factors that should be considered by employers as they plan how the organization will respond to a potential pandemic (e.g., Immuno-compromised individuals and pregnant women).
- Stockpile items such as soap, tissue, hand sanitizer, cleaning supplies and recommended personal protective equipment. When stockpiling items, be aware of each product's shelf life and storage conditions (e.g., avoid areas that are damp or have temperature extremes) and incorporate product rotation (e.g., consume oldest supplies first) into your stockpile management program.

Make sure that your disaster plan protects and supports your employees, customers and the general public. Be aware of your employees' concerns about pay, leave, safety and health. Informed employees who feel safe at work are less likely to be absent.

- Develop policies and practices that distance employees from each other, customers and the general public. Consider practices to minimize face-to-face contact between employees such as e-mail, websites and teleconferences. Policies and practices that allow employees to work from home or to stagger their work shifts may be important as absenteeism rises.
- Organize and identify a central team of people or focal point to serve as a communication source so that your employees and customers can have accurate information during the crisis.
- Work with your employees and their union(s) to address leave, pay, transportation, travel, childcare, absence and other human resource issues.
- Provide your employees and customers in your workplace with easy access to infection control supplies, such as soap, hand sanitizers, personal protective equipment (such as gloves or surgical masks), tissues, and office cleaning supplies.
- Provide training, education and informational material about business-essential job functions and employees health and safety, including proper hygiene practices and the use of any personal protective equipment to be used in the workplace. Be sure that informational material is available in a usable format for individuals with sensory disabilities and/or limited English proficiency. Encourage employees to take care of their health by eating right, getting plenty of rest and getting a seasonal flu vaccination.
- Work with your insurance companies, and state and local health agencies to provide information to employees and customers about medical care in the event of a pandemic.
- Assist employees in managing additional stressors related to the pandemic. These are likely to include distress related to personal or family illness, life disruption, grief related to loss of family, friends or coworkers, loss of routine support systems, and similar challenges. Assuring timely and accurate communication will also be important throughout the duration of the pandemic in decreasing fear or worry. Employers should provide opportunities for support, counseling, and mental health assessment and referral should these be necessary. If present, Employee Assistance Programs can offer training and provide resources and other guidance on mental health and resiliency before and during a pandemic.

Protect Employees and Customers

Educate and train employees in proper hand hygiene, cough etiquette and social distancing techniques. Understand and develop work practice and engineering controls that could provide additional protection to your employees and customers, such as: drive-through service windows, clear plastic sneeze barriers, ventilation, and the proper selection, use and disposal of personal protective equipment.

These are not comprehensive recommendations. The most important part of pandemic planning is to work with your employees, local and state agencies and other employers to develop cooperative pandemic plans to maintain your operations and keep your employees and the public safe. Share what you know, be open to ideas from your employees, then identify and share effective health practices with other employers in your community and with your local chamber of commerce.

HOW ORGANIZATIONS CAN PROTECT THEIR EMPLOYEES

For most employers, protecting their employees will depend on emphasizing proper hygiene (disinfecting hands and surfaces) and practicing social distancing (see page 26 for more information). Social distancing means reducing the frequency, proximity, and duration of contact between people (both employees and customers) to reduce the chances of spreading pandemic influenza from person-to-person. All employers should implement good hygiene and infection control practices.

Occupational safety and health professionals use a framework called the "hierarchy of controls" to select ways of dealing with workplace hazards. The hierarchy of controls prioritizes intervention strategies based on the premise that the best way to control a hazard is to systematically remove it from the workplace, rather than relying on employees to reduce their exposure. In the setting of a pandemic, this hierarchy should be used in concert with current public health recommendations. The types of measures that may be used to protect yourself, your employees, and your customers (listed from most effective to least effective) are: engineering controls, administrative controls, work practices, and personal protective equipment (PPE). Most employers will use a combination of control methods. There are advantages and disadvantages to each type of control measure when considering the ease of implementation, effectiveness, and cost. For example, hygiene and social distancing can be implemented relatively easily and with little expense, but this control method requires employees to modify and maintain their behavior, which may be difficult to sustain. On the other hand, installing clear plastic barriers or a drive-through window will be more expensive and take a longer time to implement, although in the long run may be more effective at preventing transmission during a pandemic. Employers must evaluate their particular workplace to develop a plan for protecting their employees that may combine both immediate actions as well as longer term solutions.

Here is a description of each type of control:

Work Practice and Engineering Controls—Historically, infection control professionals have relied on personal protective equipment (for example, surgical masks and gloves) to serve as a physical barrier in order to prevent the transmission of an infectious disease from one person to another. This reflects the fact that close interactions with infectious patients is an unavoidable part of many healthcare occupations. The principles of industrial hygiene demonstrate that work practice controls and engineering controls can also serve as barriers to transmission and are less reliant on employee behavior to provide protection. Work practice controls are procedures for safe and proper work that are used to reduce the duration, frequency or intensity of exposure to a hazard. When defining safe work practice controls, it is a good idea to ask your employees for their suggestions, since they have firsthand experience with the tasks. These controls should be understood and followed by managers, supervisors and employees. When work practice controls are insufficient to protect employees, some employers may also need engineering controls.

Engineering controls involve making changes to the work environment to reduce work-related hazards. These types of controls are preferred over all others because they make permanent changes that reduce exposure to hazards and do not rely on employee or customer behavior. By reducing a hazard in the workplace, engineering controls can be the most cost-effective solutions for employers to implement.

During a pandemic, engineering controls may be effective in reducing exposure to some sources of pandemic influenza and not others. For example, installing sneeze guards between customers and employees would provide a barrier to transmission. The use of barrier protections, such as sneeze guards, is common practice for both infection control and industrial hygiene. However, while the installation of sneeze guards may reduce or prevent transmission between customers and employees, transmission may still occur between coworkers. Therefore, administrative controls and public health measures should be implemented along with engineering controls.

Examples of work practice controls include:

- Providing resources and a work environment that promotes personal hygiene. For example, provide tissues, no-touch trash cans, hand soap, hand sanitizer, disinfectants and disposable towels for employees to clean their work surfaces.
- Encouraging employees to obtain a seasonal influenza vaccine (this helps to prevent illness from seasonal influenza strains that may continue to circulate).
- Providing employees with up-to-date education and training on influenza risk factors, protective behaviors, and instruction on proper behaviors (for example, cough ettiquette and care of personal protective equipment).
- Developing policies to minimize contacts between employees and between employees and clients or customers.

More information about protecting yourself, your coworkers and employees, and your family can be found at www.pandemicflu.gov.

Examples of engineering controls include:

- Installing physical barriers, such as clear plastic sneeze guards.
- Installing a drive-through window for customer service.
- In some limited healthcare settings, for aerosol generating procedures, specialized negative pressure ventilation may be indicated.

Administrative Controls—Administrative controls include controlling employees' exposure by scheduling their work tasks in ways that minimize their exposure levels. Examples of administrative controls include:

- Developing policies that encourage ill employees to stay at home without fear of any reprisals.
- The discontinuation of unessential travel to locations with high illness transmission rates.
- Consider practices to minimize face-to-face contact between employees such as e-mail, websites and teleconferences. Where possible, encourage flexible work arrangements such as telecommuting or flexible work hours to reduce the number of your employees who must be at work at one time or in one specific location.
- Consider home delivery of goods and services to reduce the number of clients or customers who must visit your workplace.
- Developing emergency communications plans. Maintain a forum for answering employees' concerns. Develop internet-based communications if feasible.

Personal Protective Equipment (PPE)—While administrative and engineering controls and proper work practices are considered to be more effective in minimizing exposure to the influenza virus, the use of PPE may also be indicated during certain exposures. If used correctly, PPE can help prevent some exposures; however, they should not take the place of other prevention interventions, such as engineering controls, cough etiquette, and hand hygiene (see www.cdc.gov/flu/protect/stopgerms.htm). Examples of personal protective equipment are gloves, goggles, face shields, surgical masks, and respirators (for example, N-95). It is important that personal protective equipment be:

- Selected based upon the hazard to the employee;
- Properly fitted and some must be periodically refitted (e.g., respirators);
- Conscientiously and properly worn;
- Regularly maintained and replaced, as necessary;
- Properly removed and disposed of to avoid contamination of self, others or the environment.

Employers are obligated to provide their employees with protective gear needed to keep them safe while performing their jobs. The types of PPE recommended for pandemic influenza will be based on the risk of contracting influenza while working and the availability of PPE. Check the www.pandemicflu.gov website for the latest guidance.

THE DIFFERENCE BETWEEN A SURGICAL MASK AND A RESPIRATOR

It is important that employers and employees understand the significant differences between these types of personal protective equipment. The decision on whether or not to require employees to use either surgical/procedure masks or respirators must be based upon a hazard analysis of the employees' specific work environment and the differing protective properties of each type of personal protective equipment. The use of surgical masks or respirators is one component of infection control practices that may reduce transmission between infected and non-infected persons.

It should be noted that there is limited information on the use of surgical masks for the control of a pandemic in settings where there is no identified source of infection. There is no information on respirator use in such scenarios since modern respirators did not exist during the last pandemic. However, respirators are now routinely used to protect employees against occupational hazards, including biological hazards such as tuberculosis, anthrax, and hantavirus. The effectiveness of surgical masks and respirators has been inferred on the basis of the mode of influenza transmission, particle size, and professional judgment.

To offer protection, both surgical masks and respirators must be worn correctly and consistently throughout the time they are being used. If used properly, surgical masks and respirators both have a role in preventing different types of exposures. During an influenza pandemic, surgical masks and respirators should be used in conjunction with interventions that are known to prevent the spread of infection, such as respiratory etiquette, hand hygiene, and avoidance of large gatherings.

Surgical Masks—Surgical masks are used as a physical barrier to protect employees from hazards such as splashes of large droplets of blood or body fluids. Surgical masks also prevent contamination by trapping large particles of body fluids that may contain bacteria or viruses when they are expelled by the wearer, thus protecting other people against infection from the person wearing the surgical mask.

Surgical/procedure masks are used for several different purposes, including the following:

- Placed on sick people to limit the spread of infectious respiratory secretions to others.
- Worn by healthcare providers to prevent accidental contamination of patients' wounds by the organisms normally present in mucus and saliva.

- Worn by employees to protect themselves from splashes or sprays of blood or body fluids; they may also have the effect of keeping contaminated fingers/hands away from the mouth and nose.

Surgical masks are not designed or certified to prevent the inhalation of small airborne contaminants. These small airborne contaminants are too little to see with the naked eye but may still be capable of causing infection. Surgical/procedure masks are not designed to seal tightly against the user's face. During inhalation, much of the potentially contaminated air passes through gaps between the face and the surgical mask, thus avoiding being pulled through the material of the mask and losing any filtration that it may provide. Their ability to filter small particles varies significantly based upon the type of material used to make the surgical mask, and so they cannot be relied upon to protect employees against airborne infectious agents. Only surgical masks that are cleared by the U.S. Food and Drug Administration and legally marketed in the United States have been tested for their ability to resist blood and body fluids.

Respirators—Respirators are designed to reduce an employee's exposure to airborne contaminants. Respirators are designed to fit the face and to provide a tight seal between the respirator's edge and the face. A proper seal between the user's face and the respirator forces inhaled air to be pulled through the respirator's filter material and not through gaps between the face and respirator. Respirators must be used in the context of a comprehensive respiratory protection program, (see OSHA standard 29 CFR 1910.134, or www.osha.gov/SLTC/respiratoryprotection/index.html). It is important to medically evaluate employees to assure that they can perform work tasks while wearing a respirator. Medical evaluation can be as simple as a questionnaire (found in Appendix C of OSHA's Respiratory Protection standard, 29 CFR 1910.134). Employers who have never before needed to consider a respiratory protection plan should note that it can take time to choose a respirator to provide to employees and to arrange for a qualified trainer and provide training, fit testing, and medical evaluation for their employees. If employers wait until an influenza pandemic actually arrives, they may be unable to provide an adequate respiratory protection program in a timely manner.

Types of Respirators

Respirators can be air supplying (e.g., the self-contained breathing apparatus worn by firefighters) or air purifying (e.g., a gas mask that filters hazards from the air). Most employees affected by pandemic influenza who are deemed to need a respirator to minimize the likelihood of exposure to the pandemic influenza virus in the workplace will use some type of air purifying respirator. They are also known as "particulate respirators" because they protect by filtering particles out of the air as you breathe. These respirators protect only against particles not gases or vapors. Since airborne biological agents such as bacteria or viruses are particles, they can be filtered by particulate respirators.

DOCUMENT APPENDIX

Air purifying respirators can be divided into several types:

- *Disposable or filtering facepiece* respirators, where the entire respirator facepiece is comprised of filter material. This type of respirator is also commonly referred to as an "N95" respirator. It is discarded when it becomes unsuitable for further use due to excessive breathing resistance (e.g., particulate clogging the filter), unacceptable contamination/soiling, or physical damage.
  - *Surgical respirators* are a type of respiratory protection that offers the combined protective properties of both a filtering facepiece respirator and a surgical mask. Surgical N95 respirators are certified by NIOSH as respirators and also cleared by FDA as medical devices which have been designed and tested and shown to be equivalent to surgical masks in certain performance characteristics (resistance to blood penetration, biocompatibility) which are not examined by NIOSH during its certification of N95 respirators.
- *Reusable or elastomeric respirators*, where the facepiece can be cleaned, repaired and reused, but the filter cartridges are discarded and replaced when they become unsuitable for further use. These respirators come in half-mask (covering the mouth and nose) and full-mask (covering mouth, nose, and eyes) types. These respirators can be used with a variety of different cartridges to protect against different hazards. These respirators can also be used with canisters or cartridges that will filter out gases and vapors.
- *Powered air purifying respirators*, (PAPRs) where a battery powered blower pulls contaminated air through filters, then moves the filtered air to the wearer's facepiece. PAPRs are significantly more expensive than other air purifying respirators but they provide higher levels of protection and may also increase the comfort for some users by reducing the physiologic burden associated with negative pressure respirators and providing a constant flow of air on the face. These respirators can also be used with canisters or cartridges that will filter out gases and vapors. It should also be noted that there are hooded PAPRs that do not require employees to be fit tested in order to use them.

All respirators used in the workplace are required to be tested and certified by the National Institute for Occupational Safety and Health (NIOSH). NIOSH-approved respirators are marked with the manufacturer's name, the part number, the protection provided by the filter (e.g., N95), and "NIOSH." This information is printed on the facepiece, exhalation valve cover, or head straps. If a respirator does not have these markings it has not been certified by NIOSH. Those respirators that are surgical N95 respirators are also cleared by the FDA and, therefore, are appropriate for circumstances in which protection from airborne and body fluid contaminants is needed.

When choosing between disposable and reusable respirators, employers should consider their work environment, the nature of pandemics, and the potential for supply chain disruptions. Each pandemic influenza outbreak could last from 6 to 8 weeks and waves of outbreaks may occur over a year or more. While disposable respirators may be more convenient and cheaper on a per unit basis, a reusable respirator may be more economical on a long-term basis and reduce the impact of disruption in supply chains or shortages of respirators.

Classifying Particulate Respirators and Particulate Filters

An N95 respirator is one of nine types of particulate respirators. Respirator filters that remove at least 95 percent of airborne particles during "worst case" testing using the "most-penetrating" size of particle are given a 95 rating. Those that filter out at least 99 percent of the particles under the same conditions receive a 99 rating, and those that filter at least 99.97 percent (essentially 100 percent) receive a 100 rating.

In addition, filters in this family are given a designation of N, R, or P to convey their ability to function in the presence of oils that are found in some work environments.

"N" if they are Not resistant to oil. (e.g., N95, N99, N100)
"R" if they are somewhat Resistant to oil. (e.g., R95, R99, R100)
"P" if they are strongly resistant (i.e., oil Proof). (e.g., P95, P99, P100)

This rating is important in work settings where oils may be present because some industrial oils can degrade the filter performance to the point that it does not filter adequately. Thus, the three filter efficiencies combined with the three oil designations lead to nine types of particulate respirator filter materials. It should be noted that any of the various types of filters listed here would be acceptable for protection against pandemic influenza in workplaces that do not contain oils, particularly if the N95 filter type was unavailable due to shortages.

Replacing Disposable Respirators

Disposable respirators are designed to be used once and are then to be properly disposed of. Once worn in the presence of an infectious patient, the respirator should be considered potentially contaminated with infectious material, and touching the outside of the device should be avoided to prevent self-inoculation (touching the contaminated respirator and then touching one's eyes, nose, or mouth). It should be noted that a once-worn respirator will also be contaminated on its inner surface by the microorganisms present in the exhaled air and oral secretions of the wearer.

If a sufficient supply of respirators is not available during a pandemic, employers and employees may consider reuse as long as the device has not been obviously soiled or damaged (e.g., creased or torn), and it retains its ability to function properly. This practice is not acceptable under normal circumstances and should only be considered under the most dire of conditions. Data on decontamination and/or reuse of respirators for infectious diseases are not available. Reuse may increase the potential for contamination; however, this risk must be

DOCUMENT APPENDIX

balanced against the need to provide respiratory protection. When preparing for a pandemic, employers who anticipate providing respiratory protection to employees for the duration of the pandemic should consider using reusable or elastomeric respirators that are designed to be cleaned, repaired and reused.

Dust or Comfort Masks

Employers and employees should be aware that there are "dust" or "comfort" masks sold at home improvement stores that look very similar to respirators. Some dust masks may even be made by a manufacturer that also produces NIOSH-certified respirators. Unless a mask has been tested and certified by NIOSH, employers do not know if the device will filter very small airborne particles. The occupational use of respirators, including those purchased at home improvement or convenience stores, are still covered by OSHA's Respiratory Protection standard.

Note: Some respirators have an exhalation valve to make it easier for the wearer to breathe. While these respirators provide the same level of particle filtration protection to the wearer, they should not be used by healthcare providers who are concerned about contaminating a sterile field, or provided to known or suspected pandemic patients as a means of limiting the spread of their body fluids to others.

Note: Additional respirator and surgical mask guidance for healthcare workers has been developed and is available at www.pandemicflu.gov/plan/healthcare/mask guidancehc.html. This document, "Interim Guidance on Planning for the Use of Surgical Masks and Respirators in Health Care Settings during an Influenza Pandemic," provides details on the differences between a surgical mask and a respirator, the state of science regarding influenza transmission, and the rationale for determining the appropriate protective device.

STEPS EVERY EMPLOYER CAN TAKE TO REDUCE THE RISK OF EXPOSURE TO PANDEMIC INFLUENZA IN THEIR WORKPLACE

The best strategy to reduce the risk of becoming infected with influenza during a pandemic is to avoid crowded settings and other situations that increase the risk of exposure to someone who may be infected. If it is absolutely necessary to be in a crowded setting, the time spent in a crowd should be as short as possible. Some basic hygiene (see www.cdc.gov/flu/protct/stopgerms.htm) and social distancing precautions that can be implemented in every workplace include the following:

- Encourage sick employees to stay at home.
- Encourage your employees to wash their hands frequently with soap and water or with hand sanitizer if there is no soap or water available. Also, encourage your employees to avoid touching their noses, mouths, and eyes.

- Encourage your employees to cover their coughs and sneezes with a tissue, or to cough and sneeze into their upper sleeves if tissues are not available. All employees should wash their hands or use a hand sanitizer after they cough, sneeze or blow their noses.
- Employees should avoid close contact with their coworkers and customers (maintain a separation of at least 6 feet). They should avoid shaking hands and always wash their hands after contact with others. Even if employees wear gloves, they should wash their hands upon removal of the gloves in case their hand(s) became contaminated during the removal process.
- Provide customers and the public with tissues and trash receptacles, and with a place to wash or disinfect their hands.
- Keep work surfaces, telephones, computer equipment and other frequently touched surfaces and office equipment clean. Be sure that any cleaner used is safe and will not harm your employees or your office equipment. Use only disinfectants registered by the U.S. Environmental Protection Agency (EPA), and follow all directions and safety precautions indicated on the label.
- Discourage your employees from using other employees' phones, desks, offices or other work tools and equipment.
- Minimize situations where groups of people are crowded together, such as in a meeting. Use e-mail, phones and text messages to communicate with each other. When meetings are necessary, avoid close contact by keeping a separation of at least 6 feet, where possible, and assure that there is proper ventilation in the meeting room.
- Reducing or eliminating unnecessary social interactions can be very effective in controlling the spread of infectious diseases. Reconsider all situations that permit or require employees, customers, and visitors (including family members) to enter the workplace. Workplaces which permit family visitors on site should consider restricting/eliminating that option during an influenza pandemic. Work sites with on-site day care should consider in advance whether these facilities will remain open or will be closed, and the impact of such decisions on employees and the business.
- Promote healthy lifestyles, including good nutrition, exercise, and smoking cessation. A person's overall health impacts their body's immune system and can affect their ability to fight off, or recover from, an infectious disease.

WORKPLACES CLASSIFIED AT LOWER EXPOSURE RISK (CAUTION) FOR PANDEMIC INFLUENZA: WHAT TO DO TO PROTECT EMPLOYEES

If your workplace does not require employees to have frequent contact with the general public, basic personal hygiene practices and social distancing can help

DOCUMENT APPENDIX

protect employees at work. Follow the general hygiene and social distancing practices previously recommended for all workplaces (see page 26). Also, try the following:

- Communicate to employees what options may be available to them for working from home.
- Communicate the office leave policies, policies for getting paid, transportation issues, and day care concerns.
- Make sure that your employees know where supplies for hand hygiene are located.
- Monitor public health communications about pandemic flu recommendations and ensure that your employees also have access to that information.
- Work with your employees to designate a person(s), website, bulletin board or other means of communicating important pandemic flu information.

More information about protecting employees and their families can be found at: www.pandemicflu.gov.

WORKPLACES CLASSIFIED AT MEDIUM EXPOSURE RISK FOR PANDEMIC INFLUENZA: WHAT TO DO TO PROTECT EMPLOYEES

Medium risk workplaces require frequent close contact between employees or with the general public (such as high-volume retail stores). If this contact cannot be avoided, there are practices to reduce the risk of infection. In addition to the basic work practices that every workplace should adopt (see page 26), medium risk occupations require employers to address enhanced safety and health precautions. Below are some of the issues that employers should address when developing plans for workplace safety and health during a pandemic.

Work Practice and Engineering Controls

- Instruct employees to avoid close contact (within 6 feet) with other employees and the general public. This can be accomplished by simply increasing the distance between the employee and the general public in order to avoid contact with large droplets from people talking, coughing or sneezing.
- Some organizations can expand internet, phone-based, drive-through window, or home delivery customer service strategies to minimize face-to-face contact. Work with your employees to identify new ways to do business that can also help to keep employees and customers safe and healthy.

- Communicate the availability of medical screening or other employee health resources (e.g., on-site nurse or employee wellness program to check for flu-like symptoms before employees enter the workplace).
- Employers also should consider installing physical barriers, such as clear plastic sneeze guards, to protect employees where possible (such as cashier stations).

Administrative Controls

- Work with your employees so that they understand the office leave policies, policies for getting paid, transportation issues, and day care concerns.
- Make sure that employees know where supplies for hand and surface hygiene are located.
- Work with your employees to designate a person(s), website, bulletin board or other means of communicating important pandemic flu information.
- Use signs to keep customers informed about symptoms of the flu, and ask sick customers to minimize contact with your employees until they are well.
- Your workplace may consider limiting access to customers and the general public, or ensuring that they can only enter certain areas of your workplace.

Personal Protective Equipment (PPE)

Employees who have high-frequency, close contact with the general population that cannot be eliminated using administrative or engineering controls, and where contact with symptomatic ill persons is not expected should use personal protective equipment to prevent sprays of potentially infected liquid droplets (from talking, coughing, or sneezing) from contacting their nose or mouth. A surgical mask will provide such barrier protection. Use of a respirator may be considered if there is an expectation of close contact with persons who have symptomatic influenza infection or if employers choose to provide protection against a risk of airborne transmission. It should be noted that wearing a respirator may be physically burdensome to employees, particularly when the use of PPE is not common practice for the work task. In the event of a shortage of surgical masks, a reusable face shield that can be decontaminated may be an acceptable method of protecting against droplet transmission of an infectious disease but will not protect against airborne transmission, to the extent that disease may spread in that manner.

    Eye protection generally is not recommended to prevent influenza infection although there are limited examples where strains of influenza have caused eye infection (conjunctivitis). At the time of a pandemic, health officials will assess whether risk of conjunctival infection or transmission exists for the specific pandemic viral strain.

    Employees should wash hands frequently with soap or sanitizing solutions to prevent hands from transferring potentially infectious material from surfaces to their mouths or noses. While employers and employees may choose to wear gloves, the exposure of concern is touching the mouth and nose with a contaminated hand and not exposure to the virus through non-intact skin (for example, cuts or scrapes). While the use of gloves may make employees more aware of potential hand contamination, there is no difference between intentional or unintentional touching of the mouth, nose or eyes with either a contaminated glove or a contaminated hand. If an employee does wear gloves, they should always wash their hands with soap or sanitizing solution immediately after removal to ensure that they did not contaminate their hand(s) while removing them.

When selecting PPE, employers should consider factors such as function, fit, ability to be decontaminated, disposal, and cost. Sometimes, when a piece of PPE will have to be used repeatedly for a long period of time, a more expensive and durable piece of PPE may be less expensive in the long run than a disposable piece of PPE. For example, in the event of a pandemic, there may be shortages of surgical masks. A reusable face shield that can be decontaminated may become the preferred method of protecting against droplet transmission in some workplaces. It should be noted that barrier protection, such as a surgical mask or face shield, will protect against droplet transmission of an infectious disease but will not protect against airborne transmission, to the extent that the disease may be spread in that manner. Each employer should select the combination of PPE that protects employees in their particular workplace. It should also be noted that wearing PPE may be physically burdensome to employees, particularly when the use of PPE is not common practice for the work task.

Educate and train employees about the protective clothing and equipment appropriate to their current duties and the duties which they may be asked to assume when others are absent. Employees may need to be fit tested and trained in the proper use and care of a respirator. Also, it is important to train employees to put on (don) and take off (doff) PPE in the proper order to avoid inadvertent self-contamination (www.osha.gov/SLTC/respiratoryprotection/index.html). During a pandemic, recommendations for PPE use in particular occupations may change, depending on geographic proximity to active cases, updated risk assessments for particular employees, and information on PPE effectiveness in preventing the spread of influenza.

WORKPLACES CLASSIFIED AT VERY HIGH OR HIGH EXPOSURE RISK FOR PANDEMIC INFLUENZA: WHAT TO DO TO PROTECT EMPLOYEES

If your workplace requires your employees to have contact with people that are known or suspected to be infected with the pandemic virus, there are many practices that can be used to reduce the risk of infection and to protect your employees. Additional guidance for very high and high exposure risk workplaces, such as health care facilities, can be found at: www.pandemicflu.gov and www.osha.gov.

Very high and high exposure risk occupations require employers to address enhanced safety and health precautions in addition to the basic work practices that every workplace should adopt (see page 26). Employers should also be aware that working in a high risk occupation can be stressful to both employees and their families. Employees in high risk occupations may have heightened concern about their own safety and possible implications for their family. Such workplaces may experience greater employee absenteeism than other lower risk workplaces. Talk to your employees about resources that can help them in the event of a pandemic crisis. Keeping the workplace safe is everyone's priority. More information about protecting employees and their families can be found at: www.pandemicflu.gov.

Work Practice and Engineering Controls

Employers should ensure that employees have adequate training and supplies to practice proper hygiene. Emergency responders and other essential personnel who may be exposed while working away from fixed facilities should be provided with hand sanitizers that do not require water so that they can decontaminate themselves in the field. Employers should work with employees to identify ways to modify work practices to promote social distancing and prevent close contact (within 6 feet), where possible. Employers should also consider offering enhanced medical monitoring of employees in very high and high risk work environments.

In certain limited circumstances ventilation is recommended for high and very high risk work environments. While proper ventilation can reduce the risk of transmission for healthcare workers in the same room as infectious patients, it cannot be relied upon as the sole protective measure. Thus, a combination of engineering controls and personal protective equipment will be needed.

- When possible, health care facilities equipped with isolation rooms should use them when performing aerosol generating procedures for patients with known or suspected pandemic influenza.
- Laboratory facilities that handle specimens for known or suspected pandemic patients will also require special precautions associated with a Bio-Safety Level 3 facility. Some recommendations can be found at: www.cdc.gov/flu/h2n2bs13.htm.

Employers should also consider installing physical barriers, such as clear plastic sneeze guards, to protect employees where possible (for example, reception or intake areas). The use of barrier protections, such as sneeze guards, is common practice for both infection control and industrial hygiene.

Administrative Controls (Isolation Precautions)

If working in a health care facility, follow existing guidelines and facility standards of practice for identifying and isolating infected individuals and for protecting employees. See the U.S. Department of Health and Human Services' pandemic influenza plan for health care facilities at: www.hhs.gov/pandemicflu/plan/sup4.html.

Personal Protective Equipment (PPE)

Those who work closely with (either in contact with or within 6 feet) people known or suspected to be infected with pandemic influenza should wear:

- Respiratory protection for protection against small droplets from talking, coughing or sneezing and also from small airborne particles of infectious material.
  - N95 or higher rated filter for most situations.
  - Supplied air respirator (SAR) or powered air purifying respirator (PAPR) for certain high risk medical or dental procedures likely to generate bioaerosols.
  - Use a surgical respirator when both respiratory protection and resistance to blood and body fluids is necessary.
- Face shields may also be worn on top of a respirator to prevent bulk contamination of the respirator. Certain respirator designs with forward protrusions (duckbill style) may be difficult to properly wear under a face shield. Ensure that the face shield does not prevent airflow through the respirator.
- Medical/surgical gowns or other disposable/decontaminable protective clothing.
- Gloves to reduce transfer of infectious material from one patient to another.
- Eye protection if splashes are anticipated.

The appropriate form of respirator will depend on the type of exposure and on the transmission pattern of the particular strain of influenza. See the National Institute for Occupational Safety and Health (NIOSH) Respirator Selection Logic at: www.cdc.gov/niosh/docs/2005-100.

Educate and train employees about the protective clothing and equipment appropriate to their current duties and the duties which they may be asked to assume when others are absent. Education and training material should be easy to understand and available in the appropriate language and literacy level for all employees. Employees need to be fit tested and trained in the proper use and care of a respirator. It is also important to train employees to put on (don) and take off (doff) PPE in the proper order to avoid inadvertent self-contamination (www.osha.gov/SLTC/respiratoryprotection/index.html). Employees who dispose of PPE and other infectious waste must also be trained and provided with appropriate PPE.

During a pandemic, recommendations for PPE use in particular occupations may change depending on geographic location, updated risk assessments for particular employees, and information on PPE effectiveness in preventing the spread of influenza. Additional respirator and surgical mask guidance for healthcare workers has been developed and is available at www.pandemicflu.gov/plan/healthcare/maskguidancehc.html. This document, Interim Guidance on Planning for the Use of Surgical Masks and Respirators in Health Care Settings during an Influenza Pandemic, provides details on the differences between a surgical mask and a respirator, the state of science regarding influenza transmission, and the rationale for determining the appropriate protective device.

WHAT EMPLOYEES LIVING ABROAD OR WHO TRAVEL INTERNATIONALLY FOR WORK SHOULD KNOW

Employees living abroad and international business travelers should note that other geographic areas have different influenza seasons and will likely be affected by a pandemic at different times than the United States. The U.S. Department of State emphasizes that, in the event of a pandemic, its ability to assist Americans traveling and residing abroad may be severely limited by restrictions on local and international movement imposed for public health reasons, either by foreign governments and/or the United States. Furthermore, American citizens should take note that the Department of State cannot provide Americans traveling or living abroad with medications or supplies even in the event of a pandemic.

In addition, the Department of State has asked its embassies and consulates to consider preparedness measures that take into consideration the fact that travel into or out of a country may not be possible, safe, or medically advisable during a pandemic. Guidance on how private citizens can prepare to shelter in place, including stocking food, water, and medical supplies, is available at the www.pandemicflu.gov website. Embassy stocks cannot be made available to private American citizens abroad, therefore, employers and employees are encouraged to prepare appropriately. It is also likely that governments will respond to a pandemic by imposing public health measures that restrict domestic and international movement, further limiting the U.S. government's ability to assist Americans in these countries. As it is possible that these measures may be implemented very quickly, it is important that employers and employees plan appropriately.

More information on pandemic influenza planning for employees living and traveling abroad can be found at: www.pandemicflu.gov/travel/index.html; www.cdc.gov/travel; www.state.gov/travelandbusiness

FOR MORE INFORMATION

Federal, state and local government agencies are your best source of information should an influenza pandemic take place. It is important to stay informed about the latest developments and recommendations since specific guidance may change based upon the characteristics of the eventual pandemic influenza strain, (for example, severity of disease, importance of various modes of transmission).

DOCUMENT APPENDIX

Below are several recommended websites that you can rely on for the most current and accurate information:

www.pandemicflu.gov
(Managed by the Department of Health and Human Services; offers one-stop access, including toll-free phone numbers, to U.S. government avian and pandemic flu information.)

www.osha.gov
(Occupational Safety and Health Administration website)

www.cdc.gov/niosh
(National Institute for Occupational Safety and Health website)

www.cdc.gov
(Centers for Disease Control and Prevention website)

www.fda.gov/cdrh/ppe/fluoutbreaks.html
(U.S. Food and Drug Administration website)

OSHA ASSISTANCE

OSHA can provide extensive help through a variety of programs, including technical assistance about effective safety and health programs, state plans, workplace consultations, voluntary protection programs, strategic partnerships, training and education, and more. An overall commitment to workplace safety and health can add value to your business, to your workplace and to your life.

Safety and Health Program Management Guidelines

Effective management of employee safety and health protection is a decisive factor in reducing the extent and severity of work-related injuries and illnesses and their related costs. In fact, an effective safety and health program forms the basis of good employee protection and can save time and money (about $4 for every dollar spent) and increase productivity and reduce employee injuries, illnesses and related workers' compensation costs.

To assist employers and employees in developing effective safety and health programs, OSHA published recommended *Safety and Health Program Management Guidelines* (*54 Federal Register* (16): 3904–3916, January 26, 1989). These voluntary guidelines apply to all places of employment covered by OSHA.

The guidelines identify four general elements critical to the development of a successful safety and health management program:

- Management leadership and employee involvement.
- Work analysis.
- Hazard prevention and control.
- Safety and health training.

The guidelines recommend specific actions, under each of these general elements, to achieve an effective safety and health program. The *Federal Register* notice is available online at www.osha.gov.

State Programs

The Occupational Safety and Health Act of 1970 (OSH Act) encourages states to develop and operate their own job safety and health plans. OSHA approves and monitors these plans. Twenty-four states, Puerto Rico and the Virgin Islands currently operate approved state plans: 22 cover both private and public (state and local government) employment; Connecticut, New Jersey, New York and the Virgin Islands cover the public sector only. States and territories with their own OSHA-approved occupational safety and health plans must adopt standards identical to, or at least as effective as, the Federal standards.

Consultation Services

Consultation assistance is available on request to employers who want help in establishing and maintaining a safe and healthful workplace. Largely funded by OSHA, the service is provided at no cost to the employer. Primarily developed for smaller employers with more hazardous operations, the consultation service is delivered by state governments employing professional safety and health consultants. Comprehensive assistance includes an appraisal of all mechanical systems, work practices and occupational safety and health hazards of the workplace and all aspects of the employer's present job safety and health program. In addition, the service offers assistance to employers in developing and implementing an effective safety and health program. No penalties are proposed or citations issued for hazards identified by the consultant. OSHA provides consultation assistance to the employer with the assurance that his or her name and firm and any information about the workplace will not be routinely reported to OSHA enforcement staff.

Under the consultation program, certain exemplary employers may request participation in OSHA's Safety and Health Achievement Recognition Program (SHARP). Eligibility for participation in SHARP includes receiving a comprehensive consultation visit, demonstrating exemplary achievements in workplace safety and health by abating all identified hazards and developing an excellent safety and health program.

Employers accepted into SHARP may receive an exemption from programmed inspections (not complaint or accident investigation inspections) for a period of one year. For more information concerning consultation assistance, see the OSHA website at www.osha.gov.

Voluntary Protection Program (VPP)

Voluntary Protection Programs and on-site consultation services, when coupled with an effective enforcement program, expand employee protection to help

DOCUMENT APPENDIX

meet the goals of the OSH Act. The three levels of VPP are Star, Merit, and Star Demonstration designed to recognize outstanding achievements by companies that have successfully incorporated comprehensive safety and health programs into their total management system. The VPPs motivate others to achieve excellent safety and health results in the same outstanding way as they establish a cooperative relationship between employers, employees and OSHA.

For additional information on VPP and how to apply, contact the OSHA regional offices listed at the end of this publication.

Strategic Partnership Program

OSHA's Strategic Partnership Program, the newest member of OSHA's cooperative programs, helps encourage, assist and recognize the efforts of partners to eliminate serious workplace hazards and achieve a high level of employee safety and health. Whereas OSHA's Consultation Program and VPP entail one-on-one relationships between OSHA and individual worksites, most strategic partnerships seek to have a broader impact by building cooperative relationships with groups of employers and employees. These partnerships are voluntary, cooperative relationships between OSHA, employers, employee representatives and others (e.g., trade unions, trade and professional associations, universities and other government agencies).

For more information on this and other cooperative programs, contact your nearest OSHA office, or visit OSHA's website at www.osha.gov.

Alliance Programs

The Alliance Program enables organizations committed to workplace safety and health to collaborate with OSHA to prevent injuries and illnesses in the workplace. OSHA and the Alliance participants work together to reach out to, educate and lead the nation's employers and their employees in improving and advancing workplace safety and health.

Groups that can form an Alliance with OSHA include employers, labor unions, trade or professional groups, educational institutions and government agencies. In some cases, organizations may be building on existing relationships with OSHA that were developed through other cooperative programs.

There are few formal program requirements for Alliances and the agreements do not include an enforcement component. However, OSHA and the participating organizations must define, implement and meet a set of short-and long-term goals that fall into three categories: training and education; outreach and communication; and promoting the national dialogue on workplace safety and health.

OSHA Training and Education

OSHA area offices offer a variety of information services, such as compliance assistance, technical advice, publications, audiovisual aids and speakers for special engagements. OSHA's Training Institute in Arlington Heights, IL, provides basic and advanced courses in safety and health for Federal and state compliance officers, state consultants, Federal agency personnel, and private sector employers, employees and their representatives.

The OSHA Training Institute also has established OSHA Training Institute Education Centers to address the increased demand for its courses from the private sector and from other Federal agencies. These centers are nonprofit colleges, universities and other organizations that have been selected after a competition for participation in the program.

OSHA also provides funds to nonprofit organizations, through grants, to conduct workplace training and education in subjects where OSHA believes there is a lack of workplace training. Grants are awarded annually. Grant recipients are expected to contribute 20 percent of the total grant cost.

For more information on grants, training and education, contact the OSHA Training Institute, Office of Training and Education, 2020 South Arlington Heights Road, Arlington Heights, IL 60005, (847) 297-4810 or see "Outreach" on OSHA's website at www.osha.gov. For further information on any OSHA program, contact your nearest OSHA area or regional office listed at the end of this publication.

Information Available Electronically

OSHA has a variety of materials and tools available on its website at www.osha.gov. These include *eTools* such as *Expert Advisors, Electronic Compliance Assistance Tools (e-cats), Technical Links;* regulations, directives and publications; videos and other information for employers and employees. OSHA's software programs and compliance assistance tools walk you through challenging safety and health issues and common problems to find the best solutions for your workplace.

A wide variety of OSHA materials, including standards, interpretations, directives, and more, can be purchased on CD-ROM from the U.S. Government Printing Office, Superintendent of Documents, phone toll-free (866) 512-1800.

OSHA Publications

OSHA has an extensive publications program. For a listing of free or sales items, visit OSHA's website at www.osha.gov or contact the OSHA Publications Office, U.S. Department of Labor, 200 Constitution Avenue, NW, N-3101, Washington, DC 20210. Telephone (202) 693-1888 or fax to (202) 693-2498.

Contacting OSHA

To report an emergency, file a complaint, or seek OSHA advice, assistance, or products, call (800) 321-OSHA or contact your nearest OSHA Regional or Area office listed at the end of this publication. The teletypewriter (TTY) number is (877) 889-5627.

DOCUMENT APPENDIX

Written correspondence can be mailed to the nearest OSHA Regional or Area Office listed at the end of this publication or to OSHA's national office at: U.S. Department of Labor, Occupational Safety and Health Administration, 200 Constitution Avenue, N.W., Washington, DC 20210.

By visiting OSHA's website at www.osha.gov, you can also:

- file a complaint online,
- submit general inquiries about workplace safety and health electronically, and
- find more information about OSHA and occupational safety and health.

OSHA REGIONAL OFFICES

Region I
(CT,* ME, MA, NH, RI, VT*)
JFK Federal Building, Room E340
Boston, MA 02203
(617) 565-9860

Region II
(NJ,* NY,* PR,* VI*)
201 Varick Street, Room 670
New York, NY 10014
(212) 337-2378

Region III
(DE, DC, MD,* PA, VA,* WV)
The Curtis Center
170 S. Independence Mall West
Suite 740 West
Philadelphia, PA 19106-3309
(215) 861-4900

Region IV
(AL, FL, GA, KY,* MS, NC,* SC,* TN*)
61 Forsyth Street, SW
Atlanta, GA 30303
(404) 562-2300

Region V
(IL, IN,* MI,* MN,* OH, WI)
230 South Dearborn Street
Room 3244
Chicago, IL 60604
(312) 353-2220

Region VI
(AR, LA, NM,* OK, TX)
525 Griffin Street, Room 602
Dallas, TX 75202
(214) 767-4731 or 4736 x 224

Region VII
(IA,* KS, MO, NE)
City Center Square
1100 Main Street, Suite 800
Kansas City, MO 64105
(816) 426-5861

Region VIII
(CO, MT, ND, SD, UT,* WY*)
1999 Broadway, Suite 1690
PO Box 46550
Denver, CO 80202-5716
(720) 264-6550

Region IX
(American Samoa, AZ,* CA,* HI,* NV,* Northern Mariana Islands)
71 Stevenson Street, Room 420
San Francisco, CA 94105
(415) 975-4310

Region X
(AK,* ID, OR,* WA*)
1111 Third Avenue, Suite 715
Seattle, WA 98101-3212
(206) 553-5930

* These states and territories operate their own OSHA-approved job safety and health programs (Connecticut, New Jersey, New York and the Virgin Islands plans cover public employees only). States with approved programs must adopt standards identical to, or at least as effective as, the Federal standards.

Note: To get contact information for OSHA Area Offices, OSHA-approved State Plans and OSHA Consultation Projects, please visit us online at www.osha.gov or call us at 1-800-321-OSHA.

INDUSTRY SELF-ASSESSMENT CHECKLIST FOR FOOD SECURITY

U.S. Department of Agriculture
Food Safety and Inspection Service

It is vital that all food slaughter and processing establishments, and all import, export, and identification warehouses take steps to ensure the security of their operations. USDA's Food Safety and Inspection Service (FSIS) created this self-assessment instrument to provide a tool for establishments to assess the extent to which they have secured their operations. The contents of the instrument are based primarily on the food security guidelines that FSIS published in 2002, *Food Security Guidelines for Food Processors*, available at www.fsis.usda.gov. Those guidelines identify security measures that establishments can adopt to enhance the security of their operations.

The checklist consists of the following (9) sections:

I. Food Security Plan Management
II. Outside Security
III. Inside Security
IV. Slaughter and Processing Security
V. Storage Security
VI. Shipping and Receiving Security
VII. Water and Ice Supply Security
VIII. Mail Handling Security
IX. Personnel Security

To use the checklist, read each question under each section and check the response that best describes the food security practice in the establishment. If a question is not applicable, check "N/A." For example, if an establishment only conducts processing activities, then questions that ask about live animals or

DOCUMENT APPENDIX

slaughter operations would not apply. Similarly, if an establishment only conducts import/export inspection activities, then questions related to processing or slaughter would not apply. A "Yes" response for every question is desirable but not expected due to the layering of certain security measures. A "No" answer on a question does not necessarily constitute a breach in security. The establishment might have other management strategies or activities conducted at different frequencies that accomplish food security goals.

A "No" should, however, trigger a critical thought process in establishment operators on whether management decisions should be made regarding additional security measures they may need to put in place at the particular operational sector/area of the establishment covered by the "No" response. In addition, Appendix 1 lists resources and websites that discuss additional security measures for establishments. Resources/websites specific to a section or question are shown at the corresponding section/question throughout the checklist for quick and easy reference. "NA" responses should be reviewed periodically to validate them against current operations.

The final outcome of this self-assessment should provide establishments with a relative measure of overall security of their operations and guide them in the development and/or revision of their food security strategies.

This checklist is one of several outreach efforts by FSIS to assist the industry to enhance the security of its regulated food products. Model food security plans have also been developed for voluntary use by the industry. The Agency is also considering additional measures that may be appropriate to ensure the safety and security of meat, poultry and egg products under certain elevated threat conditions specific to food and agriculture.

I. FOOD SECURITY PLAN MANAGEMENT

A food security plan is a written document developed using established risk management procedures and consists of specific standard operating procedures for preventing intentional product tampering and responding to threats or actual incidents of intentional product tampering.

1. Does this establishment have a written food security plan?
   ☐ Yes
   ☐ No [GO TO QUESTION 3]

2. Which of the following procedures, plans, or information are either included in the food security plan or have been put in place as a result of the food security plan? *(Check "Yes" or "No" for each item.)*

	Yes	No
Is there a designated person or team to implement and oversee the food security plan?	☐	☐
Are members of the food security management team trained in all provisions of the food security plan?	☐	☐
Are periodic drills conducted on operational elements of the food security plan?	☐	☐
Are regular food security inspections conducted to verify key provisions of the food security plan?	☐	☐
Is the security plan reviewed (and revised if necessary) periodically?	☐	☐
Are the details of food security procedures kept confidential?	☐	☐
Is the emergency contact information for local, state, and federal government homeland security authorities and public health officials included in the security plan? State contact list: www.whitehouse.gov/homeland/contactmap.html	☐	☐
Is the above contact information periodically reviewed and updated?	☐	☐
Is there an established liaison between plant officials and the local homeland security officials and other law enforcement officials?	☐	☐
Is there an established relationship between the establishment and the appropriate analytical laboratories for possible assistance in investigation of product tampering cases?	☐	☐
Are procedures for responding to threats of product tampering included in the plan?	☐	☐
Are procedures for responding to actual incidents of product tampering detailed in the plan? http://www.state.tn.us/agriculture/security/fsig.html	☐	☐
Are communication procedures for notifying law enforcement, public health officials, and FSIS inspectors in-charge when a food security threat is received or when evidence of actual product tampering is observed included in the plan?	☐	☐
Are procedures in the plan for corrective action in cases of product tampering to ensure that adulterated or potentially injurious products do not enter commerce?	☐	☐
Are procedures in the plan for safe handling and disposal of contaminated products?	☐	☐
Are employees encouraged to report signs of possible product tampering or breaks in food security system (e.g., award system)?	☐	☐
Are evacuation procedures in the security plan? Visit www.osha.gov/dep/evacmatrix/index.html for guidance material provided by the U.S. Department of Labor, Occupational Safety and Health Administration	☐	☐
Are procedures in place to restrict access to the facility during an emergency to authorized personnel only?	☐	☐
Are designated entry points for emergency personnel clearly marked?	☐	☐
Does the establishment have a documented recall plan?	☐	☐
Are procedures in the recall plan reviewed and updated as necessary?	☐	☐
Do recall procedures ensure segregation and disposition of recalled products?	☐	☐

DOCUMENT APPENDIX 623

II. OUTSIDE SECURITY

3. Which of the following security procedures does this establishment have in place for the exterior of this establishment? *(Check "Yes" or "No" for each procedure.)*

	Yes	No
Are the plant's boundaries and grounds secured to prevent entry by unauthorized persons (e.g., by locked fence or gate)?	☐	☐
Are "No Trespassing" signs posted at plant's boundaries?	☐	☐
Is there sufficient outside lighting to allow detection of unusual activities on any part of the establishment outside premises during non-daylight hours?	☐	☐
Do emergency exits have self-locking doors and/or alarms?	☐	☐
Is positive identification required to control entry of visitors to the plant (e.g., picture IDs or sign-in/sign-out at entrance)?	☐	☐
Is an updated list of establishment personnel with open or restricted access to the establishment maintained at the security office or another secure location?	☐	☐

4. Are the following secured with locks, seals, or sensors at all times to prevent entry by unauthorized persons? *(Check "Yes" or "No" for each item, or "N/A" if the item is not applicable.)*

	Yes	No	N/A
Outside doors and gates?	☐	☐	☐
Windows?	☐	☐	☐
Roof openings?	☐	☐	☐
Vent openings?	☐	☐	☐
Trailer (truck) bodies?	☐	☐	☐
Tanker truck hatches?	☐	☐	☐
Railcars?	☐	☐	☐
Bulk storage tanks?	☐	☐	☐

5. Which of the following security procedures does this establishment have in place for vehicles entering the establishment? *(Check "Yes" or "No" for each procedure.)*

	Yes	No
Are *incoming* private vehicles (e.g., employees or visitors) inspected for unusual cargo or activity?	☐	☐
Are *outgoing* private vehicles (e.g., employees or visitors) inspected for unusual cargo or activity?	☐	☐
Are *incoming* commercial vehicles (e.g., delivery trucks) inspected for unusual cargo or activity?	☐	☐
Are *outgoing* commercial vehicles (e.g., delivery trucks) inspected for unusual cargo or activity?	☐	☐

	Yes	No
Are *incoming* tanker truck shipments checked for documentation of chain of custody prior to loading?	☐	☐
Are *employee* vehicles identified using placards, decals, or some other form of visual identification?	☐	☐
Are authorized *visitor/quest* vehicles identified using placards, decals, or some other form of visual identification?	☐	☐

III. GENERAL INSIDE SECURITY

6. Which of the following security procedures does this establishment have in place within the interior of this establishment? *(Check "Yes" or "No" for each procedure, or "N/A" if the procedure is not applicable.)*

	Yes	No	N/A
Is emergency lighting provided in the establishment?	☐	☐	☐
Is active surveillance of the inside facility and operations maintained?	☐	☐	☐
Are emergency alert systems tested periodically?	☐	☐	☐
Are the locations of controls for emergency alert systems clearly marked?	☐	☐	☐
Are all restricted areas (i.e., areas where only authorized employees have access) within the plant clearly marked?	☐	☐	☐
Are visitors, guests, and other non-establishment employees (e.g., contractors, salespeople, truck drivers) restricted to non-product areas unless accompanied by an authorized establishment employee?	☐	☐	☐
Are updated plant layout schematics provided at strategic and secured locations?	☐	☐	☐
Are procedures in place to check toilets, maintenance closets, personal lockers, and storage areas for suspicious packages?	☐	☐	☐
Is an inventory of tools and utensils (e.g., knives) conducted regularly as needed to ensure security?	☐	☐	☐
Is an inventory of keys to secured areas of the facility conducted regularly to ensure security?	☐	☐	☐
Are ventilation systems constructed in a manner that provides for isolation of contaminated areas or rooms?	☐	☐	☐

7. Are the central controls for the following restricted (e.g., by locked door/gate or limiting access to designated employees) to prevent access by unauthorized persons?
   Check "Yes" or "No" for each item, or "N/A" if the item is not applicable.
   www.cdc.gov/niosh/bldvent/2002-139.html

	Yes	No	N/A
Heating, Ventilation, and Air Conditioning systems?	☐	☐	☐
Propane Gas?	☐	☐	☐
Water systems?	☐	☐	☐

DOCUMENT APPENDIX

Electricity?	☐	☐	☐
Disinfection systems?	☐	☐	☐
Clean-in-place (CIP) systems?	☐	☐	☐

8. Does this establishment collect and analyze samples in-house?
   ☐ Yes
   ☐ No [GO TO QUESTION 10]

9. Which of the following security procedures does this establishment have in place for its in-plant laboratory facilities, equipment, and operations? *(Check "Yes" or "No" for each procedure, or "N/A" if the procedure is not applicable [e.g., this establishment does not use live cultures of pathogenic bacteria].)*

	Yes	No	N/A
Is access to the in-plant laboratory facilities restricted to authorized employees? (e.g., by locked door, pass card, etc.)	☐	☐	☐
Is a procedure in place to control receipt of samples received from other establishments?	☐	☐	☐
Is a procedure in place to receive and securely store reagents?	☐	☐	☐
Is a procedure in place to control and dispose of reagents?	☐	☐	☐
Is a procedure in place to receive and securely store live cultures of pathogenic bacteria?	☐	☐	☐
Is a procedure in place to dispose of live cultures of pathogenic bacteria?	☐	☐	☐

10. Does this establishment use a computer system to monitor processing operations?
    ☐ Yes
    ☐ No [GO TO QUESTION 12]

11. Which of the following security procedures does this establishment have in place for its computer systems? *(Check "Yes" or "No" for each procedure.)*
    http://www.fiu.edu/security.guidelines.html

	Yes	No
Is the access to the system password-protected? http://www.umich.edu/~policies/pw-security.html	☐	☐
Are firewalls built into the computer network?	☐	☐
Is the system using a current virus detection system?	☐	☐

IV. SLAUGHTER AND PROCESSING SECURITY

12. Which of the following security procedures does this establishment have in place for its slaughter and processing operations? *(Check "Yes" or "No" for*

each procedure, or "N/A" if the procedure is not applicable [e.g., this establishment does not mix or batch ingredients].)

	Yes	No	N/A*
Is access to product production/slaughter and holding pen areas restricted to establishment employees and FSIS inspection personnel only?	☐	☐	☐
Is the mixing and batching of product and ingredients and other operations where large amounts of exposed product are handled continuously monitored?	☐	☐	☐
Are lines that handle and transfer products, water, oil, or other ingredients monitored to ensure integrity?	☐	☐	☐
Is the packaging integrity of ingredients examined for evidence of tampering before use?	☐	☐	☐
Is the restricted access to in-plant irradiation equipment and materials clearly marked and maintained?	☐	☐	☐
Are records maintained to ensure the capability to trace-back raw materials to suppliers?	☐	☐	☐
Are records maintained to ensure the capability to trace-forward finished products to vendors?	☐	☐	☐

*"N/A" RESPONSE POSSIBLE IN ABOVE SECTION FOR IMPORT, EXPORT, AND ID ESTABLISHMENTS.

V. STORAGE SECURITY

13. Which of the following security procedures does this establishment have in place for its storage areas? *(Check "Yes" or "No" for each procedure, or "N/A" if the procedure is not applicable [e.g., this establishment does not use restricted ingredients].)*

	Yes	No	N/A
Is access to raw product storage areas, including holding coolers restricted (e.g., by locked door/gate) to designated employees?	☐	☐	☐
Is an access log maintained for raw product storage areas?	☐	☐	☐
Is access to non-meat ingredient storage areas restricted to designated employees only?	☐	☐	☐
Is an access log maintained for ingredient storage areas?	☐	☐	☐
Is access to finished product storage areas restricted to designated employees?	☐	☐	☐
Is access to external storage facilities restricted to designated employees only?	☐	☐	☐
Are silo storage tanks for raw egg product and other bulk ingredients (syrup, oils, etc.) maintained under lock and seal?	☐	☐	☐
Are silo storage tanks for pasteurized egg product and other bulk finished products maintained under lock and seal?	☐	☐	☐
Are silo storage tanks for inedible egg product and other bulk inedible products maintained under lock and seal?	☐	☐	☐

DOCUMENT APPENDIX

	Yes	No	N/A
Are periodic security inspections of storage facilities (including temporary storage vehicles) conducted?	☐	☐	☐
Are records maintained on facility security inspections results?	☐	☐	☐
Is the inventory of restricted ingredients (i.e., nitrites, etc) reconciled against the actual use of such ingredients on a regular basis?	☐	☐	☐
Are product labels and packaging held in a secure area to prevent theft and misuse?	☐	☐	☐
Is the inventory of finished products regularly checked for unexplained additions and withdrawals from existing stock?	☐	☐	☐

14. Which of the following security procedures does this establishment have in place for the storage of hazardous materials/chemicals such as pesticides, industrial chemicals, cleaning materials, sanitizers, and disinfectants? *(Check "Yes" or "No" for each procedure.)*

	Yes	No
Is the access to inside and outside storage areas for hazardous materials/chemicals such as pesticides, industrial chemicals, cleaning materials, sanitizers, and disinfectants restricted to designated employees?	☐	☐
Are hazardous material/chemical storage areas separated from production areas of plant?	☐	☐
Is a regular inventory of hazardous materials/chemicals maintained?	☐	☐
Are discrepancies in daily inventory of hazardous materials/chemicals immediately investigated?	☐	☐
Are the storage areas for hazardous materials/chemicals constructed and safely vented in accordance with national or local building codes?	☐	☐
Is a procedure in place to receive and securely store hazardous chemicals?	☐	☐
Is a procedure in place to control disposition of hazardous chemicals?	☐	☐

VI. SHIPPING AND RECEIVING SECURITY

Visit: http://www.fsis.usda.gov/oa/topics/transportguide.htm

15. Which of the following security procedures does this establishment have in place for its shipping and receiving operations? *(Check "Yes" or "No" for each procedure, or "N/A" if the procedure is not applicable [e.g., no tanker trucks on premises].)*

	Yes	No	N/A
Are trailers on the premises maintained under lock and/or seal when not being loaded or unloaded?	☐	☐	☐
Are tanker trucks on the premises maintained under lock and seal when not being loaded or unloaded?	☐	☐	☐
Is the loading and unloading of vehicles transporting raw materials, finished products, or other materials used in food processing closely monitored?	☐	☐	☐

16. Which of the following security procedures does this establishment have in place for handling outgoing shipments? *(Check "Yes" or "No" for each procedure, or "N/A" if the procedure is not applicable [e.g., no tanker trucks on premises].)*

	Yes	No	N/A
Are outgoing shipments sealed with tamper-evident seals?	☐	☐	☐
Are the seal numbers on outgoing shipment documented on the shipping documents?	☐	☐	☐
Are tanker trucks visually inspected to detect the presence of any material, solid or liquid, in tanks prior to loading liquid products?	☐	☐	☐
Are records maintained of the above inspections of tanker trucks?	☐	☐	☐
Are chain-of-custody records maintained for tanker trucks?	☐	☐	☐

17. Which of the following security procedures does this establishment have in place for handling incoming shipments? *(Check "Yes" or "No" for each procedure, or "N/A" if the procedure is not applicable [e.g., this establishment does not receive live animals].)*

	Yes	No	N/A
Is access to loading docks controlled to avoid unverified or unauthorized deliveries?	☐	☐	☐
Is advance notification from suppliers (by phone, e-mail, or fax) required for all incoming deliveries?	☐	☐	☐
Are suspicious alterations in the shipping documents immediately investigated?	☐	☐	☐
Are all deliveries verified against the roster of scheduled deliveries?	☐	☐	☐
Are unscheduled deliveries held outside facility premises pending verification?	☐	☐	☐
Are off-hour deliveries accepted?	☐	☐	☐
If off-hour deliveries are accepted, is prior notice of the delivery required?	☐	☐	☐
If off-hour deliveries are accepted, is the presence of authorized individual to verify and receive the delivery required?	☐	☐	☐
Is the integrity of internal compartments in the truck, lot packaging, or in-transit security checks for less-than-truckload (LTL) or partial load shipments of materials verified?	☐	☐	☐
Are incoming shipments of raw product, ingredients, and finished products required to be sealed with tamper-evident or numbered seals (and documented in the shipping documents) which are verified prior to entry?	☐	☐	☐
Is the integrity of incoming shipments of raw product, ingredients, and finished products checked at receiving dock for evidence of tampering?	☐	☐	☐
Is the FSIS Public Health Veterinarian notified immediately when animals with unusual behavior and/or symptoms are received? http://www.inspection.gc.ca/english/ops/secur/livbete.shtml	☐	☐	☐

DOCUMENT APPENDIX

	Yes	No	N/A
Are the feed and drinking water supplies for live animals protected from possible intentional contamination?	☐	☐	☐
Are transportation companies selected with consideration of the procedures companies have in place to safeguard the security of product/animals being shipped?	☐	☐	☐
Are transportation companies selected with consideration of background checks conducted on drivers and other employees who have access to product/animals?	☐	☐	☐
Are ingredient suppliers selected with consideration of food security measures implemented by the suppliers?	☐	☐	☐
Are vendors of compressed gas selected with consideration of food security measures implemented by vendors?	☐	☐	☐
Are vendors of packaging materials and labels selected with consideration of food security measures implemented by vendors?	☐	☐	☐

18. Does this establishment allow returned goods, including returns of U.S. exported products, to enter the plant?
    ☐ Yes
    ☐ No [GO TO QUESTION 20]

19. Which of the following security procedures does this establishment have in place for returned goods? *(Check "Yes" or "No" for each procedure.)*

	Yes	No
Are all returned goods examined for evidence of possible tampering before salvage or use in rework?	☐	☐
Are records maintained of returned goods used in rework?	☐	☐
Are returned goods reworked/examined at a separate designated location in the establishment to prevent potential cross-contamination of products?	☐	☐
Does the establishment follow the procedures outlined in FSIS Directive 9010.1 for return of U.S. exported products? http://www.fsis.usda.gov/oppde/rdad/fsisdirectives/9010-1.pdf	☐	☐

VII. WATER AND ICE SECURITY

Visit http://cfpub.epa.gov/safewater/watersecurity/index.cfm and www.epa.gov/region1/eco/drinkwater/pdfs/drinkingH2Ofactsheet.pdf for guidance material from the U.S. Environmental Protection Agency, Water Security.

20. Which of the following security procedures does this establishment have in place for its water and ice supply? *(Check "Yes" or "No" for each procedure, or "N/A" if the procedure is not applicable.)*

	Yes	No	N/A
Is access to water wells restricted? (e.g., by locked door/gate or limiting access to designated employees)	☐	☐	☐
Is access to ice-making equipment restricted?	☐	☐	☐
Is access to ice storage facilities restricted?	☐	☐	☐
Is access to storage tanks for potable water restricted?	☐	☐	☐
Is access to water reuse systems restricted?	☐	☐	☐
Are *potable* water lines periodically inspected for possible tampering? (i.e., visual inspection for physical integrity of infrastructure etc.)?	☐	☐	☐
Are *non-potable* water lines inspected for possible tampering (i.e., visual inspection for physical integrity of infrastructure, connection to potable lines, etc.)?	☐	☐	☐
Have arrangements been made with local health officials to ensure immediate notification of the plant if the potability of the public water supply is compromised?	☐	☐	☐

VIII. MAIL HANDLING SECURITY

21. Which of the following security procedures does this establishment have in place to ensure mail handling security?

	Yes	No	N/A
Is mail handling activity conducted in a separate room or facility away from in-plant food production/processing operations?	☐	☐	☐
Are mail-handlers trained to recognize and handle suspicious pieces of mail using U.S. Postal Service guidelines? http://www.usps.com/news/2001/press/serviceupdates.htm	☐	☐	☐

IX. PERSONNEL SECURITY

22. Which of the following security procedures does this establishment have in place for ensuring that establishment personnel adhere to the security requirements?
    (Check "Yes" or "No" for each procedure, or "N/A" if the procedure is not applicable [e.g., the establishment does not use contractors].)

	Yes	No	N/A
Are background checks or selective background checks conducted for new permanent staff who will be working in sensitive operations prior to hiring?	☐	☐	☐
Are background checks or selective background checks conducted for workers in sensitive operations for new temporary, seasonal, and contract employees prior to hiring?	☐	☐	☐

DOCUMENT APPENDIX

	Yes	No	N/A
Do all plant employees receive training on security procedures as part of their orientation training?	☐	☐	☐
Are procedures in place to ensure positive identification/recognition of all establishment employees?	☐	☐	☐
Are identification procedures in place to ensure the positive identification/recognition for temporary employees and contractors (including construction workers, cleaning crews, and truck drivers) in the establishment?	☐	☐	☐
Are procedures in place to screen employees entering the plant during *working* hours?	☐	☐	☐
Are procedures in place to screen entry of employees into the plant during *non-working* hours?	☐	☐	☐
Are procedures in place to screen the entry of contractors into the plant during *working* hours?	☐	☐	☐
Are procedures in place to screen entry of employees into the plant during *non-working* hours?	☐	☐	☐
Are procedures in place to restrict temporary employees and contractors (including construction workers, cleaning crews, and truck drivers) to areas of plant relevant to their work?	☐	☐	☐
Are procedures in place to ensure clear identification of personnel with their specific functions/assignments (e.g., colored garb)?	☐	☐	☐
Is an updated shift roster, i.e., who is absent, who the replacements are, and when new employees are being integrated into the workforce, distributed to supervisors at the start of each shift?	☐	☐	☐
Is a policy in place on what personal items may and may not be allowed inside the plant and within production areas?	☐	☐	☐
Are announced and unannounced inspections of employees' lockers conducted?	☐	☐	☐
Are employees and/or visitors restricted on what they can bring (cameras, etc.) into plant?	☐	☐	☐
Are employees prohibited from removing company-provided clothing or protective gear from the premises?	☐	☐	☐

APPENDIX—LIST OF RESOURCES

These resources contain security guidelines applicable to multiple sections of the checklist that establishments can adopt to enhance their capabilities to prevent intentional product tampering and to respond to threats or actual incidents of intentional product tampering. Additional resources with guidelines that apply only to specific sections are shown at appropriate sections throughout the document for easy access and reference.

FSIS Model Food Security Plans
http://www.fsis.usda.gov/Food_Security_&_Emergency_Preparedness/Security_Guidelines/index.asp#industry

FSIS "Security Guidelines for Food Processors"
http://www.fsis.usda.gov/oa/topics/SecurityGuide.pdf

World Health Organization (WHO) — "Terrorist Threats to Food — Guidelines for Establishing and Strengthening Prevention and Response Systems" (ISBN 92 4 154584 4)
http://www.who.int/foodsafety/publications/general/terrorism/en/

U. S. Food and Drug Administration (FDA) — "Food Security, Processors, and Transporters; Food Security Preventive Measures Guidance"
http://www.cfsan.fda.gov/~dms/secguid6.html

U.S. Food and Drug Administration (FDA)—"Retail Food Stores and Food Service Establishments; Food Security Preventive Measures Guidance"
http://www.cfsan.fda.gov/~dms/secgui11.html

Center for Disease Control and Prevention (CDC), National Institute of Occupational Safety and Health (NIOSH) —"Protecting Building Environments from Airborne Chemical, Biological, or Radiological Attacks"
http://www.cdc.gov/niosh/bldvent/2002-139.html

USDA, Food and Nutrition Service (FNS) "A Biosecurity Checklist for Food Service Programs, Developing a Biosecurity Management Plan"
http://schoolmeals.nal.usda.gov/Safety/FNSFoodSafety.html

Canadian Food Inspection Agency (CFIA)—"Suggestions for Improving Security"
http://www.inspection.gc.ca/english/ops/secur/protrae.shtml

Center for Infectious Disease Research and Policy (CIDRAP), Academic Health Center, University of Minnesota
http://www.cidrap.umn.edu/cidrap/content/biosecurity/food-biosec/guidelines

County of San Diego, Department of Environmental Health, "Guidelines for Food Safety and Security"
http://www.sdcounty.ca.gov/deh/fhd/pdf/food_safety_security_217.pdf

SECURITY GUIDELINES FOR AMERICAN ENTERPRISES ABROAD

Overseas Security Advisory Council

INTRODUCTION

This appendix is a compilation of security guidelines for American private sector executives operating outside the United States. This guidance is the product of many years of experience by a cross section of American security practitioners from both the public and private sectors. Obviously, the implementation should be consistent with the level of risk in the country where you conduct business. For the most part the guidelines are for protection in high threat areas. It is recognized that the level of risk varies from country to country and time to so that you may need to choose among the suggested options or apply the concepts in a manner modified to meet your needs. Since levels of risk can change very rapidly, it is advisable to continuously monitor factors that may impact the risk level. Security precautions must be flexible and dynamic to respond effectively to changing risks. A static, inflexible security posture will almost certainly result in a lack of preparedness or unnecessary expense.

The Department of State has three threat assessment designators: High, Medium, and Low. One of these three threat designators is applied to each country where the United States has diplomatic representation. Threat assessment information is available to the American business community in countries where the United States has diplomatic representation through the Regional Security Officer or Post Security Officer at the nearest U.S. diplomatic post, i.e. Embassy or Consulate. The level assigned to a particular country is determined by an analysis of the political, terrorist, and criminal environment of that country. It is reviewed quarterly by the Department of State and changed when appropriate.

A High Threat country is one where the threat is serious and forced entries and assaults on residents are common, or where an active terrorist threat exists. A Medium Threat country is one where the threat is moderate, with some forced entries and assaults on residents occurring, or where the area has the potential for terrorist activity. A Low Threat country is one where the threat is minimal and forced entry of residences and assault of occupants is not common, and there is no known terrorist threat.

For emphasis again, the guidelines set forth in this publication are generally most appropriate for High Threat areas. One will probably want to moderate them for applications where the risk is lower; or where other considerations preclude their implementation at the level discussed here. In many situations, professional technical security assistance will be required.

These guidelines emphasize site selection and operational security. Appendices I and II are checklists which will help you determine your security needs.

SITE SELECTION GUIDELINES

Need for Security Criteria

From a security point of view, proper site selection is the most important initial step to provide adequate protection. It is the intent of this appendix to bring to the attention of all responsible personnel the wide range of security matters that should be addressed and integrated into the site selection process for new office buildings and existing buildings.

Because of car bombings there are new criteria for site selection on a worldwide basis. Regardless of the geographic process, thereby preparing for what might happen during the life of the building or its occupancy. We have all seen how quickly a benign security situation can evolve into a significant threat to facilities. It is only prudent to incorporate adequate security measures based on an evaluation of the existing threat and the potential for a higher future threat level to protect your employees and visitors for the long term. It will be evident from the factors highlighted that security considerations will impact on operational matters. The implication of this fact may be greater in some geographic regions than in others and will certainly affect some more seriously than others. Where this is the case, it is incumbent on all interested parties to evaluate potential damage while engaged in the site selection process and balance it against security requirements. If, in high threat areas, many of the suggested key criteria cannot be met the firm should consider choosing another, more secure location.

Everyone involved in site selections should be aware of the following suggested criteria for facilities.

New Office Building

Topography
Site ideally should be situated at the high point, if any, of a land tract, which makes it less vulnerable to weapons fire, makes egress/ingress more difficult and easier to detect or observe any intrusions.

Siting
Site should be located away from main thoroughfares and provide for the following:

- 100 feet minimum setback from the building to perimeter walls and vehicular entrances to the building.
- Sufficient parking space for personnel outside the compound in a secure area within sight of the building, preferably, immediately adjacent to the compound.

DOCUMENT APPENDIX

- Sufficient parking space for visitors near the site but not on the site itself.
- Sufficient space to allow for the construction of a vehicular security control checkpoint (lock-type system), which would allow vehicles to be searched, if deemed necessary, and cleared without providing direct access to the site.
- Sufficient space to allow for the construction of a pedestrian security control checkpoint (gatehouse/booth) to check identification, conduct a package check or parcel inspection or carry out visitor processing before the pedestrian is allowed further access to the site. If a need for a thorough check of purses and briefcases, as well as items carried on a person may be required, sufficient space for a Walk-Through Metal Detector (WTMD) should be considered. Walking through a WTMD is less intrusive than a personal search or even one conducted with a hand-held detector.
- Sufficient space for construction of a 9-foot outer perimeter barrier or wall.

Environmental Considerations

Site should be located in a semi-residential, semi-commercial area where local vehicular traffic flow patterns do not impede access to or from the site.

Existing Office Building

The following security considerations for high-rise buildings are listed in order of preference as the availability of local facilities dictate:

- A detached (free-standing) building and site entirely occupied and controlled by you.
- A semidetached office building that is entirely occupied by you.
- A nondetached office building that is entirely occupied and controlled by you.
- A detached (free-standing) office building in which the uppermost floors are entirely occupied and controlled by you.
- A semidetached office building in which the uppermost floors are entirely occupied and controlled by you.
- A nondetached office building in which the uppermost floors are entirely occupied and controlled by you.
- A detached (free-standing) office building in which the central floors are entirely occupied and controlled by you.
- A semidetached office building in which the central floors are entirely occupied and controlled by you.
- A nondetached office building in which the central floors are entirely occupied and controlled by you.
- A detached (free-standing) office building in which some floors are occupied and controlled by you.
- A semidetached office building in which some floors are occupied and controlled by you.

- A nondetached office building in which some floors are occupied and controlled by you.

Common Requirements

Both new and existing office buildings should be capable of accommodating these security items:

- Floor load capacity must be able to maintain the additional weight of public access control (PAC) equipment (ballistic doors, walls, windows), security containers, and disintegrators and shredders, if needed.
- Exterior walls must be smooth shell, sturdy, and protected to a height of 16 feet to prevent forced entry.
- Building must be conducive to grilling or eliminating all windows below 16 feet.

The previously listed criteria should be adopted to provide satisfactory protection for employees and visitors. If the site is found to be deficient in some areas, attempt to resolve those deficiencies by instituting security measures that will negate the deficiencies. Professional security and/or engineering assistance should be considered to address unique situations.

At a minimum, the following general security measures should be incorporated into planning designs: perimeter controls, grillwork, and shatter-resistant film for windows, public access controls, package search and check, secured area, provisions for emergency egress, and emergency alarms and emergency power.

Standards of Design for Site and Building Security

This section establishes the minimum physical security standards to be incorporated in the design of facilities.

The intent is to provide protection for assets, personnel, property, and customers; ensure that consistent security measures are used at various locations; and ensure design integrity and compatibility of all elements of security with the architecture of the site.

Labor-saving and state-of-the-art security system components and assemblies should be used in all U.S. activities operating overseas, provided they can be maintained locally and there are spare parts available locally.

For manufacturing plant and laboratory facilities, security equipment such as closed-circuit television (CCTV) cameras and monitors, intercoms, card readers, and special glass protection, should be considered. Special care should be taken to verify the vendor's references, especially as they pertain to the quality of alarms, a visit should be made to the central station to observe the professionalism of the operation. Design, purchase, and installation should be coordinated through your architect. Bear in mind, and make provisions for, the cost of maintenance

DOCUMENT APPENDIX

on your security equipment. In some locations overseas, security equipment may be less expensive and more reliable than guards who receive relatively low pay and little training.

Security Design Objectives

In designing business or activity sites, roadways, buildings, and interior space, the following functional security objectives should be achieved:

- Physical and psychological boundaries (signs, closed doors, etc.) should establish four areas with increasing security controls beginning at the property boundaries. The areas are defined as:
  - perimeter—property boundaries;
  - exterior—lobbies/docks;
  - interior—employee space; and
  - restricted—laboratories, computer rooms, etc.
- Vehicular traffic signs should clearly designate the separate entrances for trucks/deliveries and visitors and employee vehicles. Where feasible control points should be provided near the site boundaries. Sidewalks should channel pedestrians toward controlled lobbies and entrances.
- Avoid having unsecured areas where there is no one nearby with responsibility for the function of the areas.

EXTERIOR PROTECTION

Perimeter Security

Walls, Fences, Berms, etc.
The overall design for perimeter security should consider using natural barriers, fencing, landscaping, or other physical or psychological boundaries to demonstrate a security presence to all site visitors.

If the threat is considered to be high at free-standing facilities, there should be a smooth faced perimeter wall or combination wall/fence, a minimum of 9 feet tall and extending 3 feet below grade. The wall or fence may be constructed of stone, masonry, concrete, chain link, or steel grillwork. However, if space limitations and local conditions dictate the need, any newly constructed wall should be designed to prevent vehicle penetration, and should use a reinforced concrete foundation wall, 18 inches thick with an additional 1-1/2 inches of concrete covering on each side of the steel reinforcement, and extending 36 inches above the grade. This type of wall is designed to support three wall toppings: masonry, concrete, or steel picket fencing. The toppings should be securely anchored into the foundation wall. If a picket fence is used instead of a wall, the upright supports should be spaced at least 9 feet apart so that the fence, if knocked down, can not be used as a ladder. In addition, intrusion alert systems can be used to enhance perimeter security.

In cases where the above standards of construction are neither feasible, fiscally prudent, nor required by the threat, alternative methods offering comparable protection can be used. These alternatives should maximize the use of locally available materials and conditions to take advantage of existing terrain features or by the creative use of earth berms and landscaping techniques such as concrete planters.

Inside the perimeter barrier, the building should be set back on the property to provide maximum distance from that portion of the perimeter barrier which is accessible by vehicle. The desirable distance of the setback is at least 100 feet depending on the bomb resistance provided by the barrier.

At facilities with less than optimum barriers, or at locations where the terrorist threat or building location increases the vulnerability to vehicular attack, bollards[1] or cement planters can be used to strengthen the perimeter boundary. At walled or fenced facilities with insufficient setback, bollards or planters can be installed outside the perimeter to increase the setback of the buildings.

(In any event, whether at a walled facility or a nonwalled one as discussed below, the design and placement of bollards or other antivehicular devices should be considered in the early planning stages. It would prevent having impenetrable gates connected by easily penetrated walls, or necessitate relocating because local authorities forbid the construction of required barriers.)

Nonwalled Facilities Barriers

In locations without perimeter wall protection, buildings should be protected with bollards, cement planters, or any other perimeter protection device. Such devices should be placed in a manner as to allow the maximum distance between the building and the roadway and/or vehicle access area. They should be positioned to impede vehicular access to lobbies and other glassed areas that could be penetrated by a vehicle (i.e., low or no curb, glass wall or door structure between lobby and driveway). Driveways should be designed and constructed to minimize or preclude high-speed vehicular approaches to lobbies and glassed areas. (There may be local ordinances that make placement of these devices illegal or ineffective.)

A positive and concerted effort should be made to contact local host country law enforcement or governmental authorities and request that they prohibit, restrict, or impede motor vehicles from parking, stopping, or loading in front of the facility.

In high threat locations, if local conditions or government officials prohibit antivehicular perimeter security measures and your business is either the sole occupant of the building or located on the first or second floor, you should consider relocating to more secure facilities.

Building Exterior

Facade

The building exterior should be a sheer/smooth shell, devoid of footholds, decorative lattice work, ledges, or balconies. The building facade should be protected to a height of 16 feet to prevent access by intruders using basic handtools. The use of glass on the building facade should be kept to an absolute minimum, only being used for standard size or smaller windows and, possibly, main entrance doors. All glass should be protected by plastic film. Consider the use of lexan or other polycarbonate as alternatives to glass where practical.

External Doors

Local fire codes may impact on the guidance presented here. As decisions are made on these issues, local fire codes will have to be considered.

Main entrance doors may be either transparent or opaque and constructed of wood, metal, or glass. The main entrance door should be equipped with a double-cylinder dead bolt and additionally secured with crossbar or sliding dead bolts attached vertically to the top and bottom of each leaf. All doors, including interior doors, should be installed to take advantage of the doorframe strength by having the doors open toward the attack side.

All other external doors should be opaque hollow metal fire doors with no external hardware. These external doors should be single doors unless used for delivery and loading purposes.

Should double doors be required, they should be equipped with two sliding dead bolts on the active leaf and two sliding dead bolts on the inactive leaf vertically installed on the top and bottom of the doors. A local alarmed panic bar and a 180-degree peephole viewing device should be installed on the active leaf.

All external doors leading to crawl spaces or basements must be securely padlocked and regularly inspected for tampering.

Windows

The interior side of all glass surfaces should be covered with a protective plastic film that meets or exceeds the manufacturer's specifications for shatter-resistant protective film. A good standard is 4-millimeter thickness for all protective film applications. This film will keep glass shards to a minimum in the event of an explosion or if objects are thrown through the window.

Grillwork should be installed on all exterior windows and air-conditioning units that are within 16 feet of grade or are accessible from roofs, balconies, etc. The rule of thumb here is to cover all openings in excess of 100 square inches if the smallest dimension is 6 inches or larger.

Grillwork should be constructed of 1/2 inch diameter or greater steel rebar, anchored or imbedded (not bolted) into the window frame or surrounding masonry to a depth of 3 inches. Grillwork should be installed horizontally and vertically on center at no more than 8 inch intervals. However, grillwork installed in exterior window frames within the secure area should be spaced 5 inches on center, horizontally and vertically, and anchored in the manner described previously. Decorative grillwork patterns can be used for aesthetic purposes.

Grillwork that is covering windows designated as necessary for emergency escape should be hinged for easy egress. All hinged grillwork should be secured with a key operated security padlock. The key should be maintained on a cup hook in close proximity of the hinged grille, but out of reach of an intruder. These emergency escape windows should not be used in planning for fire evacuations.

Roof

The roof should be constructed of fire-resistant material. All hatches and doors leading to the roof should be securely locked with dead-bolt locks. Security measures such as barbed, concertina or tape security wire, broken glass, and walls or fences may be used to prevent access from nearby trees and/or adjoining roofs.

Vehicular Entrance and Controls

Vehicular Entrance

Vehicular entry-exit points should be kept to a minimum. Ideally, to maximize traffic flow and security, only two regularly used vehicular entry-exit points are necessary. Both should be similarly constructed and monitored. The use of one would be limited to employees' cars, while the other would be used by visitors and delivery vehicles. Depending on the size and nature of the facility, a gate for emergency vehicular and pedestrian egress should be installed at a location that is easily and safely accessible by employees. Emergency gates should be securely locked and periodically checked. All entry-exit points should be secured with a heavy duty sliding steel, iron, or heavily braced chain link gate equipped with a heavy locking device.

The primary gate should be electrically operated (with a manual back-up by a security officer situated in an adjacent booth). The gate at the vehicle entrance should be positioned to avoid a long straight approach to force approaching vehicles to slow down before reaching the gate. The general technique employed is to require a sharp turn immediately in front of the gate.

In addition to the gate, and whenever justifiable, a vehicular arrest system can be installed. An appropriate vehicle arrest system, whether active, a piece of equipment designed to stop vehicles in their tracks, or passive, a dense mass, will be able to stop or instantly disable a vehicle with a minimum gross weight of 15,000 pounds traveling 50 miles per hour.

Vehicular Control

General. All facilities should have some method of vehicle access control. Primary road entrances to all major plant, laboratory, and office locations should have a vehicle control facility capable of remote operation by security personnel with automated systems.

- At smaller facilities, vehicle access control may be provided by badge-activated gates, manual swing gates, etc.
- Site security should be able to close all secondary road entrances thereby limiting access to the primary entrance. Lighting and turn space should be provided as appropriate.

Control Features. Primary perimeter entrances to a facility should have a booth for security personnel during peak traffic periods and automated systems for remote operations during other periods.

Capabilities are:

- Electrically-operated gates to be activated by security personnel at either the booth or security control center or by a badge reader located in a convenient location for a driver;
- CCTV with the capability of displaying full-facial features of a driver and vehicle characteristics on the monitor at security control center;
- An intercom system located in a convenient location for a driver to communicate with the gatehouse and security control center;
- Bollards or other elements to protect the security booth and gates against car crash;
- Sensors to activate the gate, detect vehicles approaching and departing the gate, activate a CCTV monitor displaying the gate, sound an audio alert in the security control center;
- Lighting to illuminate the gate area and approaches to a higher level than surrounding areas;
- Signs to instruct visitors and to post property as required;
- Road surfaces to enable queuing, turnaround, and parking;
- Vehicle bypass control (i.e., gate extensions), low and dense shrubbery, fences, and walls.

Booth Construction and Operation

As noted previously, at the perimeter vehicular entry-exit a security officer booth should be constructed to control access. (At facilities not having perimeter walls, the security officer booth should be installed immediately inside the facility foyer.)

If justified by the threat level the security officer booth should be completely protected with reinforced concrete, walls, ballistic doors, and windows. The booth should be equipped with a security officer duress alarm and intercom system, both annunciating at the facility receptionist and security officer's office. This security officer would also be responsible for complete operation of the vehicle gate. If necessary, package inspection and visitor screening may be conducted just outside of the booth by an unarmed security officer equipped with walk-through and hand-held metal detectors. Provisions for the environmental comfort should be considered when designing the booth.

Parking

General. Security should be considered in the location and arrangement of parking lots. Pedestrians leaving parking lots should be channeled toward a limited number of building entrances.

All parking facilities should have an emergency communication system (intercom, telephones, etc.) installed at strategic locations to provide emergency communications directly to Security.

Parking lots should be provided with CCTV cameras capable of displaying and videotaping lot activity on a monitor in the security control center. Lighting must be of adequate level and direction to support cameras while, at the same time, giving consideration to energy efficiency and local environmental concerns.

If possible, parking on streets directly adjacent to the building should be forbidden. Wherever justifiable given the threat profile of your company, there should be no underground parking areas in the neither building basement nor ground-level parking under building overhangs.

Within Perimeter Walls/Fences. All parking within perimeter walls or fences should be restricted to employees, with spaces limited to an area as far from the building as possible. Parking for patrons and visitors, except for predesignated VIP visitors, should be restricted to outside of the perimeter wall/fences.

Garages. For those buildings having an integral parking garage or structure, a complete system for vehicle control should be provided. CCTV surveillance should be provided for employee safety and building security. If the threat of car bombing is extant, consideration must be given to prohibiting parking in the building.

Access from the garage or parking structure into the building should be limited, secure, well lighted, and have no places of concealment. Elevators, stairs, and connecting bridges serving the garage or parking structure should discharge into a staffed or fully monitored area. Convex mirrors should be mounted outside the garage elevators to reflect the area adjacent to the door openings.

Exterior Lighting

Exterior lighting should illuminate all facility entrances and exits in addition to parking areas, perimeter walls, gates, courtyards, garden areas, and shrubbery rows.

DOCUMENT APPENDIX

Lighting of building exterior and walkways should be provided where required for employee safety and security. Regarding building facades, there should be a capability to illuminate them 100% to a height of at least 6 feet.

Although sodium vapor lights are considered optimum for security purposes, the use of incandescent and florescent light fixtures is adequate. Exterior fixtures should be protected with grillwork when theft or vandalism have been identified as a problem.

For leased buildings, landlord approval of exterior lighting design requirements should be included in lease agreements.

Building Access

Building Entrances

The number of building entrances should be minimized, relative to the site, building layout, and functional requirements. A single off-hours entrance near the security control center is desirable. At large sites, additional secured entrances should be considered with provisions for monitoring and control.

Door Security Requirements

- All employee entrance doors should permit installation of controlled access system hardware. The doors, jambs, hinges and locks must be designed to resist forced entry (e.g., spreading of door frames, accessing panic hardware, shimming bolts and/or latches, fixed hinge pins). Don't forget handicap requirements when applicable.
- Minimum requirement for lock cylinders are "6-pin" pin-tumbler-type. Locks with removable core cylinders to permit periodic changing of the locking mechanism should be used.
- All exterior doors should have alarm sensors to detect unauthorized openings.
- Doors designed specifically for emergency exits need to have an alarm that is audible at the door with an additional annunciation at the security control center. These doors should have no exterior hardware on them.

Window Precautions

- For protection, large showroom type plate glass and small operable windows on the ground floor should be avoided. If, however, these types of windows are used and the building is located in a high-risk area, special consideration should be given to the use of locking and alarm devices, laminated glass, film, or polycarbonate glazing.
- For personnel protection, all windows should have shatter-resistant film.

- (For a more extensive discussion of windows and how to secure them, as well as guidance for securing windows which may be used for emergency exit, see "Windows" on page 639).

Lobby. Main entrances to buildings should have space for a receptionist during the day and a security officer at night. The security control center should be located adjacent to the main entrance lobby and should be surrounded by professionally designed protective materials.

The lobby-reception area should be a single, self-sufficient building entrance. Telephones and rest rooms to meet the needs of the public should be provided in this area without requiring entry into interior space. Rest rooms should be kept locked in high-threat environments and access controlled by the receptionist.

Consistent with existing risk level, the receptionist should not be allowed to accept small parcel or courier deliveries routinely unless they are expected by addressee.

Other Building Access Points

Other less obvious points of building entry, such as grilles, grating, manhole covers, areaways, utility tunnels, mechanical wall, and roof penetrations should be protected to impede and/or prevent entry into the building.

Permanent exterior stairs or ladders from the ground floor to the roof should not be used, nor should the building facade allow a person to climb up unaided. Exterior fire escapes should be retractable and secured in the up position.

Construction Activities

Landscaping and other outside architectural and/or aesthetic features should minimize creating any area that could conceal a person in close proximity to walkways, connecting links, buildings, and recreational spaces.

Landscaping design should include CCTV surveillance of building approaches and parking areas.

Landscape plantings around building perimeters need to be located at a minimum of 4 feet from the building wall to prevent concealing of people or objects.

INTERIOR PROTECTION

Building Layout

Building space can be divided into three categories: public areas, interior areas, and security or restricted areas requiring special security measures. These areas should be separated from one another within the building with a limited number of controlled passage points between the areas. "Controlled" in this context can allow or deny passage by any means deemed necessary (i.e., locks, security officers, etc.).

Corridors, stairwells, and other accessible areas should be arranged to avoid places for concealment.

Generally, restricted space should be located above the ground-floor level, away from exterior walls, and away from hazardous operations. Access to restricted space should be allowed only from interior space and not from exterior or public areas. Exit routes for normal or emergency egress should not transit restricted or security space.

Walls and Partitions

Public space should be separated from interior space and restricted space by slab-to-slab partitions. When the area above a hung ceiling is used as a common air return, provide appropriate modifications to walls or install alarm sensors. In shared occupancy buildings, space should be separated by slab-to-slab construction or as described previously.

Doors

Normally, interior doors do not require special features or provisions for locking.

In shared occupancy buildings, every door leading to interior space should be considered an exterior door and designed with an appropriate degree of security.

Stairway doors located in multitenant buildings must be secured from the stairwell side (local fire regulations permitting) and always operable from the office side. In the event that code prevents these doors from being secured, the floor plan should be altered to provide security to your space.

Emergency exit doors that are designed specifically for that purpose should be equipped with a local audible alarm at the door and a signal at the monitoring location.

Doors to restricted access areas should be designed to resist intrusion and accommodate controlled-access hardware and alarms.

Doors on building equipment and utility rooms, electric closets, and telephone rooms should be provided with locks having a removable core, as is provided on exterior doors. As a minimum requirement, provide 6-pin tumbler locks.

For safety reasons, door hardware on secured interior doors should permit exit by means of a single knob or panic bar.

Other Public Areas

The design of public areas should prevent concealment of unauthorized personnel and/or objects.

Ceilings in lobbies, rest rooms, and similar public areas should be made inaccessible with securely fastened or locked access panels installed where necessary to service equipment.

Public rest rooms and elevator lobbies in shared occupancy buildings should have ceilings that satisfy your security requirements.

Special Storage Requirements

Building vaults or metal safes may be required to protect cash or negotiable documents, precious metals, classified materials, etc. Vault construction should be made of reinforced concrete or masonry and be resistant to fire damage. Steel vault doors are available with various fire-related and security penetration classifications.

Elevators

All elevators should have emergency communications and emergency lighting. In shared occupancy buildings, elevators traveling to your interior space should be equipped with badge readers or other controls to prohibit unauthorized persons from direct entry into your interior space. If this is not feasible, a guard, receptionist or other means of access control may be necessary at each entry point.

Cable Runs

All cable termination points, terminal blocks, and/or junction boxes should be within your space. Where practical, enclose cable runs in steel conduit.

Cables passing through space that you do not control should be continuous and installed in conduit. You might even want to install an alarm in the conduit. Junction boxes should be minimized and fittings spot welded when warranted.

Security Monitoring

Security Control Center

If you have a security control center, it should have adequate space for security personnel and their equipment. Additional office space for technicians and managers should be available adjacent to the control center.

Your security control center should provide a fully integrated console designed to optimize the operator's ability to receive and evaluate security information and initiate appropriate response actions for (1) access control, (2) CCTV, (3) life safety, (4) intrusion and panic alarm, (5) communications, and (6) fully zoned public address system control.

The control center should have emergency power and convenient toilet facilities. Lighting should avoid glare on TV monitors and computer terminals. Sound-absorbing materials should be used on floors, walls, and ceilings. All security power should be backed up by an emergency electrical system.

The control center should be protected to the same degree as the most secure area it monitors.

Controlled Access System

This type of system, if used, should include the computer hardware, monitoring station terminals, sensors, badge readers, door control devices, and the necessary communication links (leased line, digital dialer, or radio transmission) to the computer.

In addition to the normal designated access control system's doors and/or gates, remote access control points should interface to the following systems: (1) CCTV, (2) intercom, and (3) door and/or gate release.

Alarm Systems

Sensors should be resistant to surreptitious bypass. Door contact monitor switches should be recessed wherever possible. Surface-mounted contact switches should have protective covers.

Intrusion and fire alarms for restricted areas should incorporate a backup battery power supply and be on circuits energized by normal and emergency generator power.

Control boxes, external bells, and junction boxes for all alarm systems should be secured with high-quality locks and electrically wired to cause an alarm if opened.

Alarm systems should be fully multiplexed in large installations. Alarm systems should interface with the computer-based security system and CCTV system.

Security sensors should individually register an audio-visual alarm (annunciator or computer, if provided) located at the security central monitoring location and alert the security officer. A single-CRT display should have a redundant printer or indicator light. A hard-wired audible alarm that meets common fire code standards should be activated with distinguishing characteristics for fire, intrusion, emergency exit, etc. All alarms ought to be locked in until reset manually.

Closed-Circuit TV (CCTV)

CCTV systems should permit the observation of multiple camera transmission images from one or more remote locations.

Switching equipment should be installed to permit the display of any camera on any designated monitor.

Hardware

To ensure total system reliability, only high quality security hardware should be integrated into the security system.

Stairwell Door Reentry System

In multitenant high-rise facilities, stairwell doors present a potential security problem. These doors must be continuously operable from the office side into the stairwells. Reentry should be controlled to permit only authorized access and prevent entrapment in the stairwell.

Reentry problems can be fixed if you provide locks on all stairwell doors except the doors leading to the first floor (lobby level) and approximately every fourth or fifth floor, or as required by local fire code requirements. Doors without these locks should be fitted with sensors to transmit alarms to the central security monitoring location and provide an audible alarm at the door location. Appropriate signs should be placed within the stairwells. Doors leading to roofs should be secured to the extent permitted by local fire code.

Special Functional Requirements

Facilities with unique functions may have special security requirements in addition to those stated in this booklet. These special requirements should be discussed with Corporate Security personnel or a security consultant. Typical areas with special requirements are product centers, parts distribution centers; sensitive parts storage facilities, customer centers, service exchange centers, etc.

PUBLIC ACCESS CONTROLS (PAC)

Security Officers and Watchmen

All facilities of any size in threatened locations should have manned 24 hour internal protection. Security Officers should be uniformed personnel and, if possible, placed under contract. They should be thoroughly trained, bilingual and have complete instructions in their native language clearly outlining their duties and responsibilities. These instructions should also be printed in English for the benefit of American supervisory personnel. If permitted by local law/customs, investigations or checks into the backgrounds of security officers should be conducted.

At facilities with a perimeter wall, there should be one 24 hour perimeter security officer post. If the facility maintains a separate vehicular entrance security officer post, such a post should be manned from 1 hour before to 1 hour after normal business hours and during special events. Security officers should be responsible for conducting package inspections, package check-in, and, if used, should operate the walk-through and hand-held metal detectors. Security officers should also be responsible for inspecting local and international mail delivered to the facility, both visually and with a hand-held metal detector before it is distributed. X-ray equipment for package inspection should be employed if the level of risk dictates.

At facilities with a perimeter guardhouse, the walk-through metal detector could be maintained and operated in an unsecured pass-through portion of the guardhouse. In addition, this security officer could also be responsible for conducting package inspections. When there is sufficient room to store packages at the guardhouse, checked packages should be stored here—new guardhouses should provide for such storage. If package storage at the guardhouse is not feasible, then it should be in shelves in the foyer under the direction of the foyer security officer or receptionist. Generally, security screening and package storage is carried out in the foyer.

Security Hardline

Office areas should be equipped with a "hardline" to provide physical protection from unregulated public access. Protection should be provided by a forced-entry-resistant hardline that meets ballistic protection standards. These standards can be obtained from your corporate personnel or a security consultant. When a security hardline for Public Access Control (PAC) is constructed, the following criteria should apply:

Walls

Walls comprising a PAC should be constructed of no less than 6 inches of reinforced concrete from slab to slab. The reinforcement should be of at least Number 5 rebar spaced 5 inches on center, horizontally and vertically, and anchored in both slabs. In existing buildings, the following are acceptable substitutions for 5-inch reinforced concrete hardlines:

- Solid masonry, 6 inches thick or greater, with reinforcing bars horizontally and vertically installed;
- Solid unreinforced masonry or brick, 8 inches thick or greater;
- Hollow masonry block, 4 to 8 inches thick with 1/4 inch steel backing;
- Solid masonry, at least 6 inches thick, with 1/4 inch steel backing;
- Fabricated ballistic steel wall, using two 1/4 inch layers of sheet steel separated by tubular steel studs;
- Reinforced concrete, less than 6 inches thick with 1/8 inch steel backing.

Security Doors

Either opaque or transparent security doors can be used for PAC doors. All doors should provide a 15 minute forced entry penetration delay. In addition, doors should be ballistic resistant.

The PAC door should be a local access control door, meaning a receptionist or security officer can remotely open the door.

Security Windows

Whenever a security window or teller-window is installed in the hardline, it should meet the 15 minute forced entry and standard ballistic resistance requirements.

PAC Entry Requirements

No visitor should be allowed to enter through the hardline without being visually identified by a security officer, receptionist, or other employee stationed behind the hardline. If the identity of the visitor cannot be established, the visitor must be escorted at all times while in the facility.

Alarms and Intercoms

A telephone intercom between the secure office area, the foyer security officer, and guardhouse should be installed. In facilities where deemed necessary, a central alarm and public address system should be installed to alert staff and patrons of an emergency situation. Where such a system is required, the primary control console should be located in the security control center. Keep in mind that alarms without emergency response plans may be wasted alarms. Design, implement, and practice emergency plans.

Secure Area

Every facility should be equipped with a secure area for immediate use in an emergency situation. This area is not intended to be used for prolonged periods of time. In the event of emergency, employees will vacate the premises as soon as possible. The secure area, therefore, is provided for the immediate congregation of employees at which time emergency exit plans would be implemented.

The secure area should be contained within the staff office area, behind the established hardline segregating offices from public access. An individual office will usually be designated as the secure area. Entrance into the secure area should be protected by a solid core wood or hollow metal door equipped with two sliding dead bolts.

Emergency egress from the secure area will be through an opaque 15 minute forced-entry-resistant door equipped with an alarmed panic bar or through a grilled window, hinged for emergency egress. The exit preferably will not be visible from the facility's front entrance.

EMERGENCY EXIT

All facilities should have a means of emergency escape aside from the secure area exit. Positioned appropriately throughout the building should be sufficient emergency exit points to accommodate normal facility occupancy.

All emergency doors should be hollow metal doors (fire doors where appropriate) equipped with alarmed emergency exit panic bars.

DOCUMENT APPENDIX

Emergency factors regarding windows are described on page 640.

COMMUNICATIONS

Communications Facilities

Satellite ground stations, microwave parabolic reflectors, and communications towers and supports should be located on rooftops, with limited access to the public. Where this is not possible, the equipment should be installed with fences and alarms. Closed circuit television (CCTV) with video recording capability should be considered and included where justified.

Communications

Telephone systems should incorporate an external direct line telephone link for security and life safety independent of the internal telephone network dedicated to the location. This line should feed into the secure area.

Communications considerations should provide radio transmission equipment for communications between security personnel.

Intercom systems should have the capacity to accommodate all remote access control points.

Systems Integration

Security systems in new buildings or buildings undergoing renovation should be installed with distributed wiring schemes that use local telecommunication closets as distribution points. This will provide expansion capability, future networking capability, ease of maintenance, and full function implementation of the security system. At a minimum, the communications link and interface between the sensor, output devices, and computers should include conduit, multiconductor twisted shielded cable and terminal cabinets. However, recent technology such as fiber-optic cables should be considered in planning the wiring distribution scheme. Data distribution and gathering closets used for security wiring must be secure. Where possible, integrate security wiring with other systems such as telephone, paging, energy management, etc. In every case, the design of the communications link should permit ready installation and interconnection of cameras, sensors, and other input-output devices. All life safety equipment and accessories should be Underwriters Laboratory (UL) approved.

Outlying facilities should link security systems to the nearest security control center. All new systems should be compatible with existing systems or the existing system should be replaced with the new system.

OFFICE SECURITY GUIDELINES

General Procedures

Any employee, but especially the executive, can be a target of terrorist or criminal tactics and forced entry, building occupation, kidnapping, sabotage, and even assassination. Executive offices can be protected against attacks.

The executive office should have a physical barrier such as electromagnetically operated doors, a silent trouble alarm button, with a signal terminating in the Plant Protection Department or at the secretary's desk, and close screening of visitors at the reception and security officer desks in the lobbies and again at the executive's office itself. Secretaries should not admit visitors unless positively screened in advance or known from previous visits. If the visitor is not known and/or not expected, he or she should not be admitted until satisfactory identification and a valid reason to be on site is established. In such instances, Security should be called and an officer asked to come to the scene until the visitor establishes a legitimate reason for being in the office. If the visitor cannot do so, the officer should be asked to escort the visitor out of the building.

Unusual telephone calls, particularly those in which the caller does not identify himself/herself or those in which it appears that the caller may be misrepresenting himself/herself, should not be put through to the executive. Note should be made of the circumstances involved (i.e., incoming line number, date and time, nature of call, name of caller). This information should then be provided to the Security Department for follow-up investigation.

Under no circumstances should an executive's secretary reveal to unknown callers the whereabouts of the executive, his/her home address, or telephone number.

The executive, when working alone in the evening, on weekends, or holidays, should advise Security how long he/she will be in the office and check out with Security when leaving.

Security in the Office

American enterprises, particularly those in foreign countries, have been and will continue to be the subject of controversial political and economic issues that can turn their executives and offices into targets for terrorists and criminal actions. Countermeasures against these acts can and should be implemented in the office environment. The following list describes some of the measures that may be useful in improving personal security and safety at the office.

- Avoid working alone late at night and on days when the remainder of the staff is absent.
- The office door should be locked when you vacate your office for any lengthy period, at night and on weekends. Do not permit the secretary to leave keys to the office or desk.

DOCUMENT APPENDIX

- There should be limited access to the executive office area.
- Arrange office interiors so that strange or foreign objects left in the room will be immediately recognized.
- Unescorted visitors should not be allowed admittance nor should workmen without proper identification and authorization.
- Implement a clean desk policy. Do not leave papers nor travel plans on desk tops unattended.
- Control publicity in high-risk areas. Avoid identification by photographs for news release. Maintain a low profile.
- Janitorial or maintenance activity in key offices and factory areas should be supervised by competent company employees.
- A fire extinguisher, first-aid kit, and oxygen bottle should be stored in the office area.
- The most effective physical security configuration is to have doors locked (from within) with one visitor access door to the office area.
- Where large numbers of employees are involved, use the identification badge system containing a photograph.

Advice for Secretaries

A secretary has close knowledge of schedules and company business. He/she should be instructed to maximize security, and the following precautionary measures should be reviewed with him/her:

- Be alert to strangers visiting the executive without an appointment and who are unknown to him/her.
- Be alert to strangers who loiter near the office.
- Do not reveal the executive's whereabouts to unknown callers. Even if the caller is known, the information should be on a need-to-know basis. As a standard policy, take a number where the caller can be contacted. Do not give out home telephone numbers or addresses.
- When receiving a threatening call, including a bomb threat, extortion threat, or from a mentally disturbed individual, remain calm and listen carefully. Each secretary and/or receptionist should have a threatening telephone call checklist which should be completed as soon as possible. A boiler-plate checklist is attached as Appendix III.
- Keep executive travel and managers' travel itineraries confidential. Strictly limit distribution to those with a need to know.
- Incinerate, disintegrate or shred notes, drafts, correspondence and any and all material which reveal an executive's travel plans, itineraries, home address and telephone number, invitations and responses thereto or any other data about his/her whereabouts, including information about past trips which could indicate habitual contacts and travel patterns. Do not place such material in trash cans.

- Observe caution when opening mail. A list of things to look for is included in Appendix IV. You should post this list in your mail handling facility. (All persons handling mail should be made aware of the aforementioned basic signs found in Appendix IV. The mail handlers should have available an established procedure in the event that any of the above signs are found. It is also important not to accept packages from strangers until satisfied with the individual's identity and the nature of the parcel.)

Precautions for All

Money, valuables, and important papers such as passports should not be kept in a your desk. Thefts will occur in all offices, even during working hours. Some will be solved, most could have been prevented. The following suggestions will decrease the chance of further thefts:

- Do not tempt thieves by leaving valuables or money unsecured.
- If sharing an office or suite of offices, stagger lunch hours and coffee breaks so that the office is occupied at all times.
- If the office must be left vacant, lock the door.
- Locate desks in a way that persons entering the office or suite can be observed.
- Follow a clean desk policy before leaving at night. Keep valuables and company documents in locked containers.
- Confirm work to be done or property to be removed by Maintenance, outside service personnel, or vendors.
- Do not "hide" keys to office furniture under flower pots, calendars, etc. Thieves know all the hiding places. Do not label keys except by code.

VEHICULAR AND TRAVEL SECURITY

Vehicle and Travel Security

Threats of terrorism and kidnapping are serious problems involving all aspects of security management; effective management dictates that available resources be used wisely and concentrated on security weak points. Terrorists are very quick to identify the security vulnerabilities of business, family, and pleasure travel. At their best, protection strategies dealing with vehicles and travel are perhaps the hardest to formulate, and the advantage tends to be with the terrorist. Current statistics indicate that the greatest danger from acts of terrorism occurs while the executive is traveling to or from the office and just before reaching his/her destination.

    The inherent security problems of passenger vehicle travel are many. Vehicles are easily recognized by year, make, and model, and the trained terrorist can accurately assess any protection modifications and security devices. Using adequate resources, vehicles can be discreetly followed; therefore, making possible repeated dry runs of potential attacks with very low risk of detection. Under these conditions, different methods of attack can be formulated and tested until success is ensured. While traveling in a passenger vehicle, the executive has limited protection resources upon which to rely and often is dependent on fixed security manpower. This makes it easier for terrorist groups, which are geared to mobility, to ensure numerical superiority.

The attack potential against the executive in travel rests heavily on psychological instability and human weakness. The shock of surprise attack is greatest at points of changing surroundings, crossroads, and when entering or exiting vehicles. These are situations of constant change and points of activity where the executive has a tendency to be mentally off balance. Vehicles are often left in driveways, on streets, at service centers, and other isolated areas with no form of control or protection, allowing easy access to terrorists. Through illegal entry to the vehicle, the terrorist can gain a number of attack points; sabotage with the intent to maim and injure, sabotage with the intent of execution, and sabotage to ensure the success of future attacks. These psychological factors make the vehicle the ideal place to apply scare tactics, warnings, and gain initial control of the executive.

Even though travel problems provide the greatest number of security and psychological variables, there are actions and policies that can be developed to minimize the executive's risk and complicate the terrorist's plans. The basic travel policy can be divided into three areas: (1) Normal Travel Procedures, (2) Vehicle Equipment, and (3) Vehicle Defense Strategy. The following checklists will aid in formulating and evaluating an effective travel security policy.

Normal Travel Procedures Checklist

- The avoidance of routine times and patterns of travel by executives is the least expensive security strategy that can be utilized. The selection of the route should be at the discretion of the executive, not of the chauffeur. Always restrict travel plans to a need-to-know basis.
- Avoid driving in remote areas after dark and keep to established, well-traveled roads.
- In high-risk areas or when individuals are considered attractive targets, consideration should be given to executives and drivers being trained in antiterrorism strategy and defensive driving. Establish responsibilities and develop contingency plans.
- There should be a simple duress procedure established between the executive and drivers. Any oral or visual signal will suffice (i.e., something that the executive or driver says or does only if something is amiss).
- Never overload a vehicle, and all persons should wear seat belts.
- Always park vehicles in parking areas that are either locked or watched and never park overnight on the street. Before entering vehicles, check for signs of tampering.

- When using a taxi service, vary the company. Ensure that the identification photo on the license matches the driver. If uneasy for any reason, simply take another taxi.
- When attending social functions, go with others, if possible.
- Avoid driving close behind other vehicles, especially service trucks, and be aware of activities and road conditions two to three blocks ahead.
- Keep the ignition key separate and never leave the trunk key with parking or service attendants.
- Before each trip, the vehicle should be inspected to see that (1) the hood latch is secure, (2) the fender wells are empty, (3) the exhaust pipe is not blocked, (4) no one is in the back seat or on the floor, and (5) the gas tank is at least three quarters full.
- Establish a firm policy regarding the carrying and use of firearms. Local laws may prohibit firearms.

Vehicle Equipment Checklist

- The executive vehicle designed to meet the terrorist or criminal threat in a high threat area should be a hardtop model with the following special equipment: (1) inside hood latch, (2) locked gas caps, (3) inner escape latch on trunk, (4) steel-belted radial tires with inner tire devices that permit movement even with a flat tire, (5) radiator protection, (6) disk brakes, and (7) an anti-bomb bolt through the end of the exhaust pipe.
- Positive communications can be ensured with a two-way radio or a car telephone.
- It is recommended that the executive vehicle designed to meet the terrorist or criminal threat carry the following safety equipment: (1) fire extinguisher, (2) first-aid kit, (3) flashlight, (4) two spare tires, (5) large outside mirrors, and (6) a portable high-intensity spotlight.
- For additional protection, the vehicle should have an alarm system with an independent power source (an additional battery).

Vehicle Defense Strategy Checklist

- Always be alert to possible surveillance; if followed, drive to the nearest safe location, such as police stations, fire stations, or shopping center and ask for help. Carry a mini-cassette recorder in the car to dictate details of a suspect surveillance car such as color, make, model, license plate, description of occupants, etc. It is difficult to make such detailed notes while driving.
- Where feasible, drive in the inner lanes to keep from being forced to the curb.
- Beware of minor accidents that could block traffic in suspect areas; especially crossroads because they are preferred areas for terrorist or criminal activities as crossroads offer escape advantages.

DOCUMENT APPENDIX

- If a roadblock is encountered, use shoulder or curb (hit at 30-45 degree angle) to go around, or ram the terrorist or criminal-blocking vehicle. In all cases, do not stop and never allow the executive's vehicle to be boxed in with a loss of maneuverability.
- Blocking vehicles should be rammed in a nonengine area, at 45 degree, in low gear, and at a constant moderate speed. KNOCK THE BLOCKING VEHICLE OUT OF THE WAY.
- Whenever a target vehicle veers away from the terrorist vehicle, it gives adverse maneuvering room and presents a better target to gunfire.

Travel Security Suggestions

The following are general traveling security suggestions:

- Discuss travel plans on a need-to-know basis only. Telephone operators and secretaries should not advise callers and visitors when an executive is out of town on a trip.
- Remove company logos from luggage. Luggage identification tags should be of a type that allows the information on the tag to be covered. Use the business address on the tag.
- Do not leave valuables and/or sensitive documents in the hotel room.
- When sightseeing, observe basic security precautions and refrain from walking alone in known high-crime areas.
- Always have telephone change available and know how to use the phones. Learn key emergency phrases of the country to be able to ask for police, medical, etc.
- Joggers should carry identification.
- Men should carry wallets either in an inside jacket pocket or a front pants pocket, never in a hip pocket. The less money carried the better. Credit cards can be used for most purchases.
- The telephone numbers of the U.S. Embassy or U.S. Consulate, and company employee contact numbers should be carried with employees at all times.
- Always carry the appropriate documentation for the country being visited.
- When traveling, ask for a hotel room between the second and seventh floors. Most fire department equipment does not reach higher to effect rescue and ground floor rooms are more vulnerable to terrorist or criminal activity.
- American-type hotels usually offer a higher level of safety and security inasmuch as they offer smoke alarms, fire extinguishers, safety locks, hotel security, 24 hour operators, English-speaking personnel, safety deposit boxes, and normally will not divulge a guest's room number.
- Choose taxis carefully and at random. Be sure it is a licensed taxi. Do not use independent non-licensed operators.
- Be as inconspicuous as possible in dress, social activities, and amount of money spent on food, souvenirs, gifts, etc.

- Stay in or use VIP rooms or security zones when waiting in commercial airports abroad. Minimize the amount of time spent in airports.
- Confirm arrivals at destinations with office and/or family. Use an itinerary when traveling.
- When traveling internationally, keep all medicine in original containers and take a copy of the prescription.

VISITING PERSONNEL PROTECTION

General Principles

This chapter provides guidelines regarding security procedures to be implemented during visits of company executives. Guidelines for three levels of threat (minimal threat, moderate threat and high threat) are set forth below along with the factors which determine the level of threat that may exist.

These guidelines should be viewed as tools to assist in organizing and planning visits by company executives or other key personnel. Their implementation will reduce the executive's exposure to terrorist acts, criminal activity, and potential embarrassment.

Minimal Threat-Factors and Guidelines

Minimal Threat Factors

Factors which should be used by management in determining whether in view of the local security environment a minimal threat potential exists include the following:

- A stable local government;
- Effective law enforcement;
- No significant history of terrorist acts against multinational companies or their executives;
- No previous history of criminal or terrorist acts directed against company executives;
- No significant level of criminal activity (particularly violent crimes such as robbery, kidnapping, murder, and rape);
- No current adverse publicity against the company and no local group activity protesting company policies;
- Other risk factors applicable to the local environment.

Minimal Threat Guidelines

Security Coordination. A management-level employee should be assigned as security coordinator. The coordinator's responsibilities consist of implementing the established security guidelines, coordinating all other security aspects of the visit, and serving as the visitor's main contact.

The coordinator should be present at the airport, hotel, and events during arrivals and departures. He/she should ensure adequate security precautions are taken and be present at large public functions.

Air Travel

- Travel in corporate aircraft is preferable because contact with the general public is limited, but use of commercial airlines is an acceptable alternative provided the airline involved is not considered a likely terrorist target.
- When booking reservations, you should make no reference to the visitor's position.
- Personnel should be available at the airport to handle baggage and expedite customs clearances and local airport formalities, both on arrival and departure. A VIP room should be reserved at the airport for possible use in the event of a delayed departure by the aircraft.
- Time spent at the airport should be kept to a minimum. Public areas should be avoided, if at all possible.
- Use of public transportation to and from airports is not recommended.
- Distribution of travel itineraries should be restricted.

Aircraft Security. This section applies in the event that corporate aircraft are used.

- The hiring of contract security officers at major international airports to secure the corporate aircraft during stopovers is not necessary provided that the airport has a viable security system.
- The use of contract security officers on a 24 hour basis is necessary in the event that the corporate aircraft uses a remote airfield with limited operations and minimal security or is parked in a remote area of a major airport.

Local Transportation

- The use of public transportation such as taxis, buses, and subways is not recommended.
- A four-door sedan should be available for use throughout the visit. Care should be taken to ensure that the vehicle is unobtrusive, so as not to bring undue attention to the visitor. The chauffeur or driver, if used, should be bilingual and knowledgeable of the local area and routes to be traveled.

Accommodations

- Hotel reservations should be booked at a first class hotel located in a low-crime area. Hotel management need not be contacted to provide unusual security or other arrangements for the visitor. A low-key approach is essential

to ensure anonymity. Reference to the company or the visitor's position should be avoided.
- Visitors should be preregistered to avoid being required to check in at the reception desk. The room key should be provided to the visitor immediately upon his or her arrival at the hotel or airport by personnel responsible for coordinating the visit.
- The guest room or suite should be located between the second and seventh floor of the hotel, preferably on a floor with a separate concierge. The room should be away from the public elevator lobby but near an emergency exit.
- Valuables should be stored in accordance with hotel safekeeping provisions.
- Use of a guesthouse or private residence is acceptable as long as it is not located in an isolated area.

Official Functions and Activities

- Coordinate all activities and visit sites before the visitor's attendance. The coordinator should obtain guest lists and detailed itineraries, determine emergency evacuation routes, and ascertain the purpose of the function.
- The coordinator should ensure that the function or activity does not subject the visitor to undue risk.
- Official company functions should be on an invitational basis and guests should be required to present their invitations at a reception desk staffed by company personnel before being granted access to the function. The receptionist should match the invitation to the guest list.

Liaison With Local Authorities. Prior to a visit by a VIP, you should make contact with the appropriate local authorities to advise them of the upcoming visit and to ascertain whether the current local security environment necessitates an upgraded security posture for the visit.

Background Data. An information packet should be prepared before the visit and presented to the executive upon his/her arrival. Information provided should include:

- Emergency telephone contact list, including company personnel (home and office numbers), hospital, police, fire, emergency services, and company doctor;
- Maps of the area;
- Detailed itinerary;
- Availability of company transportation;
- Brief review of current security situation including curfews, government-imposed restrictions, description of high-crime areas to be avoided, and other relevant factors; and
- Explanation of local currency (exchange rates and currency control laws or regulations).

Other

- Details of visits by VIPs should be considered company confidential and distribution limited on a need-to-know basis.
- Media coverage, unless requested by the visitor, is unwarranted.

Moderate Threat-Factors and Guidelines

Moderate Threat Factors

Factors which should be used by management in determining whether in view of the local security environment a moderate threat potential exists include the following:

- Stable local government;
- Effective law enforcement;
- Some history of terrorist attacks against multinational companies and/or their executives;
- No previous history of criminal or terrorist acts directed against company executives;
- Upswing in criminal activity, particularly violent crimes with some history of criminal kidnappings for financial gain;
- Some current adverse publicity against the company and potential for nonviolent groups to protest against company policies during the executive's visit.

Moderate Threat Guidelines

Unarmed Security Escort. In addition to the guidance set forth in "Minimal Threat Guidelines," an unarmed security escort should be used when a determination is made by management that a moderate threat exists.

High Threat-Factors and Guidelines

High Threat Factors

Factors which should be used by management in determining whether in view of the local security environment a high threat potential exists include the following:

- Unstable or unpopular local government, with terrorist groups actively attempting to bring about its overthrow;
- Ineffective or corrupt law enforcement agencies unable to reduce criminal activity and bring the terrorist problem under control;
- Significant history of terrorist attacks against multinational companies and/or their executives, including bombings, assassinations, and kidnappings;
- Recent history of criminal or terrorist acts or threats against company facilities and/or their executives;

- Widespread criminal activity reaching all elements of local society with emphasis on violent crimes;
- Considerable adverse publicity against company policies and organized local groups that have been leaning toward violence and are planning to protest company policies during the executive's visit;
- Other factors appropriate to the local environment. Asking the consulate regional security officer at the embassy is a good idea.

High Threat Guidelines

Recommendation Against Visit. High-threat potential means a significant risk to the well being of the executive. You should strongly recommend against a visit by the executive if a high risk exists. By definition, this category will apply to a limited number of locations, but might vary based on the local situation at a particular point in time. For example, a potential visit might be deemed a moderate risk one month and high risk another because of changes in the local environment.

Armed Protective Security Detail. If the executive cannot be dissuaded from visiting the high threat area, an armed protective security detail should be used.

Specific guidelines for high risk protective details are beyond the scope of this document because of the multiple and various considerations in organizing each individual protective detail. However, the use of trained security professionals is essential.

The items covered in "Minimal Threat Guidelines", will still have to be addressed when an armed protective detail is required. However, the manner in which relevant tasks are performed may be modified by guidelines issued in regard to the armed details. Some general guidelines are as follows:

- Professional bodyguards dressed in plainclothes and equipped with weapons and two-way radios should accompany the executive at all times. At least one bodyguard should remain in the direct vicinity of the executive whenever potential public contact is envisioned.
- Security personnel should conduct advance surveys of all sites to be visited and be on the scene throughout the executive's visit to the location.
- Security personnel should be assigned to the hotel or residence on a 24 hour basis to ensure that unauthorized individuals do not enter the room or suite. Cleaning staff should be escorted whenever they enter the accommodations. The room or suite should be periodically checked to ensure that contraband (such as a bomb) has not been introduced into the area.
- An escort car or cars should be used on all vehicular movements by the executive to provide a response capacity in the event of an attack or vehicular

mishap or breakdown. The escort car or cars should be staffed by at least two security professionals.
- The executive's vehicle should be driven by a security professional trained in evasive or defensive maneuvers. The vehicle should be inspected before use to ensure that explosive devices have not been installed on the vehicle or that the vehicle has not been otherwise tampered with by unauthorized individuals. Use of an armored car, if available, is recommended.
- Public exposure should be limited to the minimum necessary for the executive to complete his or her assignment.

Group Activities Guidelines

The exposure created by a number of executives gathering at a single location necessitates some degree of increased security. The following is a list of some general guidelines for use in such group activities:

- The suites should be inspected before occupancy to ensure that no contraband or unauthorized individuals are located in the rooms.
- Security should be provided for corporate aircraft overnighting at the local airport. Such security may be provided by off-duty uniformed and armed police officers or contract security guards.
- The hotel activity boards should make no reference to the company. Publicity and press coverage should be minimized. A low profile is strongly recommended. Anonymity is a powerful ally of a traveling executive.
- If possible, hotel guest rooms occupied by company personnel should be located in one section of the hotel. Consideration should be given to hiring a security officer to patrol the hallway in the vicinity of the guest rooms and function rooms during hours of darkness or even on a 24-hour basis.
- Access to functions should be controlled to prevent and unauthorized individual from gaining access to the meetings or functions. This can be handled by assigning a member of the meeting staff to serve as a receptionist outside the door. Access can be granted either by personal recognition or by checking identify cards.
- Information packets provided to participants should include the name and telephone number of the staff person responsible for security. Staff personnel should be provided with an emergency contact list, including the telephone numbers of the nearest hospital with an emergency room, ambulance service, police department, and fire department.
- Consideration should be given to leasing pagers to ensure that staff personnel can be rapidly contacted in the event of an emergency.
- Upon the conclusion of the meeting, staff personnel should inspect all guest rooms and function rooms to ensure that no documents, personal effects, or equipment have been left behind by participants.

NOTE

1. A device constructed to protect against a ramming vehicle attack. They are deployed in lines around a perimeter for anti-ram protection, or to provide supplemental control of vehicle traffic through permanent checkpoints when other means are not practical or effective.

APPENDIX I: SECURITY SURVEY CHECKLIST

General, Preparatory Data

1. Site name, address, telephone number:
   Please fill in the name of the people holding the following positions:
   Manager
   Assistant Manager
   Human Resources Manager
   Person responsible for site security
   Number of Employees
   Area Covered/Office Size
   Operating Hours
   Function
2. Survey should include review of theft reports prepared by this site for an appropriate prior period. Where appropriate, has corrective action been taken?
   Do theft reports reflect patterns, trends, or particular problems at this location?
3. What does site management regard as the most prevalent or serious security problem?
4. Does the site maintain items of value, such as works of art, paintings, wall hangings, etc.?
5. What are the site's most valuable physical assets?
6. Does the location have an employees' handbook or manual or other means of enumerating rules of conduct?
   Have employees been notified that violation of these rules are grounds for disciplinary action up to and including discharge?
   Do these rules of conduct include theft of company, customer, and employee property, including information?
7. Identify off-site locations that should be included in survey, to include warehouses, offices, storage facilities, etc.
   Are these locations protected against vandalism and theft?
8. What is the police agency having jurisdiction over the site?
   Does the plant have dedicated telephone line to this agency?
   Have they been called for assistance in the recent past?
   What has been their response?

Do they normally include any of our perimeter in their patrols?
If requested, would they?
9. Are police emergency numbers readily available to personnel who should have this information?
10. Is information readily available on how to reach the proper agency for assistance with illegal narcotics, bomb threats, obscene calls, etc.?
Do you have a policy of reporting identifiable items of stolen property to the local police for addition to their files, indexes?
11. Some police agencies have a Crime Prevention Unit that responds to invitations to speak on various topics (drugs, rape, etc.) or that may conduct limited security surveys.
Is this service available?
If so, have you taken advantage of it?

Perimeter Security

Lighting Evaluation

12. Is the perimeter adequately lighted?
13. Does lighting aid or inhibit guards in the performance of their duties?
14. Is lighting compatible with closed-circuit television (CCTV)?
Does it cause monitor to "bloom"?
15. Is the power supply adequately protected?
16. Is lighting properly maintained and cleaned?
17. Are sensitive areas (parking lots, computer areas, stores, storage rooms, shipping/receiving areas) adequately lighted?
18. If an emergency occurred, is the site adequately lighted?
Is the fenceline adequately lighted?
In appropriate areas, is glare projection lighting used?

Security Force

19. Proprietary?
Contract?
If contract, name of agency and telephone number if proprietary, what is method and source of selection of personnel?
20. Are perimeter patrols conducted? Frequency?
21. Is an incident log, including alarms/responses maintained? Reviewed daily?
By Whom?
22. Are security personnel used for nonsecurity related duties?
If yes, what duties?
23. Does site use photo ID cards?
Compatible with access control system?
Who administers it?

24. Are all employees required to show photo ID card upon entry? Is duplicate copy kept on file?
25. Are parking decals or other methods of registering employee vehicles used? Are privately owned vehicles permitted to park on site? If so, can an individual reach a vehicle without passing a guard?
26. Does the site have a receptionist in place at all times? Are visitors required to register? Are they provided with an identifying badge, and are non-company employees escorted while on the site? Is visitor identification verified (e.g., vending company ID, etc.)?

Perimeter Protection

27. If outside building walls form part of the perimeter, are all doors and windows secured against surreptitious entry? Can entry be achieved via the roof? Can hinge pins be removed from doors? Are all entry/egress points controlled when opened?

Internal Security

Lock/Key Control

28. With whom does physical and administrative key control rest?
29. Is a master key system in use?
    How many grandmaster/master keys have been issued?
    Is adequate control exercised over these keys?
30. Is a cross-control system (name versus key number) in use?
    What type of numbering system is in use?
    Is the entire system, including blanks, inventories on a regular basis?
    Are they stamped "Do Not Duplicate"?
31. What level of management authorization (written) is required for issuance of keys?
32. Identify personnel who are permitted to have keys to perimeter fence, doors.
33. Are office/facility keys, particularly masters, permitted to be taken home? Are keys signed in/out in a daily log?
34. Are locks rotated?
35. How long has the present lock/key system been in use?
    Have keys been reported lost?
    What level key?
    What is the policy when this happens?
36. Is a record of locations of safes and their combinations maintained?
    Are combinations routinely changed annually and when an individual who knows one no longer has that need to know? (separation, transfer, retirement)
    Are safe combinations, if written, maintained in a secure place?

DOCUMENT APPENDIX

Alarms and Electronics

37. What type, if any, electronic security system is in use here?
Do alarms terminate at the site or at an outside central station?
Has service/response been satisfactory?
38. List alarms such as burglar (doors, windows, space (motion)), duress (receptionist, cashier, nurse), other (card access, CCTV, etc.)

Theft Control Procedures

39. Does the site have a policy of marking items susceptible to theft (calculators, office equipment, hand tools, microwave ovens, TV monitors, VCRs, etc.) so they can be identified as company property?
Describe the extent of the program.
Does it include die stamping or etching and painting?
40. Are serial numbers of all items bearing them recorded?
In the event of theft, is this information related to the police for inclusion in stolen property indexes, and for identification and return in case of subsequent recovery?
41. Are trash receptacles periodically inspected to determine whether items of value may be removed from the site via them?
42. Are all store/office supplies, etc., attended when open?
What is the procedure for drawing supplies when no attendant is present?
43. Are telephone records properly safeguarded to prevent unauthorized destruction?
Is access to telephone switching equipment (the "frame room") restricted?
44. Who performs custodial services—proprietary or contract janitorial people?
Is access limited to the office area only?
Are they bonded?
Are they required to wear ID badges?
Are they checked during the performance of their duties?
Are they inspected by guards as they leave?
Are the janitors' vehicles inspected on the way off the property?
How is trash removed from the site?
Are the vehicles used to remove trash inspected on the way off the property?
Do the janitors have access to restricted or sensitive areas?
Are they given office keys (masters?)?
Are they permitted to take these keys off the site with them?
45. How much cash is kept on site? Is it handled at more than one location?
How is cash supply replenished?
Where is it kept during working hours?
Where is it kept after hours?
Where are blank payroll checks kept?
Where are blank disbursement checks kept?

Considering the neighborhood the site is located in, and the amount of cash on site, how do you assess your vulnerability to armed robbery or burglary?

Proprietary/Limited Information

46. Is there Proprietary and/or Limited data on site?
    If so, in what form?
    Is it properly marked?
    Is it stored in a secure location?
    Are the following locked at the end of the day:
    a. Offices?
    b. Filing Cabinets?
    c. Desks?
47. What are office destruction procedures and file purging for Proprietary data?
48. Does the site have a clean desk policy?

Personnel Security

49. Are any background checks conducted prior to employment? Are previous employment dates verified? Are personnel medical records properly safeguarded? Is security included in the new hire orientation? Is company property (credit cards, ID keys) retrieved during exit interviews?

Emergency Procedures

50. Do you have a current bomb threat procedure?
    Who implements it? (searches areas)
    Does the procedure include a checklist for the switchboard operator?
51. Is there a contingency plan for acts of violence?
    A disaster plan?
52. If personnel are required to work alone, are they periodically checked by someone to ascertain their well-being?
    What means do they have of calling for help in an emergency?

Computer Security

53. Are terminated employees immediately separated from the EDP function?
54. Is access to the data center controlled physically, electronically?
    Locked when not in use?
55. Is output distributed via user controlled lock boxes? Is tape library maintained physically separate from machine room?

Threat Information

56. Has liaison been established by your office with the American Embassy Regional Security Officer (RSO)?

DOCUMENT APPENDIX

Is the RSO able to notify you of security threats concerning known terrorist groups active in the area?
Any groups that harbor hatred for U.S. corporations, your company, its manager, and employees?
Anniversary dates that local population or terrorist groups celebrate?
What tactics and activities are practiced or adopted by local terrorist groups that might affect your company, Its managers and employees?

57. Do you have sources that will inform you of any political controversy or labor disputes that might impact your operations?
58. Will you provide Security with copies of information, that may be detrimental to the company, received as a result of your contracts with the RSO, as well as other sources, including newspaper articles?

APPENDIX II: FACILITY QUESTIONNAIRE

1. Are there any known groups that harbor hatred for U.S. businesses, managers and employees?
   Identify:
2. What terrorist groups are known to be active in the area?
   What tactics have these groups been known to use?
   What is the possibility of a change in these tactics?
3. Are there known groups that vocally oppose foreign capitalism or imperialism in the area?
   Identify:
4. Are there any known groups that vocally or actively oppose the local government that the United States supports?
   Identify:
5. Is there any current political controversy or labor dispute that we should be aware of?
6. Are there any upcoming anniversary dates that the local population or terrorist groups celebrate? Identify:
7. Have there been any previous hostage taking or kidnapping incidents, bombings, assassinations, strikes against U.S. businesses or the government, demonstrations, assaults, sabotage against corporate facilities or products, or occupation of corporate facilities in the area?
   Identify:
8. If there have been previous hostage taking or kidnapping incidents,
   a. How were the victims seized?
   b. What was the fate of the hostages?
   c. How much ransom was demanded?
   d. Was it paid?
   e. How were the negotiations handled?

9. Does the host country prohibit negotiating with hostage takers or prohibit the payment of ransom?
10. Do you consider the local police and intelligence services effective?
11. What are the aims of the local criminals or terrorist groups?
    What tactics or type of activity by these groups would best further those aims?
12. What is the identified groups' capability of carrying out planned activities such as ambush, hostage taking, kidnapping, execution, bombing, etc.?
13. In the event of terrorist activity, which organizations, businesses, groups, or individuals would be the most likely targets?

APPENDIX III: THREATENING PHONE CALL CHECKLIST

PLACE THIS UNDER YOUR TELEPHONE — BOMB THREAT!
QUESTIONS TO ASK:

1. When is bomb going to explode?
2. Where is it right now?
3. What does it look like?
4. What kind of bomb is it?
5. What will cause it to explode?
6. Did you place the bomb?
7. Why?
8. What is your address?
9. What is your name?

EXACT WORDING OF THE THREAT:

Sex of caller:
Race:
Age:
Length of Call:
Number at which call is received:
Time: Date: /   /

CALLER'S VOICE:

Calm	Nasal
Angry	Stutter
Excited	Lisp
Slow	Raspy
Rapid	Deep
Soft	Ragged
Loud	Clearing throat
Laughter	Deep breathing
Crying	Cracking voice

DOCUMENT APPENDIX

Normal   Disguised
District   Accent
Slurred   Familiar

If voice is familiar, who did it sound like?

BACKGROUND SOUNDS:

Voices   Street noises
Crockery   Animal noises
Clear   PA System
Static   Music
Local   House noises
Booth   Long distance
Motor   Other
Factory machinery
Office machinery

THREAT LANGUAGE:

Well spoken   Incoherent
(educated)   Taped
Foul   Irrational
Message read by threat maker

REMARKS:

Report call immediately to:
Phone number
Date   /   /
Name
Position
Phone number

HOSTAGE!
QUESTIONS TO ASK:

2. Who is this?
3. Where are you calling from?
4. Is this a prank?
5. How do I know this is not a prank?
6. May I talk to the hostage?
7. Is the hostage all right?
8. What do you want?

VERY IMPORTANT:

9. Will you call back in 15 minutes?
10. How can I contact you if I have trouble meeting your demands?

EXACT WORDING OF DEMAND:

Sex of Caller: Race:
Age: Length of call:
Number at which call is received:
Time: Date: /   /

APPENDIX IV: LETTER AND PARCEL BOMB RECOGNITION POINTS

WARNING!
LETTER AND PARCEL BOMB RECOGNITION POINTS

- Foreign Mail, Air Mail, and Special Delivery
- Restrictive Markings, such as Confidential, Personal, Etc.
- Excessive Postage
- Hand Written or Poorly Typed Addresses
- Incorrect Titles
- Titles but No Names
- Misspellings of Common Words
- Oily Stains or Discolorations
- No Return Address
- Excessive Weight
- Rigid Envelope
- Lopsided or Uneven Envelope
- Protruding Wires or Tinfoil
- Excessive Securing Material, such as Masking Tape, String, etc.
- Visual Distractions

PERSONAL SECURITY GUIDELINES FOR THE AMERICAN BUSINESS TRAVELER OVERSEAS

Overseas Security Advisory Council

INTRODUCTION

This appendix was developed to inform and make the American business traveler aware of the potentially hostile overseas environment in which they may be traveling or working. The information contained in this appendix will familiarize the traveler with personal security guidelines for traveling overseas. The potential hazards and vulnerabilities that are inherent in protecting/carrying sensitive or proprietary information while traveling are described, as are surveillance and/or targeting recognition, personal conduct abroad, hostage/hijacking survival and fire safety.

There are several scenarios to traveling abroad that are addressed: first, the actual getting from point A to B; second, the airport; third, the hotel or temporary quarters; fourth, traveling within a foreign country; and, lastly, the office or workplace. Each of these five situations presents different potential security problems.

The most effective means of protecting yourself and your property is the liberal use of common sense reinforced with a high state of security awareness. Do not give anyone the opportunity to exploit vulnerabilities. Stay alert and exercise good judgment.

TRAVEL PREPARATION AND PLANNING

Travel Itinerary

DO NOT publicize your travel plans, but limit that knowledge to those who need to know. Leave a full itinerary of your travel schedule, hotel phone numbers and business appointments with your office and with a family member or friend.

Passport

Is it valid? Are the visas current for the country of destination? If not, you and everything in your possession may be looked at in-depth by host government authorities. If you are carrying documents that are sensitive or proprietary, they will be examined in detail to see if there is anything that would be of interest. If there is, you can bet that copies will be made, and there is not much that you will be able to do about it.

Make photocopies of your passport, visa and other important documents that you will be traveling with. Put copies in both your carry on and checked luggage. This makes it easier to replace your identification documents should anything happen. (Also, it is a good idea to leave a photocopy with someone at home.)

Visas

Is a visa required for any of the countries that you are visiting and do you have the appropriate visa(s)? Is the information on your visa application true and correct? In some countries, falsifying information on a visa application can result in an unexpected vacation in the local bastilles.

Some countries are sensitive to which visa you obtain. If you are traveling on business, a business visa should be obtained; otherwise a tourist visa is acceptable.

Medical

Take plenty of any prescription medication with you, as well as an extra set of eyeglasses or contact lenses. Also, take a copy of your prescription should you need to have glasses, contacts or medication replaced. Keep an inoculation record and update it before each trip as each country has different requirements.

- Carry with you a list with your blood type, allergies, medical conditions and special requirements. It is a good idea to have a medical alert bracelet if you have a special medical condition.
- Inoculations—Does the country to be visited require any specific inoculations? This information is available from the embassy or consulate. Be sure to carry your international shot record, just in case.
- If you do not have comprehensive medical coverage, consider enrolling in an international health program. Hospitals in foreign countries do not take credit cards and most will not honor U.S. based medical insurance plans.

Miscellaneous

Keep your personal affairs up to date. If possible, leave a power of attorney with a family member or friend should anything happen to you.

- Do research on the country you will be traveling to before you go. Talk with friends, family or business associates who have visited the country. They can usually give you some good tips for your trip. Also, for any travel warnings or other conditions that you should be aware of, check with the U.S. State Department, Bureau of Consular Affairs.
- Travelers should discuss with their travel agents, which airlines, hotels and car rental companies are recommended.

DOCUMENT APPENDIX

- Carry in your wallet/pocketbook only the documents you will need. Take only the credit cards you plan to use on your trip.
- If you plan to rent a car, check to see if you must obtain an international drivers permit for the country you plan to visit.
- Obtain information from U.S. Customs regarding any special requirements for the country you are visiting.

Local Import Restrictions

Request from the embassy of the country you plan to visit a copy of any list or pamphlet describing customs restrictions or banned materials. This is a hint designed to minimize the possibility of an encounter with the local authorities.

Leave all expensive and heirloom jewelry at home.

Luggage

DO NOT pack sensitive or proprietary information in your checked luggage. Double envelope the material and hand carry it. Be sure that your luggage is tagged with covered tags that protect your address from open observation. Put your name and address inside each piece of luggage and be sure that all luggage is locked or secured in some fashion.

Luggage Locks

The locks on your luggage are not that secure when it comes to the professional thief or manipulator and are really no more than a deterrent. But, if time is of the essence to the perpetrator, and it usually is when a crime is involved, there are a couple of suggestions that might deter surreptitious entry and/or theft.

- For added security on all luggage, run a strip of nylon filament tape around the suitcase to preclude its opening accidentally if dropped or mistreated by baggage handlers.
- For luggage and briefcases with two combination locks, reset the combination locks from the factory combination (000) to different combinations on each of the right and left locks.
- For luggage with single locks, set the lock on each piece of luggage with a different combination.
- DO NOT pack extra glasses or necessary daily medication in your luggage. Carry it in your briefcase, purse or pocket. If you are the victim of a hijacking you may need these items—if they are in your luggage, you probably will not be able to get to them.
- On your luggage use your business address and telephone number. If possible, use a closed name tag with a cover. Do not use a laminated business card on your luggage, and avoid putting the company name or any logos on your luggage.
- Check with the airline and/or your personal insurance company regarding any lost luggage coverage.

- Make sure you use sturdy luggage. Do not over pack as the luggage could open if dropped. Bind the luggage with strapping so that it will remain intact.
- Never place your valuables (jewelry, money and travelers checks) in your checked luggage. Never leave your bags unattended.
- Consider obtaining a modest amount of foreign currency before you leave your home country. Criminals often watch for and target international travelers purchasing large amounts of foreign currency at airport banks and currency exchange windows.

Airline Security and Seat Selection

- Try to book a non-stop flight, as these have fewer takeoffs and landings.
- Choose an airline with a good safety and on-time record.
- Try to make your stopovers in airports that have a high security standard and good security screening.
- Try to fly wide body planes. Hijackers tend to avoid these as having too many passengers.
- Most travelers prefer an aisle seat. Choose a window or center seat. This will keep you away from the hijackers and any action that may be happening in the aisle.

AT THE AIRPORT

To diminish the risks of becoming an innocent bystander victim of a terrorist attack and reduce your exposure to the criminal threat, there are a number of things that you should remember when checking into an airport.

- In the event of a disturbance of any kind, go in the opposite direction. DO NOT GET INVOLVED!
- Plan to check in early for your flight to avoid long lines at the ticket counter.
- Go directly to the gate or secure area after checking your luggage. (Secure Zone—Area between security/immigration and the departure gate.) Avoid waiting rooms and shopping areas outside the secure areas.
- Stay away from glass wall areas and airport coffee shops which are open to the concourse or public waiting areas.
- From the time you pack your luggage until you check it with the carrier at the airport maintain positive control of all items, both hand carried and checked.
- At many airports security personnel, following FAA protocol, will ask you questions about control of your luggage. Know what items you are carrying and be able to describe any/all electrical items.
- When going through the pre-board screening process cooperate with security personnel and remember that they are there to help ensure that your travel is safe.

- When arriving at or departing from an airport it is a good idea not to be exchanging items between bags while waiting in line for security screening or immigration/customs processing. Complete all packing before entering such areas.
- If a conflict should arise while undergoing the screening process, cooperate. Obtain the names of the screeners involved, and then discuss the matter with a supervisor from the appropriate air carrier.
- Remember that x-ray will not damage film, videos or computer equipment. Many times such items can be cleared using x-ray which means that they will not have to be handled by the screener.
- Consider being transported to/from the airport by a hotel vehicle. Generally the cost is not prohibitive, and arrangements can be made in advance by your travel agent.
- Declare all currency and negotiable instruments as required by law.
- NEVER leave your luggage or briefcase unattended, even while checking in or once in the secure zone. In some countries, the police or security forces assume that an unattended bag is a bomb, and your luggage could be forcefully opened or even destroyed.
- Always be aware of where you are in conjunction with where you are going. If an incident occurs, you need to know how to avoid it and either get out of the area or to your boarding area.
- Dress casually when traveling, as this will keep any undue attention from you. Once aboard the flight, remove your shoes for better circulation. Walk around the flight cabin to keep your blood circulating and swelling down.
- Avoid last minute dashes to the airport.
- Eat moderately, avoid alcoholic beverages and drink plenty of water as this will help to avoid dehydration.
- If possible, before you leave make an effort to adjust your sleep patterns.
- Sleep as much as possible during the flight.
- Carry airsickness medication with you. Even the best traveler sometimes experiences airsickness.
- Avoid a demanding schedule upon arrival. Give yourself a chance to adjust to your surroundings.

SELECTING A SECURE HOTEL

Many U.S. corporations have hotels abroad that are owned by local businessmen and staffed by local workers but managed by first class U.S. hoteliers. You usually can expect levels of safety and security that are consistent with U.S. standards.

- Ask the corporate travel agent for a list of recommended hotels.

- Check with the Regional Security Officer at the U.S. Embassy for a list of hotels utilized by officials visiting the area.

Making Reservations

Make your own reservations when practical and consistent with company policies. The fewer people that become involved in your travel and lodging arrangements, the better.

- If traveling abroad, especially in politically sensitive areas, consider making reservations using your employer's street address, without identifying the company, and using your personal credit card. Again, the less known about your travel itinerary, and whom you represent, the better.
- If arriving after 6:00 P.M., ensure that reservations are guaranteed.
- Request information about parking arrangements if anticipating renting an automobile.
- Be aware that credit card information has been compromised in the past. Always audit monthly credit card statements to ensure that unauthorized use has not been made of your account.
- It is advisable to join frequent travelers' programs available with many lodging companies. These programs enable upgrades to executive or concierge floors where available. Be sure to advise the person taking reservations that you are a member and request an upgrade.

Arriving at or Departing From the Hotel

The most vulnerable part of your journey is traveling between the point of debarkation/embarkation and the hotel. Do not linger or wander unnecessarily in the parking lot, indoor garage or public space around the hotel—be alert for suspicious persons and behavior. Watch for distractions that are intentionally staged to setup a pickpocket, luggage theft or purse snatch.

- Stay with your luggage until it is brought into the lobby, or placed into the taxi or limo.
- Consider using the bellman. Luggage in the "care, custody and control" of the hotel causes the hotel to be liable for your property. Protect claim checks; they are your evidence!
- Keep in mind though that there are limits of liability created by states and countries to protect hoteliers. Personal travel documents, lap tops, jewelry, and other valuables and sensitive documents in excess of $1,000 in value should be hand carried and personally protected.
- If you arrive by auto, park as close to a hotel access point as possible, and park in a lighted area. Remove all property from the car interior and place it in the trunk. Avoid leaving valuables or personal documents in the glove compartment. Prior to leaving the security of the vehicle, note any suspicious persons or behavior.

DOCUMENT APPENDIX 679

- If using valet service, leave only the ignition key, and take trunk, house, or office keys with you. Often, valets are not employees of the hotel and work for contract firms.
- Parking garages are difficult to secure. Avoid dimly lit garages that are not patrolled and do not have security telephones or intercoms.
- Female travelers should consider asking for an escort to their vehicles whether parked in the lot or garage.

Registration

In some countries, your passport may be temporarily held by the hotel for review by the police or other authorities, obtain its return at the earliest possible time.

- Be aware of persons in the hotel lobby who may have unusual interest in your arrival.
- If carrying your luggage, keep it within view or touch. One recommendation is to position luggage against your leg during registration but place a briefcase or a purse on the desk or counter in front of you.
- Ground floor rooms, which open to a pool area or beach with sliding glass doors and window access, are considered vulnerable. Depending upon the situation, area, and security coverage, exercise a higher level of security if assigned a first floor room.
- It is suggested that female travelers request rooms that are away from the elevator landing and stairwells. This is to avoid being caught by surprise by persons exiting the elevator with you or hiding in the stairwell.
- Always accept bellman assistance upon check-in. Allow the bellman to open the room, turn lights on, and check the room to ensure that it is vacant and ready for your stay. Before dismissing the bellman, always inspect the door lock, locks on sliding glass doors, optical viewer, privacy latch or chain, guest room safes, dead bolt lock on interconnecting suite door, and telephone. If a discrepancy is found, request a room change.
- Ask where the nearest fire stairwell is located. Make a mental note which direction you must turn and approximately how many steps there are to the closest fire stairwell. In the event of a fire, there is frequently dense smoke and no lighting.
- Also observe where the nearest house telephone is located in case of an emergency. Determine if the telephone is configured in such a manner that anyone can dial a guest room directly, or whether the phone is connected to the switchboard. Most security-conscious hotels require a caller to identify whom they are attempting to telephone rather than providing a room number.
- Note how hotel staff are uniformed and identified. Many "pretext" crimes occur by persons misrepresenting themselves as hotel employees on house telephones to gain access to guest rooms. Avoid permitting a person into the

guest room unless you have confirmed that the person is authorized to enter. This can be verified by using the optical viewer and by calling the front desk.

IN YOUR HOTEL

All hotel rooms abroad are bugged for audio and visual surveillance. This statement, of course, is NOT TRUE, but that is the premise under which you must operate to maintain an adequate level of security awareness while conducting business abroad. Many hotel rooms overseas are under surveillance. In those countries where the intelligence services are very active, if you are a business person working for an American company of interest to the government or government sponsored competitor, everything that you do in that hotel room may be recorded and analyzed for possible vulnerabilities or for any useful information that can be derived from your conversation.

With the basic premise established above, here are some security tips that will minimize the potential risks.

Hotel Room Key

Keep it with you at all times. The two most common ways that thieves and others use to determine if a person is in their hotel room is to look at the hotel room mail slot or key board or call the room on the house phone. If you do not answer the phone that is one thing, but, if your room key is there, you are obviously out and the coast is clear for a thief or anyone else who is interested in searching your room and luggage.

Upon Arrival

Invest in a good map of the city. Mark significant points on a map such as your hotel, embassies and police stations. Study the map and make a mental note of alternative routes to your hotel or local office should your map become lost or stolen.

- Be aware of your surroundings. Look up and down the street before exiting a building.
- Learn how to place a telephone call and how to use the coin telephones. Make sure you always have extra coins for the telephone.
- Avoid jogging or walking in cities you are not familiar with. If you must jog, be aware of the traffic patterns when crossing public streets. (Joggers have been seriously injured by failing to understand local traffic conditions.)

Valuables

Valuables should normally be left at home. The rule of thumb is, if you neither want nor can afford to lose them, DO NOT TAKE THEM! However, if you must carry valuables, the best way to protect them is to secure them in your local

DOCUMENT APPENDIX

offices. If that is not possible, the next best course of action is to seal any valuables by double enveloping, initialing across seams and taping all edges and seams before depositing them in the hotel's safe deposit box or safe.

Luggage

Keep it locked whenever you are out of the room. It will not stop the professional thief or intelligence agent but it will keep the curious maid honest.

Passport

Keep your passport with you at all times. The only time that you should relinquish it is:

- To the hotel if required by law when registering.
- If you are required to identify yourself to local authorities for any reason.

At night, lock your passport and your other valuables in your luggage. This eliminates their mysterious disappearance while you are asleep or in the shower.

Utilize a portable or improvised burglar alarm while asleep. Two ash trays and a water glass are quite effective as an alarm when placed on the floor in front of the entry door into your room. Place a water glass in one ashtray and put the second ashtray on top of the glass. If a straight chair is available, place it next to the door and put the ash tray/water glass alarm on the edge of the chair where it will fall with enough racket to wake you.

GUEST ROOM AS A "SAFE HAVEN"

Hotels are required to provide reasonable care to ensure that guests have a safe and secure stay. Hotels are not required to guarantee guest security. You are responsible for your personal security and property.

- While in the room, keep the door closed and engage the dead bolt and privacy latch or chain. A limited number of hotel emergency keys can override the dead bolt locks. To ensure privacy use the latch or chain!
- Hoteliers provide guest room "safes" for the convenience of guests. However, these containers are not as durable as bank safes and can be breached. Furthermore, the Housekeepers Liability Laws provide that if guest property is not in the "care, custody and control of the hotel," the hotel is not liable. Guests should always place money or valuables in the safe deposit box at the front desk of the hotel.
- When leaving the guest room, ensure that the door properly closes and is secured. Make a mental note of how your property was left; avoid leaving valuables in plain view or in an unorganized manner. A number of hotel employees enter the room each day to clean, repair and restock the room.

Although most hotel employees are honest and hardworking, a few succumb to the temptation of cash or jewelry left unprotected.
- If you determine that an item is missing, conduct a thorough search prior to reporting the incident to hotel security. Do not expect to receive a copy of the security report, as it is an internal document. The incident should be reported to the local police, the Regional Security and Consular Officers at the U.S. Embassy, and your insurance carrier. Hotel security can provide a letter verifying that you reported property missing.
- Prior to traveling, it is recommended that you copy all credit cards, passport, air tickets and other documents to facilitate reporting loss and replacing them. While traveling abroad, secure these documents in the room safe deposit box and carry copies of your passport and visa.
- Request housekeeping make up your room while you are at breakfast, rather than leave a "Please Service This Room" sign on the door knob. This sign is a signal to criminals that the room is unoccupied.
- If you are required to use parking stickers in your auto, be sure that it does not indicate your name or room number.

AROUND THE HOTEL

Most first class international hotels have spent a considerable sum to ensure your safety and security. Fire safety equipment, CCTVs, and security patrols are often part of the hotel's security plan. Regardless of the level of security provided by the hotel, you need to become familiar with certain aspects of the security profile of the hotel. This will take on increased significance when you may be forced to stay at the only hotel at a particular location.

- Vary the time and route by which you leave and return to the hotel. Be alert for persons watching your movements.
- Note if hotel security locks certain access points after dark. Plan to use the main entrance upon return to the property.
- Speak with the bellman, concierge and front desk regarding safe areas around the city in which to jog, dine or sightsee. Ask about local customs and which taxi companies to use or avoid.
- Do not take valuables to the spa or work out room. Note if there are house phones available in the event of a confrontation or emergency.
- Be cautious when entering rest rooms in the hotel. On occasion, unauthorized persons use these facilities to deal drugs or engage in prostitution or theft. Female travelers should be alert to placing purses on hangers on the inside of the lavatory doors, or on the floor in stalls—two frequent locations for grab and run thefts.
- Criminals often use areas around public telephones to stage pickpocket activity or theft. Keep briefcases and purses in view or "in touch" while using phones. Caution is urged in safeguarding telephone credit card numbers.

Criminals wait for callers to announce credit card numbers on public phones and then sell the numbers for unauthorized use.
- Purse snatchers and briefcase thieves are known to work hotel bars and restaurants waiting for unknowing guests to drape these items on chairs or under tables only to discover them missing as they are departing. Keep items in view or "in touch". Be alert to scams involving an unknown person spilling a drink or food on your clothing. An accomplice may be preparing to steal your wallet, briefcase or purse.
- The pool or beach area is a fertile area for thieves to take advantage of guests enjoying recreation. Leave valuables in the hotel. Safeguard your room key and camera. Sign for food and beverages on your room bill rather than carry cash.
- Prostitutes take advantage of travelers around the world through various ploys, use of "knock out" drugs, and theft from the victim's room. Avoid engaging persons who you do not know and refrain from inviting them to your guest room.

FIRE SAFETY FOR THE TRAVELER

Fire safety at home and abroad is a matter of thinking ahead, knowing what to do, and keeping your fear under control. Panic and smoke are the most dangerous threats in the case of a fire. To minimize the risk of a fire, the traveler should remember the precautions listed below and where feasible:

- Stay only at hotels, which have smoke detectors and/or sprinklers installed in all rooms and provide information about fire/safety procedures.
- Request a room between the second and seventh floor. Most fire departments do not have the capability to rescue people above the seventh floor level with external rescue equipment (i.e., ladders).
- Inquire as to how guests are notified if there is an emergency.

Your Hotel Room

- Note the location of the fire exits (stairs) on your floor. Count the number of doors between your room and the exit. If there is a fire, you may have to crawl there in the dark.
- Check exit doors to be sure that they are unlocked and that stairwells are clear of obstructions.
- Note the location of fire alarms, extinguishers and hoses and read any fire safety information available in your room.
- Check outside your room window to ascertain if there is a possible escape route that would be feasible in an extreme emergency.

In Case of a Fire

- KEEP CALM — DO NOT PANIC.
- Call the front desk and notify them of the location of the fire.

- Check your door by placing your palm on the door and then on the doorknob. If either feels hot, DO NOT OPEN THE DOOR.
- If it is safe to exit from your room, head for the stairs. TAKE YOUR ROOM KEY WITH YOU; YOU MAY HAVE TO RETURN TO YOUR ROOM.
- If the corridor is full of smoke, crawl to the exit and again check the door before opening it to see if it is hot. The fire could be in the stairwell.
- DO NOT USE THE ELEVATOR!
- If you can not leave your room or the stairwells are unsafe and you must return to your room:
  - Notify the front desk that you are in your room awaiting rescue.
  - Open a window for fresh air. Do not break the window as you may need to close it again if smoke starts to enter from the outside.
  - Fill the tub and sink with water. Soak towels and blankets as necessary to block vents and openings around doors to keep the smoke and fumes out.
  - Attempt to keep the walls, doors and towels covering vents and cracks cool and wet.
  - A wet towel swung around the room will help clear the room of smoke.
  - Cover your mouth and nose with a wet cloth.
  - Stay low, but alert to any signs of rescue from the street or the halls. Let the firemen know where you are by waving a towel or sheet out the window.

IN THE WORK PLACE

The work place, your home away from home. Here you are safe and secure in the one place where you no longer have to worry about what you do or say. WRONG! You can be just as vulnerable here as anywhere else in the country. You probably are safer, but there are still some precautions that should be taken.

- Safeguard all sensitive or proprietary papers and documents; do not leave them lying around in the office or on top of a desk.
- Guard your conversations so that unauthorized personnel are not able to eavesdrop on discussions pertaining to proprietary information, personnel issues or management planning or problems. In many countries, employees are debriefed by the local intelligence or security services in an effort to learn as much as possible about activities of American companies and their personnel.
- Be careful of all communications. Be aware that the monitoring of telephone, telegraph and international mail is not uncommon in some countries.

TRAVELING BY TRAIN

In many countries, railroads continue to offer a safe, reliable and comfortable means of travel between major metropolitan areas. Other countries, however, operate rail systems that use antiquated equipment, are often over crowded and seldom run on time. As a general rule, the more advanced (socially and economically)

DOCUMENT APPENDIX

a country is, the more modern and reliable will be its rail service. Frequently, rail travel provides a more economical method of travel than other modes of transportation, and frequently it is the only available transportation to smaller cities and towns. However, rail travel can present some security risks to the traveler, just like other means of travel.

Railroads are "soft" targets for several types of criminal or terrorist attacks. They operate over open ground and are easily accessible to the public. The tracks on which the trains operate are in the open for most of the distance they cover. This easy accessibility provides an inviting target for bombings and other forms of sabotage.

The railroad terminals and stations are like self-contained cities, open to the public, frequently for 24 hours a day. They provide a fertile ground for pickpockets, purse snatchers, baggage thieves, bombers and other criminals to operate.

Likewise, trains themselves offer similar opportunities to criminals and terrorists. A train is like a hotel on wheels, offering temporary accommodations, such as restaurants, sleeping space, bars and lounges. All of these can be, and often times are, subject to criminal activities including robbery, thievery, bombing and even, albeit rarely, hostage taking.

Security Risks

Generally, railroad terminals and trains are easy targets for the following types of attacks:

- Bombing and other forms of sabotage to railroad tracks, terminals and trains;
- Robberies and burglaries;
- Theft of unattended baggage on board trains and in rail terminals; and
- Thefts from sleeping compartments.
- Just as air travel calls for planning and preparation to lessen the risks of unfortunate experiences while traveling, rail travel also requires certain preventive measures in order to lessen the likelihood of the traveler becoming a victim. Some of these simple, yet effective, precautions can help make a rail trip a comfortable and convenient means of moving between or within many countries of the world.

Some Precautionary Measures

Prior to Departure:

- It should be noted that many cities have more than one railroad station. Travelers should confirm in advance from which station your train will depart. Make certain that you use the right one.
- Make reservations in advance so that you do not have to stand in the frequently long lines at the rail station ticket counters. This is where pickpockets, baggage thieves and purse-snatchers like to operate. Your

hotel concierge can assist in making your reservations and picking up your ticket.
- Travel light and always keep your luggage under your control. In the time it takes to set down your luggage to check a timetable, a baggage thief can make off with it.
- Watch your tickets. Keep them in an inside pocket or purse to lessen the chance that they can be stolen.
- Do not discard your train ticket until completion of your trip and you have left the arrival area. In some countries you will be required to show your ticket at the exit of the arrival station. If you do not have it, you may be required to purchase another one. Hold on to your ticket, whether or not a conductor checks it.
- Make certain that you board the right car and that it is going to your intended destination.
- Find out in advance if your car will have to be switched to another train en route, when and where this will occur, and the name of the stop just prior to the switching point; be prepared accordingly.
- If you have to transfer to another train to reach your destination, determine this in advance and know where you will make the transfer, the time of transfer, and the train number and departure time of your connecting train (and the track number if possible).
- Learn how to tell if you are in the correct car and if it goes to your destination. Name boards on the side of the car will tell you this.

For example, a name board that appears like this:

VENEZIA

Bologna - Firenze

ROMA

shows that the car began in Venice, stops in Bologna and Florence, and terminates in Rome. Next to the steps leading into the car you should see the numeral "1" or "2," or both. The "1" indicates First Class; the "2" indicates Second Class; and "1" at one end of the car and "2" at the other indicates one part of the car is First Class and the other is Second Class.

- Make certain you know how to spell and pronounce the name of your destination city so you can recognize it when announced.
- Be alert to train splitting. This occurs when part of the train is split off and attached to another train while the remainder of the original train then continues on its way. Check with the ticket agent or on-board conductor to determine this.
- Try not to schedule a late night or early morning arrival. You might find yourself stranded at a rail station with no public transportation.
- Arrange to be met at your arrival point whenever possible.

DOCUMENT APPENDIX

On Board the Train

- If possible, check unneeded luggage into the baggage car.
- Keep your luggage with you at all times. If you must leave your seat, either take the luggage with you or secure it to your seat or the baggage rack with a strong cable-lock.
- Try to get a window seat. This provides a quick means of escape in the event of an accident.
- Have necessary international documents, including your passport, handy and ready for inspection by immigration officials at each border crossing.
- Always keep your camera and other valuables with you at all times.
- If you have a private compartment, keep the door locked and identify anyone wishing to gain access. Know the names of your porters and ask them to identify themselves whenever entering your compartment.
- When in your compartment, be aware that some train thieves will spray chemicals inside to render the occupant(s) unconscious in order to enter and steal valuables. A locked door will at least keep them out.
- If you become suspicious of anyone, or someone bothers you, notify the conductor or other train personnel.
- If you feel you must leave the train temporarily at a stop other than your destination, make certain that you are not left behind.
- An understanding of military time (the so-called 24-hour clock) will make it easier for you to understand the train schedule.
- Make certain you have currency from each of the countries through which you will be traveling. In some lesser-developed countries (and on some trains) it may be advisable to carry your own food and water.

Upon Arrival

- Make certain that you depart from the train at the correct location.
- Use only authorized taxis for transportation to your hotel or other destination.
- Be alert to criminals such as pickpockets, baggage thieves and/or unauthorized taxi drivers/guides.
- If you do not have a hotel reservation, go to the in-station hotel services and reservations desk for help in obtaining a hotel room.

DRIVING ABROAD

Obtain an International Drivers Permit (IDP). This can be purchased through your AAA Club. Have your passport photos and a completed application. There will be a fee involved. Carry both your IDP and your state driver's license with you at all times.

- Some countries have a minimum and maximum driving age. Check the laws before you drive in any country.

- Always "buckle up." Some countries have penalties for people who violate this law.
- If you rent a car, always purchase the liability insurance. If you do not, this could lead to financial disaster.
- As many countries have different driving rules, obtain a copy of them before you begin driving in that country.
- If the drivers in the country you are visiting drive on the opposite side of the road than in the U.S., practice driving in a less populated area before attempting to drive during the heavy traffic part of the day.
- Be aware of the countryside you will be driving in. Many countries require you to honk your horn before going around a sharp corner or to flash your lights before passing.
- Find out before you start your journey that has the right of way in a traffic circle.
- Always know the route you will be traveling. Have a copy of a good road map, and chart your course before beginning.
- Do not pick up hitchhikers or strangers.
- When entering your vehicle, be aware of your surroundings.

PERSONAL CONDUCT OVERSEAS

A hostile or even friendly intelligence organization is always on the lookout for sources who are vulnerable to coercion, addictions, greed or emotional manipulation. To eliminate, or at least diminish, the possibility of your doing something inadvertent that would bring your activities to the special attention of one of these agencies, here are some
   DO NOT's to remember:

- DO NOT do anything which might be misconstrued or reflect poorly on your personal judgment, professional demeanor, or embarrassing to you and/or your company.
- DO NOT gossip about character flaws, financial problems, emotional relationships or marital difficulties of anyone working for the company, including yourself. This type of information is eagerly sought after by those who would like to exploit you or another employee.
- DO NOT carry, use or purchase any narcotics, marijuana, or other abused drugs. Some countries have very stringent laws covering the import or use of medications and other substances. If you are using a prescribed medication that contains any narcotic substance or other medication that is subject to abuse, such as amphetamines or tranquilizers, carry a copy of the doctor's prescription for all medications and check your local restrictions and

requirements prior to departure. Some countries may require additional documentation/certification from your doctor.
- DO NOT let a friendly ambiance and alcohol override your good sense and capacity when it comes to social drinking. In some countries, heavy drinking in the form of toasting is quite common, and very few westerners can keep up with a local national when it comes to drinking the national brew. An intoxicated or hung over business negotiator could, if they are not careful, prove to be very embarrassing to themselves and expensive to the company. In these situations, prudence is essential.
- DO NOT engage in "Black Market" activities such as the illegal exchange of currency, or the purchase of religious icons or other local antiquities.
- DO NOT accept or deliver letters, packages or anything else from anyone unknown to you. You have no way of knowing what you are carrying and it could result in your being arrested for illegally exporting a prohibited item.
- DO NOT engage in any type of political or religious activity, or carry any political or religious tracts or brochures, or publications likely to be offensive in the host country, such as pornography or mercenary/weapons.
- DO NOT photograph anything that appears to be associated with the military or internal security of the country, including airports, ports, or restricted areas such as military installations. If in doubt, DO NOT.
- DO NOT purchase items that are illegal to import such as endangered species or agricultural products.

I'VE BEEN ARRESTED!—WHAT DO I DO NOW?

Foreign police and intelligence agencies detain persons for a myriad of reasons or for no other reason than suspicion or curiosity. The best advice is to exercise good judgement, be professional in your demeanor and remember the suggestions and hints that are listed in this booklet. But, if you are detained or arrested for some reason, here are some points to remember:

- DO ask to contact the nearest embassy or consulate representing your country. As a citizen of another country, you have this right; but that does not mean that your hosts will allow you to exercise that right. If you are refused or just ignored, continue to make the request periodically until they accede and let you contact your embassy or consulate.
- DO stay calm, maintain your dignity and do not do anything to provoke the arresting officer(s).
- DO NOT admit anything or volunteer any information.
- DO NOT sign anything. Often, part of the detention procedure is to ask or tell the detainee to sign a written report. Decline politely until such time as the document is examined by an attorney or an embassy/consulate representative.

- DO NOT accept anyone on face value. When the representative from the embassy or consulate arrives, request some identification before discussing your situation.
- DO NOT fall for the ruse of helping the ones who are detaining you in return for your release. They can be very imaginative in their proposals on how you can be of assistance to them. Do not sell yourself out by agreeing to anything. If they will not take no for an answer, do not make a firm commitment or sign anything. Tell them that you will think it over and let them know. Once out of their hands, contact the affiliate or your embassy for protection and assistance in getting out of the country.

TARGETING RECOGNITION

Any person traveling abroad on business should be aware of the fact that they could be targeted by an intelligence agency, security service or, for that matter, a competitor if they are knowledgeable of, or carrying, sensitive or proprietary information. In the course of doing business abroad, there are certain indicators that may occur which should be recognized as potential hazards and indicative of unwarranted interest in your activities. These situations should be closely scrutinized and avoided if at all possible. A few of the most common scenarios that have been utilized by intelligence/security services and have led to successful targeting and acquisition of information are listed below:

- Repeated contacts with a local or third country national who is not involved in your business interests or the purpose of your visit, but as a result of invitations to social or business functions, appears at each function. This individual's demeanor may indicate more than just a passing interest in you and your business activities.
- A close personal social relationship with a foreign national of a hostile host government is often unavoidable for business reasons. In these instances, be cautious and do not allow the relationship to develop any further than the strictly business level.
- Be suspicious of the accidental encounter with an unknown local national who strikes up a conversation and wants to:
  - Practice English or other language.
  - Talk about your country of origin or your employment.
  - Buy you a drink because they have taken a liking to you.
  - Talk to you about politics.
  - Use a myriad of other excuses to begin a "friendly" relationship.

If any of the above or anything else occurs which just does not ring true, BE SUSPICIOUS!! It may be innocent but, exercise prudence and good judgment.

DOCUMENT APPENDIX

SURVEILLANCE RECOGNITION

The subject of surveillance is extremely important to anyone conducting business abroad. Surveillance could be indicative of targeting for reasons other than interest by a foreign intelligence or security service. Terrorists and criminals also use surveillance for operational preparation prior to committing other terrorist or criminal acts. It should be noted, however, that the normal business traveler, who only spends a few days in each city and has a low profile, is not really a viable target for terrorists and the risk is very low.

The real terrorist threat to a traveler is that of being at the wrong place at the wrong time and becoming an inadvertent victim of a terrorist act.

Surveillance is an assessment of vulnerabilities in an attempt to determine any information available, from any source, about you or your activities, such as lifestyle or behavior that can be used against you. If the intended target recognizes the fact that he or she is under surveillance, preventive measures can be taken that will hopefully deter further interest. As an example, if the surveillant(s) realizes that he or she has been spotted, then the assumption must be that the operation has been compromised and that the police have been notified or other preventive measures have been taken. On the other hand, if a traveler is being scrutinized by a foreign intelligence or security agency, the surveillance may well continue.

Surveillance takes many forms, from static, such as an observer physically or electronically watching or monitoring your activities in your hotel room or office, to mobile surveillance where the individual being watched is actually followed either on foot or by vehicle.

How do you recognize surveillance? There is only one way: be ALERT to your surroundings. As a traveler, you probably will not be at any one location long enough to know what the norm is in your surroundings, and this puts you at a disadvantage. You will not realize that the person sitting in the car across the street is a stranger and should not be there, whereas a resident would immediately become suspicious.

Be observant and pay attention to your sixth sense. If you get the funny feeling that something is not right or that you are being watched, PAY ATTENTION! That sixth sense is trying to tell you something, and more often than not it will be right.

In any event, report your suspicions or any information to the general manager of the local affiliate or your embassy or consulate just in case something does occur. If there is any question about what actions should be taken, and guidance is not available from the affiliate, contact your embassy or consulate and they will advise you as to what you should do and whether or not the information should be reported to the local authorities. But, the most important thing you should do is making sure that your demeanor is professional and everything you do is above board and not subject to compromise.

If you have reason to believe that you are under surveillance, here is what you should NOT do:

- DO NOT try to slip away or lose the followers as this will probably alert them and belie the fact that you are just a businessperson or tourist going about your business.
- In your hotel room, assume that the room and telephone are being monitored. DO NOT try to play investigator and start looking for electronic listening devices. This again could send the wrong signals to the surveillant. Just make sure that you do not say or do anything in your hotel room that you would not want to see printed on the front page of the *New York Times*.

Response to Targeting

If you have any reason to believe that you are targeted by an intelligence or security service, there is really only one course of action to follow. Report your suspicions to the affiliate or embassy or consulate and follow their guidance.

HOSTAGE SURVIVAL

Any traveler could become a hostage. The odds of that happening are extremely low when the number of travelers is compared to the number of people that have actually become a hostage. However, there is always that slim chance that a traveler could end up being in the wrong place at the wrong time. With this in mind, the traveler should make sure that his/her affairs are in order before they travel abroad. Items of particular importance to an individual in a hostage situation are the currentness of an up-to-date will, insurance policy and a power of attorney for the spouse. If these items have been taken care of before departure, the employee will not have to worry about the family's welfare and the hostage can focus all of his/her efforts on the one thing of paramount importance and that is SURVIVAL!!

To survive, travelers should realize that there are certain dynamics involved in a hijacking or a kidnapping, and, to increase their ability to survive, they must understand how these interacting forces affect the end result. Each individual involved in an incident of this type will have an impact on the eventual outcome. One wrong move by either a victim or a perpetrator could easily result in a disaster rather than a peaceful conclusion to the incident.

The first thing that a traveler should remember is that he or she is not the only one that is scared and nervous. Everyone involved is in the same emotional state, including the perpetrators. Fear can trigger a disaster, and it does not take much for some individuals to set off a defensive spate of violence. Whether it is a demonstration of violence to reinforce a demand or to incite fear in the minds of the hostages, the violence will be motivated by fanaticism and/or fear and that violence will be directed at the person(s) who are perceived to be a threat or a nuisance to the hijackers.

To minimize the possibility of being selected for special attention by the perpetrators and to maximize your ability to survive a hostage situation, here are some guidelines to remember:

Hijacking Survival Guidelines

The physical takeover of the aircraft by the hijackers may be characterized by noise, commotion, and possibly shooting and yelling, or it may be quiet and methodical with little more than an announcement by a crew member. These first few minutes of the hijacking are crucial:

- Stay calm, and encourage others around you to do the same.
- Remember that the hijackers are extremely nervous and are possibly scared.
- Comply with your captor(s) directions.
- If shooting occurs, keep your head down or drop to the floor.
- Remain alert.

Once the takeover of the aircraft has occurred, you may be separated by citizenship, sex, race, etc. Your passport may be confiscated and your carry-on luggage ransacked. The aircraft may be diverted to another country. The hijackers may enter into a negotiation phase, which could last indefinitely, and/or the crew may be forced to fly the aircraft to yet another destination. During this phase passengers may be used as a bargaining tool in negotiations, lives may be threatened, or a number of passengers may be released in exchange for fuel, landing/departure rights, food, etc. This will be the longest phase of the hijacking:

- If you are told to keep your head down or maintain another body position, talk yourself into relaxing into the position; you may need to stay that way for some time.
- Prepare yourself mentally and emotionally for a long ordeal.
- Do not attempt to hide your passport or belongings.
- If addressed by the hijackers, respond in a regulated tone of voice.
- Use your time wisely by observing the characteristics and behavior of the hijackers, mentally attach nicknames to each one and notice their dress, facial features and temperaments.
- If you or a nearby passenger are in need of assistance due to illness or discomfort, solicit the assistance of a crew member first—do not attempt to approach a hijacker unless similar assistance has been rendered by them for other passengers.
- If the hijackers single you out, be responsive but do not volunteer information.

The last phase of the hijacking is resolution, be it by use of a hostage rescue team or resolution through negotiation. In the latter instance, the hijackers may simply surrender to authorities or abandon the aircraft, crew and passengers. In the case of a hostage rescue operation to resolve the hijacking:

- The characteristics of a hostage rescue force introduction into the aircraft will be similar to the hijacker's takeover—noise, chaos, possibly shooting—the rescue force is re-taking control of the aircraft.

- If you hear shots fired inside or outside the aircraft, immediately take a protective position—put your head down or drop to the floor.
- If instructed by a rescue force to move, do so quickly, putting your hands up in the air or behind your head; make no sudden movements.
- If fire or smoke appears, attempt to get emergency exits open, and use the inflatable slides or exit onto the wing.
- Once you are on the tarmac, follow the instructions of the rescue force or local authorities; if neither are there to guide you, move as quickly as possible away from the aircraft and eventually move towards the terminal or control tower area.
- Expect to be treated as a hijacker or co-conspirator by the rescue force; initially you will be treated roughly until it is determined by the rescue force that you are not part of the hijacking team.
- Cooperate with local authorities and members of the U.S. Embassy, Consulate or other U.S. agencies in relating information about the hijacking.
- Onward travel and contact with family members will be arranged by U.S. authorities as soon as possible.

KIDNAPPING SURVIVAL GUIDELINES

Kidnapping can take place in public areas where someone may quietly force you, by gunpoint, into a vehicle. They can also take place at a hotel or residence, again by using a weapon to force your cooperation in leaving the premises and entering a vehicle. The initial phase of kidnapping is a critical one because it provides one of the best opportunities to escape.

- If you are in a public area at the time of abduction, make as much commotion as possible to draw attention to the situation.
- If the abduction takes place at your hotel room, make noise, attempt to arouse the suspicion or concern of hotel employees or of those in neighboring rooms—minimally, the fact that an abduction has taken place will be brought to the attention of authorities and the process of notification and search can begin. Otherwise, it could be hours or days before your absence is reported.
- Once you have been forced into a vehicle, you may be blindfolded, physically attacked (to cause unconsciousness), drugged, or forced to lie face down on the floor of the vehicle. In some instances, hostages have been forced into trunks or specially built compartments for transporting contraband.
- Do not struggle in your confined state; calm yourself mentally, concentrate on surviving.
- Employ your mind by attempting to visualize the route being taken, take note of turns, street noise, smells, etc. Try to keep track of the amount of time spent between points.

DOCUMENT APPENDIX

- Once you have arrived at your destination, you may be placed in a temporary holding area before being moved again to a more permanent detention site. If you are interrogated:
  - Retain a sense of pride but be cooperative.
  - Divulge only information that cannot be used against you.
  - Do not antagonize your interrogator with obstinate behavior.
  - Concentrate on surviving; if you are to be used as a bargaining tool or to obtain ransom, you will be kept alive.

After reaching what you may presume to be your permanent detention site (you may be moved several more times), quickly settle into the situation:

- Be observant—Notice the details of the room, the sounds of activity in the building and determine the layout of the building by studying what is visible to you. Listen for sounds through walls, windows or out in the streets, and try to distinguish between smells.
- Stay mentally active by memorizing the aforementioned details. Exercise your memory and practice retention.
- Keep track of time. Devise a way to track the day, date and the time, and use it to devise a daily schedule of activities for yourself.
- Know your captors. Memorize their schedule, look for patterns of behavior to be used to your advantage, and identify weaknesses or vulnerabilities.
- Use all of the above information to seek opportunities to escape.
- Remain cooperative. Attempt to establish rapport with your captors or guards. Once a level of communication is achieved, try asking for items that will increase your personal comfort. Make them aware of your needs.
- Stay physically active even if your movement is extremely limited. Use isometric and flexing exercises to keep your muscles toned.
- If you detect the presence of other hostages in the same building, devise ways to communicate.
- DO NOT be uncooperative, antagonistic, or hostile towards your captors. It is a fact that hostages who display this type of behavior are kept captive longer or are singled out for torture or punishment.
- Watch for signs of Stockholm Syndrome, which occurs when the captive, due to the close proximity and the constant pressures involved, begins to relate to, and empathize with, the captors. In some cases, this relationship has resulted in the hostage become empathetic to the point that he/she actively participates in the activities of the group. You should attempt to establish a friendly rapport with your captors, but maintain your personal dignity and do not compromise your integrity.
- If you are able to escape, attempt to get first to a U.S. Embassy or Consulate to seek protection. If you cannot reach either, go to a host government or friendly government entity.

CONCLUSION

It is no wonder that most U.S. business people consider business travel hard work —and one of the most stressful aspects of their job. The running, waiting, and anxiety associated with travel can take its toll on the mind and body. Add an unfamiliar location, a foreign language, and a different culture to the situation and you have the potential for all sorts of problems.

As pointed out in this publication, the keys to safe travel are planning and sound security practices. Proper planning ensures your logistical plan is in place and you have the necessary background information to support your itinerary. Incorporating sound security practices into your travel routine will reduce the likelihood of problems. Together, these keys allow you to get on with the real purpose of your trip.

GLOSSARY

Abridged Security Policy Manual is a simplified manual that tells users what they need to know in order to do their jobs securely. It is important that it be short, relevant, easy to read, but with pointers to the full policies.

Access is the ability to use an information system or enter a physical location.

Access Authority is the person responsible for monitoring and granting access rights to other people.

Access Control is the process of deciding who is given or denied access to an information system or a physical location.

Access Control List is a register of people who have been granted access and the level of access each is allowed.

Actionable Activities are behaviors that create legal liabilities.

Active Competitive Intelligence involves seeking information about strategy, tactics, targets, and technology.

Agroterrorism refers to terrorist attacks on agricultural targets.

American Chemistry Council (**ACC**) is a trade group for the chemical industry that was founded in 1872. It provides information about the industry and campaigns on issues critical to the chemical industry. It created the Responsible Care© program in 1988 to reduce accidents and emissions.

Americans with Disabilities Act (**ADA**) is a federal civil rights law prohibiting the exclusion of people with disabilities from everyday activities, and it includes a requirement that the safety and evacuation special needs of all employees are met.

Antivirus Software scans files for infections. An infection is detected by discovering patterns associated with an infection, which means a program can find only viruses it knows.

Area Challenge involves asking people you do not recognize who they are and why they are in a particular area.

Asset's Vulnerability is determined by the strength of its mitigating controls against the capabilities of a threat.

Attacks on an Organization are crises that involve premeditated actions intended to harm the organization and/or its personnel.

Attack Signature is a sign indicating an attempt at unauthorized access.

Audit Trail records who has accessed an information system and what users did during their access.

Automated External Defibrillator (AED) is a computerized medical device. It can check an individual's heart rhythm, recognize a rhythm that requires a shock, and advise the rescuer when it is necessary to deliver an electric shock to a victim of sudden cardiac arrest to restore normal rhythm.

Automated Notification Systems, also known as mass notification systems, are designed to deliver a large volume of text, voice, or data messages to a potentially large audience and in an extremely short amount of time. Normally, the system can send messages through multiple communication channels—not only telephone but also e-mail, pager, fax, instant messenger, PDA, and other channels. To facilitate two-way communication, the notification system should be able to receive an active response, such as a keypad entry, to confirm successful delivery of a message.

Backup is term for copying a file or program.

Badge Challenge involves asking people who they are and why they are in a location if they do not have identification badges.

Baselines, or **Minimum Security Baselines (MSBs),** are operating system specific and provide extensive and minute details of OS configuration settings. The minimum level of assurance that you can have will be your baseline.

Bell-LaPadula Model is an information flow security model. The security clearance and the need to know of each subject and the classification of every object are stored in an authentication database. The security clearance and the need to know of a subject are compared to the classification of the object, if a request for access is initiated. Access is permitted only if the access is in accordance with the stated security policy.

Biba Integrity Model is based on access control rules that ensure data integrity. This model organizes the subjects and objects into groups described by a data integrity level. The subject is restricted from writing to data in a higher data integrity level than its own and cannot be corrupted by data written in a lower level than its own. This model is characterized by the no write up and no read down.

Biometric Information is the electronic storage of physiological and behavioral data. Typical biometric information includes fingerprints, iris recognition, retina recognition, facial recognition, hand geometry, voice recognition, signature verification, and keystroke dynamics.

Biometric Sample is the physiological and behavioral data a user provides to be examined to verify a person's identity and determine access.

Biometric System is an automated system that stores biometric information, compares the stored information, determines whether there is a match, and decides whether the identity has been verified.

Biometric Template is the biometric information about a person that is stored in the biometric system and used as a comparison point to later samples to determine whether or not a person is granted access.

Biometrics use body parts (physiological characteristics) and actions taken by a person (behavioral characteristics) to determine or verify identity.

Black Blocs is the spontaneous collection of anarchists dressed in black.

Brewer and Nash Model provides controls such that there can be no conflict of interest between the subject and the object. A conflict of interest occurs when a trusted subject

GLOSSARY

has competing professional and personal interests. A conflict of interest does not imply that anything improper or unethical has occurred; just that it will be difficult for the subject to make a completely impartial decision. In many cases, third-party verification is required.

Business Continuity can be all the initiatives taken to assure the survival, growth, and resilience of the enterprise.

Business Continuity Plan is an extension of that fundamental business plan that enables a company to achieve its strategic goals under extraordinary conditions.

Business Impact Analysis (BIA) determines systematically how (various) disasters might disrupt processes and what such disruptions would "cost" your business.

Challenges occur when a group accuses the organization of acting improperly or unethically.

Chemical Facility is any plant or warehouse where chemicals are used, manufactured, or stored.

Citizen Corps was created by the Department of Homeland Security and encourages people to take personal responsibility for preparedness by getting trained in first aid and emergency skills and to volunteer to support local emergency management.

Clark-Wilson Integrity Model is designed to formalize information integrity, which is maintained by ensuring unauthorized subjects do not corrupt data in error or with malicious intent. The model defines both certification rules and enforcement rules. Users must use a program to modify the data. This additional layer of protection helps to ensure the integrity of the data. Users are given authorization to use only programs they are allowed to use. The programs permit the user to do only certain things. A third piece of the model revolves around auditing. The model requires the tracking of information that is received from outside the trusted system.

Cloaking refers to efforts to screen an organization's information from the eyes of competitors.

Common Vulnerability Scoring System (CVSS) is an evolving standard for rating software vulnerabilities. The National Infrastructure Advisory Council (NIAC) is responsible for developing the CVSS.

Community Emergency Response Team (CERT) is a program that educates people about emergency preparedness. Community members are trained for hazards that are likely to affect their areas and in basic emergency response skills. Those skills can include first aid, search and rescue, and fire safety.

Competitive Intelligence is the timely, relevant, accurate, and unbiased intelligence on potential threats to an organization's competitive position.

Compromise is when information is given to an unauthorized user or when a security policy is violated, allowing unauthorized or unintentional release of information.

Computer Security Incident is when a security policy or accepted use policy is violated or a violation is attempted.

Controller manages and directs an exercise.

Corporate or Industrial Espionage is the illegal or unethical efforts to collect information for commercial gain rather than national interests.

Corporate Social Responsibility is the management of actions designed to affect an organization's impacts on society. It becomes the actions a company takes to further the social good.

Corruption can be the abuse of commercial position for personal gain. This is called private/private corruption.

Countermeasures are the efforts designed to reduce the vulnerability of an information system or a physical location.

Countersurveillance is a process of collecting and analyzing data. Countersurveillance seeks data on potential attacks.

Criminal History search uses the Social Security number to construct a comprehensive criminal record search. County records are searched for each address connected to the Social Security number along with a search of federal court records.

Crisis is an unpredictable event that poses a significant threat of harm to an organization, industry, or stakeholders if it is handled improperly.

Crisis Command, or **Control Center,** is the physical location where the crisis team meets.

Crisis Management is a set of factors designed to combat crises and to lessen the actual damage inflicted by a crisis.

Crisis Management Plan is a carefully arranged selection of information that can aid a crisis team.

Crisis Management Team is assigned to handle the crisis response and to develop crisis preparation. The team is cross-function, meaning it is comprised of people from different areas of the organization. They meet in a designated area during a crisis and lead the crisis management effort.

Crisis Portfolio/Families are groupings of similar crises used to develop a set of crisis management plans.

Crisis Sensing is the process of identifying and disarming potential crises.

Crisis Sensing Mechanism is a systematic approach to collecting and analyzing prodromal information/crisis warning signs.

Crisis Spokespersons are the organization representatives that speak for the organization during a crisis. Their primary role is to speak with the news media.

Customs-Trade Partnership Against Terrorism (C-TPAT) is a voluntary partnership between the private and public sectors intended to reduce the threat of terrorism while also facilitating the speed of international trade.

Cyberslacking is when employees use the Internet, including e-mail, for personal use and waste time while at work.

Data Mirroring refers to the real-time backup of data.

Decryption is the process of converting ciphertext to plaintext.

Defamation is a false statement that hurts someone's character or good name.

Defensive Competitive Intelligence is preventing competitors from collecting information about your organization. It tries to make it more difficult for competitors to find useful information about your company.

Dictionary Attacks are attempts to break password protect by simply guessing words from the dictionary.

GLOSSARY

Digital Signature provides verification that an e-mail is really from the person claiming to have sent it and that the e-mail has not been altered.

Digital Video Surveillance is networked IP-based. The video surveillance is a component of the IT network. Each camera has an IP address and is controlled centrally through a software application. The video surveillance transition to digital is another illustration of the convergence of physical and IT security.

Distributive Justice involves perceptions with the fairness of outcomes. Employees determine whether the effort they put into their jobs results in equal rewards from their jobs.

Document Management System tracks a document's versioning/history, maintaining a record of revisions that could prove crucial if verifying when a specific requirement of a policy took effect were to become necessary. It also helps with archiving, retention, distribution, and work flow.

Drill is a supervised, coordinated activity used to test a specific operation of part of a response plan. The idea is to work on smaller parts of the response plan before integrating these components into a large drill that tests them all.

Drug Screening or **Testing** uses samples of urine, oral fluid, or hair to determine whether a person has used certain substances.

Drug Tests are processes used to detect the presence of drugs.

Due Care is the care that a reasonable person or company would exercise to secure its data, physical security, or other protective actions in an organization.

Due Diligence determines whether due care has actually occurred.

Dumpster Diving involves digging through trash for information.

Ecodefense are strategies to defend the environment.

Ecoterrorism involves attacks by environmental groups on individuals or property.

Education Verification verifies all degrees claimed by the applicant. The information obtained includes the name of the institution, date of graduation, dates of attendance, degree obtained, and type or field of study. Applicants who are currently students can be checked for degree progress, field of study, and planned graduation date.

Emergency is any unplanned event that can cause deaths or significant injuries to employees, customers or the public; or that can shut down your business, disrupt operations, cause physical or environmental damage, or threaten the facility's financial standing or public reputation.

Emergency Operations Center is a base of operations for those charged with managing emergencies, directing emergency response teams, assessing and controlling physical damage, coordinating with public safety officials, and providing status reports to management.

Emergency Planning & Community Right-to-Know Act (**EPCRA**) is part of Title III of the 1986 Superfund Amendments and Reauthorization Act (SARA). EPCRA seeks to identify the amounts of chemicals located at, or released from, facilities; to understand the potential problems that hazardous materials pose to the surrounding communities and environment; and to provide information to the public/local community and the local emergency planning and response organizations. EPCRA has four sections that cover emergency planning, emergency notification, community right-to-know hazardous chemical reporting, and community right-to-know toxic chemical release inventory reporting.

Emergency Preparedness and Response centers on a plan of action to commence during or immediately after a disaster to prevent loss of life and minimize injury and property damage. It includes developing emergency response procedures and establishing training for all employees on the proper actions to take in response to emergency and disaster situations and training for employee emergency response teams (emergency response teams, fire safety teams) on their roles and responsibilities. Also included is the acquisition and maintenance of life safety systems and emergency supplies and equipment.

Emergency Response Teams (ERTs) are the people responsible for executing an organization's emergency response and preparing for that response. The team receives training on the overall emergency management program and its specific roles and responsibilities.

Employee Background Check or **Investigation** is an inquiry into a person's character, personal characteristics, or general reputation.

Employee Special Needs Form is a means of identifying employees with special needs in emergency situations. The forms are distributed and employees are invited, not required, to complete and return the form.

Employer Reference Check verifies past employment by checking with the human resource departments of previous employers. The search can check dates of employment and job titles. Applicants must grant approval for this contact.

Encryption is a coded message, an excellent choice for sensitive information. Unless the person has the key for decoding the message, all that will be seen is a random series of characters, letters, and numbers. It is a process of converting plaintext into ciphertext using an encryption program.

End-to-End Encryption involves the process of encrypting a message and sending it through a system.

Ethical Conduct Audit is a systematic measurement of the people, perceptions, behaviors, decisions, and processes in the context of their work activities using a reliable and valid method to discern with greater certainty the status of ethics and compliance within the company.

Ethical Misconduct Disasters (EMDs) are specific, unexpected, and nonroutine unethical events or a series of unethical events that creates significant operational disruptions and threatens, or is perceived to threaten, an organization's continuity of operations.

Ethics and Compliance Programs cover behaviors related to compromising the following: customer or marketplace trust, shareholder or organizational trust, employee trust, supplier trust, and public or community trust.

E-vaulting is a general term describing a number of different data backup methods.

Exercise is a focused practice activity that puts organizational personnel into a simulated situation requiring them to act as they are expected to in a real event.

Expected Actions are the actions or choices that you want players in an exercise to carry out in order to demonstrate their competence. In short, it is what you want people to do so you can evaluate their proficiency.

GLOSSARY

Facilitation Bribes (Grease) are small amounts of money that speed up legitimate transactions.

False Acceptance is when a biometric system wrongly identifies a person or incorrectly verifies an impostor.

False Acceptance Rate is the probability that a biometric system will make an error by incorrectly identifying a user or allowing entry to an impostor.

False Positive is created when a person asks a question that when answered provides a sense of comfort that does not reflect the actual state of affairs.

False Rejection is when a biometric system does not identify an applicant or fails to verify a true user.

False Rejection Rate is the probability that a biometric system fails to properly identify a legitimate user of a system.

Financial Risk is comprised of multiple areas of risk such as credit risk, interest rate risk, and market risk, to name just a few.

Firewalls can be hardware or software designed to protect your computer from attackers.

Food Safety Inspection Service (**FSIS**) has resources to help small and very small companies address food security issues. The FSIS provides model plans for a variety of food industry establishments that can be adapted to fit the security and budgetary needs of a company.

Food Security guards against intentional acts of contamination or tampering.

Foreign Corrupt Practices Act requires any publicly traded company to maintain accurate records of all transactions and to have an adequate set of internal accounting controls. The act is designed to prevent corruption by making it difficult to hide money used for bribes.

Full-Scale Exercise simulates an actual event as closely as possible. Equipment is used, people mobilized, and simulated victims appear. The organization often coordinates the full-scale exercise with local emergency responders. These responders get practice, and the organization better understands how to coordinate responses with these units. Your people are on the scene moving and using equipment as they would in the actual event, as well as coping with the simulated victims.

Functional Exercise tests the capabilities of the organization to respond to a simulated event. As a simulation, it is interactive and time pressured. Events and information unfold in real time. The functional exercise allows an organization to examine the coordination, interaction, and integration of its roles, policies, procedures, and responsibilities.

Functions in exercises are actions or operations required during a response.

General Crisis Management Skills are knowledge, skills, and traits that are unique to a crisis management team. The two key skills are decision making and listening.

Graham-Denning Model addresses those issues involving granting rights to users and how the users can use those rights on objects. The model uses eight basic protections.

Guidelines are written to the same level of technical detail that standards are. The difference is that guidelines are sets of procedures that are suggestions, not mandates.

Hazard is the technical or scientific assessment of risk based on likelihood and impact.

Hazard Vulnerability assesses the factors that can cause the most damage to your facility and your operations.

Hidden Mode prevents other users from discovering your wireless connection. Only paired devices can find one another in this mode.

Hot Site is the term used in business continuity to indicate a facility that is fully operational and can be used if an existing facility is damaged or cannot be accessed for some reason.

Inappropriate Usage is when someone violates accepted computing practices.

Incident is a violation or threat of a violation of information security policies or accepted use policies.

Indirect Bribe is when an organization pays some agent or intermediary who helps negotiate a deal and employs bribes as part of its negotiations.

Information Security (**Cyber Security**) describes efforts to prevent, detect, and respond to attacks on a company's information with a focus on computer systems.

Information Security Awareness is a communication effort designed to make users aware of information security concerns and the need to address them.

Information Security Awareness and Training Program is a communication program designed to teach users about information security policies and procedures.

Information System is a defined set of information resources that serves to collect, process, maintain, use, share, and disseminate information.

Insider Threat is when employees, contract workers, or former employees use their knowledge of an organization's information system to violate information security.

Instant Background Check is a computer search of a rather limited database. Such checks are fast but not very accurate.

Integrity represents efforts that seek to protect information from improper modification or destruction.

Integrity Management includes creating formal and informal systems to ensure that employees will act in ways that are legally compliant as well as enact the corporate code of conduct goals in consistently professional, ethical, and desirable ways. Integrity management is intertwined with managing the larger corporate culture and informal reward/motivation processes that impact employee decisions and behaviors in ways that transcend policies printed in a written code of conduct.

Interactional Justice involves perceptions of the quality of interpersonal treatment during procedural justice episodes. Did the people in charge of the procedural justice treat those involved with dignity and respect?

Intrusion Detection System is software that scans for suspicious activity and warns the administrator of these actions.

Intrusion Prevention System is a system that detects an intrusion and takes actions to prevent it.

Issues Management is a systematic approach intended to shape how an issue, a type of problem whose resolution can impact the organization, develops and is resolved.

Key Exchange is when people exchange public keys so that they can exchange secure messages.

GLOSSARY

Key Pair is a set of keys that allows a message to be encrypted and then decrypted using the other key. The pair is usually a public key and its matching private key.

Keys are used to create a digital signature and each signature has two keys, a private key and a public key. A private key is used by the person sending the e-mail message and is password protected. A public key is made available to the receiver of the message and is used to verify the signature.

Learnable Risk is one that management can make less uncertain if it can commit the time and resources to learn more about it.

Least Privilege Principle is when each employee is given only the minimum access required to complete his or her job.

Legal Risks are the uncertainties resulting from legal actions or ambiguity in the applicability or interpretation of laws, contracts, or regulations.

Libel is written defamation.

Logic Bomb is a form of malicious code that is embedded in a system and timed to activate at a later date.

Low-Hanging Fruit is the easiest target for social engineers.

Management Misconduct is when management has intentionally placed stakeholders at risk or violated legal or regulatory statues.

Military Record Verification contacts military branches to verify dates served and type of discharge.

Mitigation involves actions taken to eliminate or reduce a hazard and the risk it poses to an organization.

Monkeywrenching involves violent forms of ecodefense including arson, tree spiking, billboard vandalism, road reclamation, and ecotage (eco-sabotage).

Motor Vehicle Records Check provides information on an applicant's driving history. The data would include speeding or moving violations, chargeable accidents, DUIs/DWIs, suspensions or revocations, and accumulation of points.

Multilevel Security Policy is one in which the computer or database contains information that has different security classifications (top secret, secret, confidential, unclassified).

Muster Area is a designated area where people are to gather during emergencies.

Narrative briefly describes the events that occurred right before the exercise began. The narrative sets the mood for the exercise as well as the stage for later actions by providing the initial information players use to make choices and take actions.

Negligent Hiring occurs when a company fails to screen employees properly and the hiring results in injuries.

Negligent Retention occurs when a company keeps an employee on staff after learning this employee is unsuitable and injuries occur.

Negligent Supervision involves a failure to provide the proper oversight to ensure that employees perform their jobs properly.

Noncompete Agreements are part of an employment contract or separate agreements that prohibit an employee from working in the same business for a specified length of time. The idea is to prevent employees from using a company's confidential information against it.

Nondisclosure Agreements (NDAs) are contracts that require parties to protect the confidentiality of secret information that they learn during employment or through some other business transaction.

Noninterference Model states that transactions at one level of a system will not affect the state of the system at a lower level of the system. If a transaction that has occurred at a level above me changes my state of the system, then I might be able to deduce what that transaction was.

Non-Repudiation is the guarantee that the information sent has been delivered and the receiver has proof of the sender's identity.

Object is the thing that is requested in an information system.

Operational Risks include a shortage in the workforce or equipment breakdown or other events that may impact businesses.

Organizational Culture is the way things are done in an organization. It represents the beliefs, values, norms, and practices that are taken for granted in an organization and guide actions in the organization.

Orientation Session introduces the plan and process. The purpose is to familiarize people with their roles, plans, procedures, and equipment. It can also help them understand how to coordinate activities and clarify their responsibilities.

Outrage is the emotional and subjective reaction to risk.

Outrage Management is a type of risk communication intended to help people to understand that a risk is not as bad as they think it is.

Outside Threat refers to unauthorized entry by a person who is not a member of a secure domain.

Outsourcing is when management shifts noncore operations from internal production to an external entity that specializes in that operation.

Packet Sniffer is software that records and watches network traffic.

Pandemic is a global disease outbreak.

Passive RFID Tags have no power source and are read by scanners from short range.

Passwords limit computers to authorized users. They are one form of authentication. Strong passwords include a combination of letters (upper and lower case), special characters, and numbers.

Patches are updates to software programs that prevent the exploitation of known vulnerabilities. Using patches reduces the risk of attackers inflicting harm by exploiting the vulnerability.

Personal Identification Number, or **PIN,** is a password composed of decimal digits only.

Petty Corruption involves small amounts of money.

Phishing Attacks purport to originate from a financial institution and ask (though sometimes warn) users to update their personal account information by clicking on the attached link and providing what is required.

Physical Security includes efforts to monitor and control the facility's exterior perimeter and interior space.

Piggyback is a way of getting around swipe cards. When a legitimate person is entering, another person claims to be late for a meeting and says he or she left his or her badge

GLOSSARY

on the desk or offers some other excuse. The person then gains entry without a valid swipe card.

Policies are guiding principles that establish management's authority and responsibility to create a secure business environment, outline acceptable and unacceptable behaviors and activities, and present specific direction toward the basic goal of protecting the organization's people, facilities, physical assets, and information assets. A policy is a formal statement of rules that people who are given access to an organization's technology and information assets must follow.

Political Risks include changes in leadership, civil unrest, and war or other political events that may have an impact on a company's ability to operate.

Precaution Advocacy is a type of risk communication designed to get people to be concerned and to take the risk seriously when people are not concerned enough about a risk.

Preemployment Evaluation Report Credit Report, or **PEER,** uses the national credit bureau database to provide the applicant's national credit history. The report will reveal any bankruptcies, tax liens, foreclosures, or repossessions.

Prime Movers are powerful figures in the organization that others look to for guidance or rely on as important sources of information.

Principle of Separation of Duties states that each transaction is divided into a number of smaller transactions. These transactions are then assigned to separate individuals to accomplish.

Procedural Justice involves perceptions of the fairness of procedures. Are procedures implemented consistently, without bias, and do they consider the interests of all parties and provide a means for correcting errors?

Procedures are documents that provide the step-by-step instructions necessary to reach a desired end state. They are specific operational steps that individuals must take to achieve goals that are often stated in policies.

Processes are activities, tasks, and procedures typically performed across multiple organizations to implement company policies and standards.

Prodromes is a name used to refer to the warning signs of a crisis.

Product Harm is when a product that an organization makes can hurt consumers in some way.

Product Tampering is when an individual or group alters a product to cause harm or for its own financial gain.

Professional License Verification verifies an applicant's professional license or certificate through the accrediting agency or professional association. This information includes the license number, expiration date, type of license or certificate issued, date of issue, and whether there have been any disciplinary actions or sanctions against the license or certificate holder.

Propaganda of Deed uses direct actions against organizations to inspire revolution or change.

*-Property (**read star-property**). This is the no write-down rule. It states that a subject with a secret classification cannot write into a document with a classification of confidential or unclassified.

Public Key is the half of the key that is available to verify signatures or encrypted information.

Purpose Statement is a broad articulation of the exercise goal that guides the entire exercise. The purpose statement limits the objectives and clarifies to participants why the exercise is being conducted.

Radio Frequency Identity (**RFID**) is a system of technology components used to track "things." Its primary use is to follow the movement of items through the supply chain. As such it can be an important component in supply chain security. The central component of the RFID is a wireless radio frequency device known as a tag.

Random Risks are such that no amount of analysis of causes or drivers can make them less uncertain.

Recovery Planning involves making provisions for first aid, search and rescue, building evacuation, and emergency communications; coping with fires and hazardous materials; and general personnel training in all of the above.

Remote Access is when a user accesses an information system from outside of the security parameters of the information system.

Remote Video Auditing (**RVA**) is the use of digital cameras to observe specific employee behaviors such as customer service or areas that involve regulatory compliance activities. The video is reviewed and critiqued by auditors.

Reputation (of an organization) is an aggregate evaluation stakeholders make about how well an organization is meeting stakeholder expectations based on its past behaviors. In other words, a reputation is an evaluation of an organization based on stakeholder perceptions of an organization.

Rescreening of Employees typically involves drug testing, random or follow-up, and periodic background rechecks.

Residual Risk is the potential risk that exists after efforts have been taken to reduce the risk.

Resume Fraud is when job applicants place false information on their resumes.

Risk is the likelihood that something bad is going to happen and the associated impact.

Risk Analysis is a formal review of the risks in a system that involves assessing the likelihood of the risk, the impact of the risk, and efforts to mitigate the risk.

Risk Assessment is the process of evaluating the likelihood that a negative event is going to occur and the estimated magnitude of loss.

Risk Bearers are people who must live with the possible consequences of a risk, such as those living near a chemical facility.

Risk Communication is a conversation about risk between an organization and its stakeholders, typically the community members living near a facility. The focus of risk communication is helping people to understand/evaluate the risk and to manage it.

Risk Management is the process of identifying, assessing, and addressing the risk.

Risk Tolerance is the amount of risk an organization can accept in efforts to achieve its objectives.

Rumors are untrue information about your organization that is publicly circulating.

Scope sets limits to an exercise.

Secure State Machine Model is when you can show that after every possible state transition action, the machine ends up in a secure state.

Security Architecture is the framework of the organization's security system.

Security Documentation defines the scope of security in the organization and determines what is to be protected and the extent of that protection. It also explains what is expected from employees and what the consequences are of noncompliance.

Security Model is a scheme or framework for specifying and enforcing the organization's security policies.

Sensitive Information includes mission critical data, private customer information, private employee information, proprietary information, data used to calculate the organization's financial performance, and any information that would damage an organization if it were to be exposed.

Shelter-in-Place means that people should stay inside of buildings and seal the buildings off from outside air.

Shoulder Surfing is simply watching as the employee logs in and then exploiting the network from home later.

Silo Effect is when business units seem to operate almost autonomously with little understanding of what other departments are doing and how they impact one another.

Silos are self-contained departments or other business units that struggle when communicating or working with other departments or silos.

Simple Security Property states that subjects cannot read up. Subjects cannot read an object with a higher classification than their security clearance.

Simulators in exercises provide messages to participants in a carefully planned sequence that mimics the actual event.

Slander is spoken defamation.

Smishing uses a text message sent to cell phone users, usually informing them that they have enrolled in a service of some sort and will be charged unless they visit a listed web site to unsubscribe. Visiting the site loads malware on the victim's computer.

Social Engineers are individuals who use charm, guile, and wit to secure information—or perhaps even direct system access—in order to achieve their goals.

Social Security Number Trace determines whether the Social Security number belongs to someone who is dead and lists all the names and addresses associated with that Social Security number.

Spear Phishing involves sending the fraudulent e-mail only to customers of the institution in question, in contrast to a standard phish e-mail that is sent to as many e-mail addresses as the scammer can find.

Spyware, or **Adware,** is used by advertisers to control material people see when they are online. Spyware can send you pop-ups, redirect your browser to a web site, or track your Internet activities.

Stakeholder Churn is when stakeholders are angry at an organization. Angry stakeholders can make it harder for an organization to operate.

Stakeholders are any group that can affect or be affected by the behavior of an organization. Common stakeholders include customers, suppliers, investors and financial analysts, community members, employees, the news media, government agents, and activist groups.

Standards are more issue specific than policies, usually focus on one technology, and have a shorter life span. They are sets of rules for implementing policy.

State Transition Actions are those processes that will alter the state of the information system.

Strategic Partnership Program Agroterrorism (SPPA) is a collaborative initiative designed to protect the food supply of the United States. The collaboration includes the Department of Homeland Security (DHS), U.S. Department of Agriculture (USDA), Food and Drug Administration (FDA), and Federal Bureau of Investigation (FBI), along with private industry, trade associations, and the states.

Subjects are any entities that request access in an information system.

Supply Chain is the series of steps from the extraction of raw materials to the finished product. These connections require a coordinated system of organizations, people, activities, information, and resources to move a product or service from supplier to customer. Every organization that comes into contact with a product is part of its supply chain.

Suspected Terrorist Watch List Search determines whether the applicant is on a suspected terrorist watch list.

Tabletop Exercises are designed to be low-stress analyses of problematic situations an organization is likely to face. The tabletop, led by a facilitator, is not an attempt to simulate the event or use equipment, but is just an analysis of the situation. Participants improve their critical thinking skills by analyzing and resolving problems using plans and procedures. The key is to have participants identify and analyze the problem areas.

Tags are small devices for RFID that have a transponder and an antenna. The tags transmit data signals when powered/queried by an RFID on the tag's frequency.

Technical Error Industrial Accidents refer to situations in which the cause is beyond the reasonable control of a person or the organization.

Template is the biometric data stored by a system and used for comparison purposes for verification.

Terrorism involves the unlawful use of force and violence against people or property that is intended to intimidate or coerce some group.

Threat is anything that can have a harmful effect on an organization.

Ticketing System is a procedure that shows who is responsible for an action, such as correcting the vulnerabilities, and the deadlines for the responsible person to take the action.

Transparency is a principle that seeks to make the operations of an organization visible to stakeholders. It typically involves the reporting of facts and figures along with the process that organizations use to make decisions.

Trojan Horses are software that pretends to be something else. They claim to do one thing but perform other nefarious operations in the background.

Trusted Subject is one that has proven to be trustworthy. Trustworthy means that subjects always manually perform functions within the security policy. Trusted subjects can transfer information from a higher-classified object to a lower-classified object.

GLOSSARY

Unauthorized Access is when a person gets into an information system without permission or when users access information they are not supposed to access.

Unauthorized Disclosure occurs when unauthorized people are given access to information.

U.S. Chemical Safety Board is an independent federal agency that is responsible for investigating chemical accidents. The focus of its work is on identifying the root cause of chemical accidents at fixed industrial facilities.

U.S. Customs and Border Protection (CBP) is part of the United States Department of Homeland Security. Its actions cover enforcing U.S. trade law, regulating and facilitating international trade, and collecting import duties. The CBP also has responsibilities for preventing terrorists from entering the United States, preventing illegal drugs from entering the United States, and protecting American businesses from the theft of intellectual property.

USDA's Agriculture Marketing Service (AMS) works with the Cotton, Dairy, Fruit and Vegetable, Livestock and Seed, Poultry, and Tobacco commodity programs. AMS provide standardization, grading, and marketing news for these six commodities. AMS also enforces federal laws including the Perishable Agricultural Commodities Act and the Federal Seed Act.

United States Department of Homeland Security was created by the National Strategy for Homeland Security and the Homeland Security Act of 2002. Its goal is to create a large network with a unified core that seeks to improve the security of the United States.

User is a person who has authorized access to an information system.

User Registration is when a person is granted access to a secure domain.

Vendor Vulnerability Identification and Control refers to establishing minimum acceptable criteria for vendor vulnerability identification and control methodologies, as these methodologies relate to vendor business continuity programs and plans, and the ability of the vendor to integrate its methodologies on a sustainable basis with the client's business continuity management strategy.

Verification is a process that determines whether the claimed identity is correct. The process involves comparing the claim to some stored information for the user.

Video Surveillance Policy lets employees know where and how they will be observed by video cameras.

Virtual Teams are when not all team members are in the same physical space with face-to-face contact. Various communication technologies are used to connect the members to the team.

Viruses are the codes we hear the most about in the news. To get a virus, you must actually do something that allows the code to infect your computer, such as open an infected file.

Vishing leverages voice-over IP (VoIP) technology to dial victims with a message that their credit card has had recent fraudulent activity. The victim is asked to type in his or her credit card information.

Vulnerability is a weakness that could be exploited or create a harmful event such as a crisis or security breach.

Vulnerability Assessment is a formal analysis and evaluation of vulnerabilities in an organization.

Vulnerability Discovery involves the procurement, placement, and scheduling of a vulnerability scanning tool.

Warm Site is term used in business continuity to refer to a facility that is partially equipped and can be used to cover some functions when there is a significant business disruption.

Workers' Compensation Record Search is a search of an applicant's accident dates and nature or type of injuries involved. Such searches can be done only in states that permit the dissemination of workers' compensation claim history and must be used in compliance with the Americans with Disabilities Act guidelines. The State Workers' Compensation Commission of the Industrial Relations Board will provide this data.

Workplace Aggression consists of efforts by current or past employees to harm current employees or the organization.

Workplace Aggression Tolerance Questionnaire (WATQ) is a twenty-eight-item instrument used to assess how willing people are to accept a variety of workplace aggression behaviors.

Workplace Violence involves someone in the workplace becoming a victim of a violent act. The violence may be perpetrated by a random customer, another employee, a former employee, or by someone else the person knows, such as an enraged spouse.

Worms are malicious programs that can move through systems without any help from a user. The worm uses weaknesses in software to infect a computer. Worms can also be spread via e-mail and web sites but are self-propagating.

INDEX

Abraxas Corporation, 291
Abridged Security Policy Manual, definition, 697
ACC. *See* American Chemistry Council
Acceptable risk, definition, 525
Access, definition, 697
Access Authority, definition, 697
Access control, 324; definition, 697
Access Control List, definition, 697
Accidents, 170–75
Accreditation, definition, 525
Actionable Activities, definition, 697
Active analysis methodology, 144–46
Active Competitive Intelligence, definition, 697
ActiveX, 3
ADA. *See* Americans with Disabilities Act
Adelphia, 231
ADMS. *See* Advanced Disaster Management Simulator
ADT, 291
Advanced Disaster Management Simulator (ADMS), 103–4
Adware, definition, 70
AED. *See* Automated external defibrillator
AFA. *See* American Family Association (AFA)
Agriculture Marketing Service (AMS), 75; definition, 711
Agroterrorism, 65–72, 166; CARVER + shock, 71; defense against, 71; definition, 697; effects, 67; elements of event, 69; food tampering, 66; ORM and, 71; potential agroterrorists, 69–70; risks, 68; Strategic Partnership Program Agroterrorism, 73–74; types of attacks, 70–71; vulnerabilities, 68–69
Airports, 336, 676–77
Alarm systems, 647, 650
ALF. *See* Animal Liberation Front
American Chemistry Council (ACC), 78, 697
American Family Association (AFA), 174–75
American Society for the Prevention of Cruelty to Animals (ASPCA), 166
Americans with Disabilities Act (ADA), 94, 697
AMS. *See* Agriculture Marketing Service
Animal Liberation Front (ALF), 63
Animal Rights Militia, 169
Anticorruption policies, components, 395–96
Anti-Phishing Working Group, 45
Antivirus software, 4, 697
Apgar, David, 141
Area Challenge, definition, 697
Arrest, 689–90
The Art of Deception (Mitnick), 45
ASPCA. *See* American Society for the Prevention of Cruelty to Animals
Assets: definition, 525; misappropriation, 243; protection, 110–11
Asset's Vulnerability, definition, 697
Associated standards, 24–25
Association of Certified Fraud Examiners, 285
Attacks. *See also* World Trade Center: 2001, 60–61, 162–63; economic effect, 67; external, 2; insider, 2; malicious code, 2; on organizations, 164–69; quarry, 64–65

Attack Signature, definition, 697
Attacks on an Organization, definition, 697
Audit Trail, definition, 697
Authorize Processing, definition, 525
Automated external defibrillator (AED), 92; definition, 698
Automated notification systems: definition, 698; during pandemics, 320
Availability Protection, definition, 525
Avian influenza. *See* "Bird flu"
Awareness, Training, and Education, definition, 525

Background screening, 280–87; cost savings and, 284; lack of use, 286–87; legal exposure and, 282–83; resume fraud, 284–85; security and, 284; types, 281–82
Backup, definition, 698
Badge Challenge, definition, 698
Bandura, Albert, 274
Baron, Robert, 275
Basel II, 23
Baselines, 21; definition, 698
BCPs. *See* Business continuity plans
Bell, Charlie, 172
Bell-LaPadula Model, 9–10; definition, 698
Benetton, 293
Best Corporate Citizen, 249
Better Business Bureau, 44
BIA. *See* Business impact analysis; Business impact assessment
Biba Integrity Model, 10; definition, 698
Biometrics, 294–95; definition, 698
Biometric Sample, definition, 698
Biometric System, definition, 698
Biometric Templante, definition, 698
"Bird flu," 308–10; web sites, 420–21
Biszick-Lockwood, bar, 321
BlackBerry, 32
Black Blocs, definition, 698
BLS. *See* Bureau of Labor Statistics
Bluetooth device, 39
BP, 179–80
Breen, Edward, 178
Brewer and Nash Model, 11–12; definition, 698–99
Bribery, 210–12. *See also* Corruption

Brink's Business Security, 291
Bureau of Labor Statistics (BLS), 257, 258
Burger King, 172–73
Business continuity, 120–32; business impact assessment phase, 125–30; definition, 120, 126, 699; evolution, 122–25; plan development, 130–31; strategy evaluation and selection, 130; tactics, 121–22; validation and maintenance, 131; web sites, 411–13
Business continuity plans (BCPs), 124, 329; definition, 699
Business impact analysis (BIA), 108–9; business continuity, 121; definition, 699; for pandemics, 311; potential events, 143; vulnerability assessment team and, 142
Business impact assessment (BIA), 125–30
Buss, Arnold, 273

Cadbury Schweppes, 177–78
Cameras. *See* Closed circuit television cameras
Cantalupo, Jim, 172
Cars, 37
CARVER + shock, 71
Case studies: American Family Association, 174–75; BP, 179–80; Cadbury Schweppes, 177–78; Chi-Chi's, 173–74; ChoicePoint, 167–68; corporate espionage, 339–40; Ford, 174–75; Lockheed Martin, 168–69; Pom Wonderful, 169; Saybolt International, 211–12; Sharp Electronics, 175; Tyco, 178–79
CASPIAN. *See* Consumers Against Supermarket Privacy Invasion and Numbering
CBP. *See* U.S. Customs and Border Protection
CBT. *See* Computer-based training
CCTV. *See* Closed circuit television cameras
CDC. *See* Centers for Disease Control
Centers for Disease Control (CDC), 67, 257, 304, 312. *See also* Pandemics
CEO. *See* Chief executive officer
CERT. *See* Community Emergency Response Team
Certification, definition, 525

INDEX   715

CGM. *See* Consumer generated media
Challenges, definition, 699
Chemical facilities: definition, 699; governmental actions, 78–80; industry actions, 78; terrorism and, 77–81
Chi-Chi's, 173–74
Chief executive officer (CEO), 172; crisis communications team and, 182–83
Chinese Wall Model, 12
ChoicePoint, 167–68
CI/KR. *See* Critical infrastructure/key resources
Citizen Corps, definition, 699
Clark-Wilson Integrity Model, 10–11; definition, 699
Classification, 24
Clients, workplace violence and, 258
Cloaking, definition, 699
Closed circuit television cameras (CCTV), 251–53, 300, 647
CMP. *See* Crisis management plan
COBIT. *See* Control Objectives for Information and related Technology
Common vulnerability scoring system (CVSS), 18; definition, 699
Communications: automated emergency notification systems, 117–18; during a crisis, 180–87; crisis planning, 115–19; delivering and receiving, 401; efficiency, 401–2; emergency notification system, 478, 549; emergency system, 93; with family, 549–50; messages, 116; pandemics and, 312–21; during postcrisis phase, 157, 158–60; risk communication, 136–41; technnology disruptions, 116–17; virtual crisis management teams, 202
Community Emergency Response Team (CERT), 98–99; definition, 699
Community outreach, 555–58
Competitive intelligence, 217–23; defensive, 220–22; definition, 699; process, 217–18; sensitive information, 222; sources, 219–20; success/failure, 218–19
Compromise, definition, 699
Computer-based training (CBT), 49
Computer Fraud and Abuse Act, 33
Computer hacking, 164–65

Computer Security Incident, definition, 699
Conflicts of interest, 12
Consumer generated media (CGM), 161
Consumers Against Supermarket Privacy Invasion and Numbering (CASPIAN), 293
Contingincy planning, 548–49
Continuity of Operations Plan (COOP), 121
Controller, definition, 699
Control Objectives for Information and related Technology (COBIT), 23
Controls, 509–10
Conveyance security, 326
COOP. *See* Continuity of Operations Plan
Copyright infringement, 33
Corporate Social Responsibility, definition, 700
Corruption, 209–17; anticorruption policies for companies, 215–16; bribery and, 210–12; case study, 211–12; definition, 210–12, 700; reforms, 214–15; as a risk/business security concern, 213–14; social costs, 213; vulnerability to, 212; web sites, 413
Council of Europe, 214
Counterintelligence, web sites, 413–14
Countermeasures, definition, 700
Countersurveillance, 290–92. *See also* Video surveillance; analysis, 291; definition, 291, 700; web sites, 414
Coworkers, workplace violence and, 258
CPR, 92
Credit risk, 150
Criminal History, definition, 700
Criminals: agroterrorism and, 69–70; disasters and, 144; history search, 281; record check, 281; workplace violence and, 258
Crises. *See also* Emergency management; Threats: crisis management, 152–64; definition, 152–53, 700; emergency management, 85–90; postcrisis evaluation survey tool, 159–60; postcrisis phase, 156–60; reputation formation and damage from, 245–47; scanning, 190; sensing mechanism, 393–94; stages, 155–60; steps for constructing a mechanism, 393–94; types, 164–80

Crisis Command/Control Center, definition, 700
Crisis communications, 180–87; internal audiences and special needs, 186–87; media: monitoring, 185–86; spokesperson, 184; message: delivery, 185; management, 184–85; team, 182–84; twenty-four-hour news cycle, 182
Crisis management, 152–64; concerns about, 160–63; definition, 153–54, 700; plan elements, 398–400; postcrisis evaluation survey tool, 159–60; postcrisis phase, 156–60; spokesperson tasks, 397; stages of a crisis, 155–60; victims and, 157
Crisis management plan (CMP), 154–55, 194–200; components, 197–98; control/command center, 198; crisis appendix, 199; stakeholder network, 195–96; time and, 197
Crisis management team, 200–203; definition, 700; demands, 200–201; selection, 201–2; training, 202; traits, 201; virtual teams, 202
Crisis Portfolio/Families, definition, 700
Crisis sensing, 187–94; basics, 187–89; building blocks, 189–94; crisis scanning, 190; definition, 700; prodromal information, 189–92; steps for constructing, 192–94
Crisis Sensing Mechanism, definition, 700
Crisis spokesperson, 203–7; definition, 700; delivery, 204; preparation, 206; question handling, 204–6; role, 203–4; tasks, 204
Critical incident, 126
Critical infrastructure/key resources (CI/KR), 73
CRM. *See* Customer relationship management
CSO Magazine, 121
C-TPAT. *See* Customs-Trade Partnership Against Terrorism
Culture. *See* Integrity, culture of; Organizational culture, managing
Customer relationship management (CRM), 121
Customs-Trade Partnership Against Terrorism (C-TPAT), 81–84; benefits, 82; definition, 700

CVSS. *See* Common vulnerability scoring system
Cyber activities, suspicious, 42–43
Cyber Security, definition, 704
Cyber Security Alerts, 54
Cyber Security Bulletins, 54
Cyberslacking, 32–33; definition, 700

DAC. *See* Discretionary access control
Data backup, 51–53; in-house and third-party options, 52–53; storage options, 52; vendors, 422
Data mirroring, 52; definition, 700
Dates, 24
Day, Nick, 339
DEA. *See* U.S. Drug Enforcement Administration
Decryption, definition, 700
Defamation, 33; definition, 700
Defensive Competitive Intelligence, definition, 700
DHS. *See* U.S. Department of Homeland Security
Dictionary Attacks, definition, 700
Digital IP surveillance, 288–90
Digital signatures, 4, 5; definition, 701
Digital Video Surveillance, definition, 701
Diligence Inc., 339–40
"Disaster Prepardness" (Surmacz), 121
Disaster recovery management, 104–20; business impact analysis, 108–9; crisis communication planning, 115–19; disaster threats, 105–7; ongoing planning, 119–20; personnel security needs, 111–15; protection of assets, obligations, and operations, 110–11; risk assessment, 107–8; web sites, 411–13
Disasters: corporate image/brand, 107; data/information loss, 106; definition, 86; ethical misconduct, 230–32; facilities/equipment, 106; human error, 106; industrial accidents, 106; IT impact, 107; malevolent acts, 106; natural, 106; operations/systems, 107; personnel, 106–7
DISCOVER project, 104
Discretionary access control (DAC), 9
Discrimination, 33

INDEX

Distributive Justice, definition, 701
Documentation. *See* Security documentation
Documentation processing security, 325
Document Management System, definition, 701
Documents: Emergency Management Guide for Business and Industry, 529–81; FEMA: Protecting Your Business from Disasters, 588–90; Guidance on Preparing Workplaces for an Influenza Pandemic, 591–620; Industry Self-Assessment Checklist for Food Security, 620–32; Personal Security Guidelines for the American Business Traveler Overseas, 673–96; Ready Business: Sample Emergency Plan, 582–88; Security Guidelines for American Enterprises Abroad, 633–72; Security Self-Assessment Guide for Information Technology Systems (Swanson), 425–528
DOD. *See* U.S. Department of Defense
Domestic violence, workplace violence and, 258
Drill, definition, 701
Driving, 687–88
Drug testing, 285–87; definition, 701; guidance, 286; web sites, 414
Due care, 13; definition, 701
Due diligence, 13; definition, 701
Dumpster Diving, definition, 701
Dunn & Bradstreet, 219
Durden, Douglas, 289–90

E. coli, 67, 69, 153, 170
Earth First!, 64
Earth Liberation Front (ELF), 63–64
Earthquakes, 568–70
Ecodefense, definition, 701
Economic disasters, 144
Economic Espionage Act (1996), 33
Ecoterrorism, 62–65; definition, 62, 701; web sites, 415
Edelman Public Relations, 246
Education: awareness, 324; for pandemics, 314; verification in hiring, 282, 701
EEA. *See* Essential element of analysis
Effect/danger ratio, 274–75

800 number, 93
EITI. *See* Extractive Industries Transparency Initiative
Electronic product code (EPC), 293
ELF. *See* Earth Liberation Front
E-mail, 32–38; actionable activities, 33; damage resources and information, 34; as damage to the organization's reputation, 34; employee misuse, 34; monitoring vendors, 422; scams, 44; surveillance policies, 36; as time waster, 32–33; use policies, 34–37; enforcment, 36; organization's right to monitor, 37
EMDs. *See* Ethical misconduct disasters
Emergency: definition, 530, 701; sample plan, 582–88
Emergency management: benefits, 85–90; definition, 530; mitigation benefits, 87; preparedness benefits, 87–88; recovery benefits, 89; response benefits, 88–89
Emergency Management Group (EMG), 546
Emergency Management Guide for Business and Industry, 529–81
Emergency notification systems, 117–18; initiation, 400–403
Emergency operations center (EOC), 547; definition, 701; in pandemics, 316
Emergency Planning and Community Right-to-Know Act (EPCRA), 136–41; definition, 701–2
Emergency preparedness, 90–98; components of the response program, 91–92; definition, 90; emergency communication system, 93; emergency equipment and supplies, 94; emergency response teams, 92; employee awareness and training, 93; employee protection, 90–91; employee special needs and skills, 94–95; evaluation, 379–81; personal preparedness, 97–98; readiness, 95–97
Emergency Preparedness and Response centers, definition, 702
Emergency response, types, 139–40
Emergency response teams (ERTs), 91, 92, 379–81; definition, 702
Emergency response training and testing, 99–104; human element, 101;

investment in, 100; role playing, 102; scenarios, 102; simulations, 103; tabletop exercises, 102–3; testing as validation, 101–2; virtual reality, 103–4
EMG. *See* Emergency Management Group
Employee Background Check/Investigation, definition, 702
Employees: awareness and training, 93; background screening and drug testing, 280–87; web sites, 415; collaboration, 352; coworker violence, 259; during disasters, 106–7; employer protection of, 600–620; ethics and, 228; exposure to pandemic influence at work, 595–96; living abroad, 614–15; negligent hiring, 284; personnel disasters, 144; protection, 90–91, 111; secretaries, 653–54; security survey, 270–71; special needs and skills, 94–95; support, 559–60; theft and fraud, 242–44; training, 544; workplace violence and, 258
Employee Special Needs Form, definition, 702
Employer reference check, 282; definition, 702
Encryption, 4, 5; definition, 702
End-to-End Encryption, definition, 702
End users, 27
Enright, Guy, 339
Enron, 175–76, 230, 231, 242
Environmental Protection Agency (EPA), 77
EOC. *See* Emergency Operations Center; Emergency operations center
EPA. *See* Environmental Protection Agency
EPC. *See* Electronic product code
EPCRA. *See* Emergency Planning and Community Right-to-Know Act
E-policy, 34–36; communicating, 36–37
Equipment and supplies, 94
ERTs. *See* Emergency response teams
Espionage, corporate/industrial, 338–41; case study, 339–40; definition, 699
Essential element of analysis (EEA), 330–32
Ethical conduct audit, 227–42; commitment to integrity contnuity, 235; considerations, 236–38; definition, 702; ethical misconduct disasters, 230–32; expectations for ethical orientations, 232–33; integrity lapses, 229–30; overview, 239–41; proactive integrity management, 227, 238
Ethical misconduct disasters (EMDs), 230–32; categories, 231; definition, 702
Ethics, 223–27; compliance programs and, 225–27, 395; misconduct and, 224–25; range of behaviors for ethics and compliance program, 223–24
Ethics and Compliance Programs: components, 395; definition, 702; web sites, 415
Evacuation: assembly areas and accountability, 551–52; devices, 94; emergency preparedness, 114; in large and multiple tenant buildings, 298–300; planning, 551; routes and exits, 51
E-vaulting, 52. *See also* Data backup; definition, 702
Exercise and training basics, 361–75; definition, 702; design team, 374–75; exercise comparison, 364; exercise documents, 374–75; overview, 361; process, 368–74; drill, 363, 365; establishing the base, 368–69; evaluation and critique, 373; exercise conduct, 372–73; exercise design, 369–72; follow-up, 373–74; orientation session, 363; progressive exercising, 361–62; rationale, 362; types, 363–68; vendors, 423
Exits, 650–51
Expected Actions, definition, 702
Expiration date, 24
Extractive Industries Transparency Initiative (EITI), 214
Exxon-Mobil, 162

FAA. *See* Federal Aviation Authority
Facilitation Bribes (Grease), definition, 703
Facilities/equipment disasters, 106
False Acceptance, definition, 703
False Acceptance Rate, definition, 703
False Positive, definition, 703
False Rejection, definition, 703
False Rejection Rate, definition, 703
Family: communications, 549–50; preparedness, 552
FBI. *See* Federal Bureau of Investigation

INDEX

FDA. *See* Food and Drug Administration
FDIC, 44
Federal Aviation Authority (FAA), 336
Federal Bureau of Investigation (FBI), 62, 73, 165
Federal Emergency Management Agency (FEMA), 61, 88; Emergency Management Guide for Business and Industry, 529–81; emergency management offices, 571–72; exercise and training basics, 370–72; FEMA: Protecting Your Business from Disasters, 588–90; state emergency management agencies, 572–79
Federal information technology, 510–22
FEMA. *See* Federal Emergency Management Agency
FEMA: Protecting Your Business from Disasters, 588–90
Finance director, 183
Financial risk, 150; definition, 703
Fingerprints, 295
Fink, Steven, 155
Fire, 562–63
Fire extinguishers, 92
Firewalls, 4; definition, 703
First aid, 92; emergency preparedness, 113–14
Floods and flash floods, 564–66
Fombrun, Charles, 247–48
Food: packaging, 68–69; security, 75–77; Industry Self-Assessment Checklist for Food Security, 620–32; web sites, 415; tampering, 66
Food and Drug Administration (FDA), 68, 73, 173–74
Food Safety Inspection Service (FSIS), 75, 620–32; definition, 703; goals, 76; planning process, 75–76; security plans, 75
Food Security, definition, 703
Ford Motor Company, 174–75, 176
Foreign agents, 340–41
Foreign Corrupt Practices Act, 214, 216; definition, 703
Foreman, Dave, 64
Framing, definition, 317
Fraud, 242–44
FSIS. *See* Food Safety Inspection Service

Full-scale exercises, 367–68; definition, 703
Functional exercises, 366–67; definition, 703
Functions, definition, 703
Futureproofing, 144

General Crisis Management Skills, definition, 703
General Support System, definition, 525
Geography, 27
Global Crossing, 231
Global security: corporate/industrial espionage, 338–41; outsourcing and security, 321–22; pandemics, 303–12; communication, 312–21; Personal Security Guidelines for the American Business Traveler Overseas, 673–96; supply chain continuity, 327–35; supply chain security, 323–26; travel and, 335–38
Graham-Denning Model, 11; definition, 703
Graves, Samuel, 249
Guardsmark, 259
Guidance on Preparing Workplaces for an Influenza Pandemic, 591–620
Guidelines, 21; definition, 703

Hazard, definition, 703
Hazardous materials, 563
Hazard Vulnerability, definition, 704
HGA. *See* Hypothetical Government Agency's questionnaire
Hidden Mode, definition, 704
Hijacking, 693–94
Hiring, negligent, 284
H5N1 influenza virus. *See* "Bird flu"
Hoover's, 219
Hostages, 692–94
Hotels, 336–37, 677–82
Hot Site, definition, 704
HR. *See* Human resources
Human error, 106
Human influenza virus, 309. *See also* Pandemics
Human resources (HR), 128; crisis communications and, 183
Hurricanes, 566
Hypothetical Government Agency's (HGA) questionnaire, 523–24

ICS. *See* Incident Command System
Identity theft, 133
IDS/IPS. *See* Intrusion detection/prevention program
Immelt, Jeffrey R., 142
Inappropriate Usage, definition, 704
Incident, definition, 704
Incident Command System (ICS), 546–47
Incident management and investigation, 325
Indirect Bribe, definition, 704
Individual Accountability, definition, 526
Industrial accidents, 106, 170–71; definition, 710
Industry Self-Assessment Checklist for Food Security, 620–32
Influenza pandemic, 307–8, 592–94. *See also* Pandemics; Guidance on Preparing Workplaces for an Influenza Pandemic, 591–620; web sites, 420–21
Information: classification, 221; data loss disaster, 106; disasters and, 144; dissemination in pandemics, 315–16; misinformation, 318–19; portable device security, 39; security, 1–6, 325, 391–92; definition, 704; guidance, 398; prevention, 2–5; threats, 2; web sites, 416–19; sensitive, 709; storage options, 52
Information cyber protection: data backup, 51–53; information security, 1–6; information security policies and standards, 20–31; insider threats, 40–42; Internet and e-mail use policies, 32–38; portable device security, 38–40; security documentation, 13–19; security models, 6–12; social engineering, 43–51; suspicious cyber activities, 42–43; United States Computer Emergency Readiness Team, 54–57
Information Owner, definition, 526
Information Security Awareness, definition, 704
Information Security Awareness and Training program, definition, 704
Information System, definition, 704
Information technology (IT), 16, 26; disasters, 107; protection of assets, obligations, and operation, 110–11

Insider threats, 40–42; agroterrorism and, 69; characteristics of attacks, 40–41; definition, 704; lessons from previous attacks, 41; nature, 40; organization and, 41; web sites, 419–20
Instant Background Check, definition, 704
Insurance, 559
Integrity. *See also* Ethical conduct audit: consistent, 229; continuity, 227; culture of, 359–60; definition, 704; lapses, 229–30; organizaational, 229; proactive management, 238
Integrity Management, definition, 704
Intelligence, 219–20. *See also* Competitive intelligence
Intentional Inducement of Copyright Infringement Act (2004), 33
Interactional Justice, definition, 704
Interest rate risk, 150
Internal audiences, 186–87
International Chamber of Commerce, 214
International Standard for Information Technology—Code of Practice for Information Security Management (ISO/IEC 17799), 23
Internet, 32–38; actionable activities, 33; competitive intelligence and, 219; damage resources and information, 34; as damage to the organization's reputation, 34; employee misuse, 34; espionage and, 340–41; legislation, 32; liabilities, 33; surveillance policies, 36; as time waster, 32–33; use policies, 34–37; enforcement, 36; e-policies and, 36–37; organization's right to monitor, 37; vendors, 422
Intrusion detection/prevention (IDS/IPS) program, 16, 18, 19; definition, 704
Intrusion Prevention System, definition, 704
IPAC, 259
ISO/IEC 17799. *See* International Standard for Information Technology—Code of Practice for Information Security Management
ISO reference, 24
Issues management, 188; definition, 704
IT. *See* Information technology

INDEX

Jargon, 351
Java, 3
JavaScript, 3
Johnson, James A., 60

Kaiser Permanente, 186–87
Key Exchange, definition, 704
Key Pair, definition, 705
Keys, definition, 705
Kickbacks, 210, 216. *See also* Corruption
Kidnapping, 694–95
KLD Research & Analytics, 249
Klitgaard, Robert, 212
Kozlowski, Dennis, 178–79
KPMG, 224–25, 339–40
Kresevich, Millie, 359–60
Kryptonite, 246

Landreth, Jeff, 259
Last review date, 24
LDRPS. *See* Living Disaster Recovery Planning System
Learnable Risk, definition, 705
Least Privilege Principle, definition, 705
Legal advisers, 183
Legal risk, 150–51; definition, 705
Legislation, political risk and, 151
LEPC. *See* Local emergency planning committee
LexisNexis, 219
Libel, 33; definition, 705
License, verification, 282
Living Disaster Recovery Planning System (LDRPS), 121
LMS. *See* Logical Management Systems Corp.
Local emergency planning committee (LEPC), 138
Lockheed Martin, 168–69
Logical Management Systems Corp. (LMS), 144–46
Logic Bomb, definition, 705
Low-Hanging Fruit, definition, 705
Luggage, 675–76, 681

MAC. *See* Mandatory access control
Major Application, definition, 526
Malicious code, 2

Managed security service provider (MSSP), 17
Management: misconduct, 175–80; definition, 705
Management and the Activity Trap (Odiorne), 127
Mandatory access control (MAC), 9
Marintek, 104
Market risk, 150
Martin, Lynn, 177
Mass notification vendors, 422
Material Weakness, definition, 526
McDonald's, 172
Mead, David, 211–12
Media, 77, 557–58; consumer generated, 161; crisis spokesperson and, 203–4; during disaster recovery, 118–19; monitoring, 185–86; public relations director, 183; risk communication and, 140–41; social, 161; spokesperson, 184; twenty-four-hour news cycle, 182
Media Reputation Index, 247–48
Merrill Lynch, 231
Messages, 116, 401; crisis management plan and, 197; delivery, 185; management, 184–85; maps and, 319–20; meta-messages and framing, 316–18; organizational culture and, 358; during pandemics, 316
Military record, verification, 282; definition, 705
Minimum security baselines (MSBs), 21; definition, 698
Mission statement, 532
Mitigation, definition, 153, 705
Mitnick, Kevin, 45
Mitroff, Ian, 152
Mitsubishi Motors, 231
MLS. *See* Multilevel security policy
Monkeywrenching, definition, 705
"Most Admired Companies," 247
Motor vehicle records, background check and, 281–82; definition, 705
MSBs. *See* Minimum security baselines
MSSP. *See* Managed security service provider
MT63, 10
Multilevel security (MLS) policy, 9; definition, 705
Muster Area, definition, 705

NAPBS. *See* National Association of Professional Background Screeners
Narrative, definition, 705
National Association of Professional Background Screeners (NAPBS), 287
National Cargo Security Council (NCSC), 327
National Incident Management System (NIMS), 88, 121
National Infrastructure Protection Plan (NIPP), 73
National Institute for Occupational Safety and Health (NIOSH), 257, 296
National Notification Network, 320
National Transportation Safety Board (NTSB), 171
Natural Resources Defense Council (NRDC), 191
NCSC. *See* National Cargo Security Council
NDAs. *See* Nondisclosure agreements
Negligent Hiring, definition, 705
Negligent Retention, definition, 705
Negligent Supervision, definition, 705
Networks, definition, 526
Neuman, Joel, 275
NGOs. *See* Nongovernmental organizations
NIMS. *See* National Incident Management System
9/11 attack, 60–61, 162–63
NIOSH. *See* National Institute for Occupational Safety and Health
NIPP. *See* National Infrastructure Protection Plan
Nokia, 259
Noncompete Agreements, definition, 705
Noncompliance/penalties, 25
Nondisclosure agreements (NDAs), 221; definition, 706
Nongovernmental organizations (NGOs), 214
Noninterference Model, definition, 706
Non-Repudiation, definition, 706
No-write-down rule, 9
NRDC. *See* Natural Resources Defense Council
NTSB. *See* National Transportation Safety Board

Object, definition, 706
Occupational Safety and Health Administration (OSHA), 157–58; offices, 619–20; safety and health program management guidelines, 615–18; training and education, 617–18
OCILLA. *See* Online Copyright Infringement Liability Limitation Act
Odiorne, George S., 127
Oklahoma City bombing (1995), 61
Ollila, Jorma, 259
Online Copyright Infringement Liability Limitation Act (OCILLA), 33
Operational Controls, definition, 526
Operational manager, 183
Operational risk, 151; definition, 706
Operations/systems, disasters, 107
Organisation for Economic Cooperation and Development, 214
Organizational culture, managing, 356–59; approach to management, 358–59; assessment, 357; definition, 356, 706; skepticism in, 357–58
Organizational transformation (OT), 238
Organizations: attacks on, 164–69; case studies, 167–69; integrity and, 229; management misconduct, 175–80
Orientation Session, definition, 706
OSHA. *See* Occupational Safety and Health Administration
OT. *See* Organizational transformation
Outrage: definition, 706; evaluation, 396–97
Outrage Management, definition, 706
Outside Threat, definition, 706
Outsourcing, security and, 321–22; definition, 706
Owner, 24
Ozymandias Sabotage Handbook, 64–65

Packet Sniffer, definition, 706
Palmer, Walter, 290
Pandemics, 303–12; Asian influenza, 307–8; "bird flu," 308–10, 420–21; communication, 312–21; information dissemination, 315–16; message characteristics, 316; message maps, 319–20; meta-messages and framing, 316–18; planning, 313–14;

INDEX

readiness, 320–21; risk communication, 314–15; role of automated mass notification systems and, 320; rumors, misinformation, and errors, 318–19; consequences, 305–7, 310; definition, 706; development of a disaster plan, 598–600; Guidance on Preparing Workplaces for an Influenza Pandemic, 591–620; Hong Kong influenza, 308; immunity, 306; influenza, 307–8; maintaining operations during, 596–98; periods of, 304–5; planning for, 594–95; prepandemic education, 314; preparation for, 310–11; preparedness evaluation, 386–91; risks, 307–10; Spanish influenza pandemic, 307; web sites, 420–21
Passive RFID Tags, definition, 706
Passports, 673–74, 681
Passwords, 3; definition, 706; guidelines for use, 3, 397
Patches, 5; definition, 706
Payment Card Industry (PCI), 12
PCI. *See* Payment Card Industry
PDA. *See* Personal digital assistant
PEER. *See* Preemployment evaluation credit report
Penalties, 25
People for the Ethical Treatment of Animals (PETA), 161, 169, 172–73
Personal digital assistant (PDA), 38, 339–40. *See also* Portable device security
Personal Identification Number (PIN), definition, 706
Personal protective equipment (PPE), 610–11, 613–14
Personal Security Guidelines for the American Business Traveler Overseas, 673–96
Personnel. *See* Employees
Personnel security, 111–15, 324; during a crisis, 156
Pesticides, 66
PETA. *See* People for the Ethical Treatment of Animals
Petty Corruption, definition, 706
"Phishing," 44
Phishing Attacks, definition, 706

Physical disasters, 144
Physical protection: biometrics, 294–95; countersurveillance, 290–92; employee background screening and drug testing, 280–87; evacuation in large and multiple tenant buildings, 298–300; integration of physical and information security, 300–302; physical security, 251–54; radio frequency identity, 292–94; security guards/officers, 255–56; shelter-in-place, 296–98; video surveillance, 288–90; workplace aggression, 272–78; contributors to, 278–80; workplace violence prevention and policies, 257–72
Physical security, 251–54, 323; areas and security measures, 252–54; definition, 706; employees and, 253–54; evolution, 254; resources, 406; suspicious activity, 253
Piggyback, definition, 706–7
PIN. *See* Personal Identification Number
PKI. *See* Public key infrastructure
Plumrose USA, 289
Policies, 13–14, 20–31; anticorruption, 215–16, 395–96; content, 23–25; definition, 526, 707; development, 25–28; exceptions, 28–30, 392; information security content area evaluation, 391–92; organization, 23; policy manual, 30–31; for social engineering, 48–49; structure, 22
Political risk, 151; definition, 707
Pom Wonderful, 169
Pornography, 33
Portable device security, 38–40; information security, 39; physical security, 38; wireless concerns, 39
Posttraumatic stress, 114–15; symptoms, 115
PPE. *See* Personal protective equipment
Precaution Advocacy, definition, 707
Preemployment evaluation credit report (PEER), 281; definition, 707
Preparedness: benefits, 87; emergency response and, 90–98
Prime Movers, definition, 707
Principle of Separation of Duties, definition, 707
Privacy, radio frequency identity and, 293–94
Procedural Justice, definition, 707

Procedural security, 324–25
Procedures, 22; definition, 526, 707
Processes, definition, 707
Procurement: incident management, 333–34; planning, 332–33
Prodromes, 188. *See also* Crisis sensing mechanism; definition, 707
Product Harm, definition, 707
Product tampering, 166, 169; definition, 707
Professional License Verification, definition, 707
Propaganda of Deed, definition, 707
Property: protection, 552
*-Property, 9–10; definition, 707
Protesters, agroterrorism and, 70
Publication date, 24
Public information, 557
Public key infrastructure (PKI), 7; definition, 708
Public relations director, 183
Purpose Statement, definition, 708

Quarry attacks, 64–65
Quest, 231
Quest Consultants International, 291

Radio frequency identity (RFID), 292–94; basics, 292–93; definition, 708; privacy concerns, 293–94
Random Risks, definition, 708
Raytheon, 176
Ready Business: Sample Emergency Plan, 582–88
Recovery Planning, definition, 708
Red Cross, 94
Remediation, 18
Remote Access, definition, 708
Remote video auditing (RVA), 289; definition, 708
RepTrak, 248
Reputation: assessment, 381–83; background testing and, 280–81; benefits, 245; corporate image/brand disasters, 107; corporate social responsibility, 248–49; damage to organization, 34; definition, 244–45, 708; disasters and, 145; evaluation, 381–83; favorable, 189; formation and damage from crises, 245–47; management, 244–50; measurement, 247–48
Reputation Institute, 247–48
Reputation Quotient, 247
Rescreening of Employees, definition, 708
Residual Risk, definition, 708
Resume Fraud, definition, 708
RFID. *See* Radio frequency identity
Rifkin, Stanley Mark, 46–47
Risk Analysis, definition, 708
Risk assessment, 189; definition, 132, 708
Risk bearers, 136–37; definition, 708
Risk communication, 136–41; definition, 708; EPCRA, 137–38; focus, 137; news media and, 140–41; outrage and, 138–40, 396–97; in pandemics, 314–15; types, 137
Risk Intelligence: Learning to Manage What We Don't Know (Apgar), 141
Risk management, 132–35; definition, 132–33, 526, 708; evaluation and measurement, 133–34
Risks: definition, 132, 526, 708; financial, 150; legal, 150–51; operational, 151; political, 151; types, 150–52
Risk Tolerance, definition, 708
Roman Catholic Church, 231
Rozzano, Mike, 289
Rules of Behavior, definition, 526–27
Rumors, 165–66; definition, 708; during pandemics, 318–19
RVA. *See* Remote video auditing

Safety: business continuity, 120–32; Community Emergency Response Team, 98–99; crisis communication, 180–87; crisis management, 152–64; crisis management plan, 194–200; crisis management team, 200–203; crisis sensing mechanism, 187–94; crisis spokesperson, 203–7; disaster recovery management, 104–20; emergency management benefits, 85–90; emergency preparedness and response, 90–98; emergency response training and testing, 99–104; risk communication, 136–41; risk management, 132–35; types of crises, 164–80; types of risks, 150–52; vulnerability assessment team, 141–49

INDEX

SANS Institute, 27
SANS Security Policy Project, 27
SARA. *See* Superfund Amendments and Reauthorization Act
SARA Title III, 136
Sarbanes-Oxley Act (2002), 225, 321
Saybolt International, 211–12
Scandals. *See* Ethical misconduct disasters
SCM. *See* Supply chain management
Scope, 25; definition, 708
Sears, 231
SEC. *See* Securities and Exchange Commission
Secretaries, 653–54
Sector Specific Agencies (SSA), 73
Secure State Machine Model, definition, 709
Securities and Exchange Commission (SEC), 220
Security: acceptance of, 353–56; communication channels, 355; examples, 355; resistance to change, 354–55; starting point, 354; outsourcing and, 321–22; physical, 251–54; resources, 406; vendors, 422–23; web sites, 415–16
Security Architecture, definition, 709
Security clearance, 9
Security documentation, 13–19; creation, 13–14; baseline, 14; guidelines, 14; policies, 13–14; procedures, 14; process, 14; standard, 14; definition, 709; goals, 14; proof, 16–17; requirements, 14–15; steps, 15–16; vulnerability management, 17–19; customizable reporting, 18–19; prioritization, 17–18; remediation, 18; verification, 18; vulnerability discovery, 17
Security guards/officers, 255–56; hiring, 255–56; types, 255
Security Guidelines for American Enterprises Abroad, 633–72
Security Magazine, 221, 222
Security manual, 30–31
Security models, 6–12; definition, 6, 709; purpose, 6–7; secure state machine models, 8–12; security planning process, 8; uniqueness, 7–8
Security Pacific Nation Bank of Los Angeles, 46–47

Security Self-Assessment Guide for Information Technology Systems (Swanson), 425–528
Sensitive Information, definition, 527, 709
Sensitivity, definition, 527
SERC. *See* State emergency response commission
Sexual harassment, 33
Shadowcrew, 45
Sharp Electronics, 175
Shelter-in-place, 296–98, 552; definition, 709; how to, 394–95; web sites, 421
Shigella dyseneriae, 66
Shoulder Surfing, definition, 709
Sigificant Weakness, definition, 526
Silo effect, 346–53; avoiding, 349–53; dangers of, 347; definition, 709; efforts to protect organizations and, 347–48; evaluation, 348–49; existence of, 346–47
Silos: definition, 709; evaluation, 378–79
Simple Security Property, definition, 709
Simulators, definition, 709
Slander, 33; definition, 709
Smishing, 44; definition, 709
Smith Barney Citigroup, 231
Social engineering, 43–51, 392–93; close to home, 44–45; combating, 47–48; definition, 709; evaluation, 392–93; physical attacks, 46; policies, 48–49; psychological attacks, 46–47; real-world training, 49–50; web sites, 421
Social media, 161
Social responsibility: corporate, 248–49; definition, 700
Social Security number, 281; definition, 709
Society of Human Resource Managers, 285
Software, blocking, 36
Spear phishing, 44; definition, 709
Special needs, 186–87
Spokesperson. *See* Crisis spokesperson: tasks for crisis management, 397
SPPA. *See* Strategic Partnership Program Agroterrorism
Spyware, 4; definition, 709; symptoms, 4–5, 397–98
SSA. *See* Sector Specific Agencies
Stakeholder churn, 161; definition, 709

Stakeholders: definition, 161, 709–10; network, 195–96; reputation and, 245; security breaches and, 246
Standards, 21; definition, 710
Starr Report, 33
State emergency response commission (SERC), 138
State Transition Actions, definition, 710
Steganography, 10
Stewart, Martha, 231
Storms, 568
Strategic Partnership Program Agroterrorism (SPPA), 73–74; definition, 710
Subjects, definition, 710
Superfund Amendments and Reauthorization Act (SARA), 136
Supply Chain, definition, 710
Supply chain continuity, 327–35; assessment, 328–32; vendor continuity capability assessment, 330–32; vendor continuity capability questionnaire, 329–30; elements, 327; procurement incident management considerations, 333–34; procurement planning considerations, 332–33; vulnerability drivers, 328–29
Supply chain management (SCM), 121
Supply chain security, 323–26; access control, 324; conveyance security, 326; definition, 23; documentation processing security, 325; education and training awareness, 324; incident management and investigation, 325; information security, 325; personnel security, 324; physical security, 323; procedural security, 324–25; trading partner security, 325–26
Surmacz, Jon, 121
Surveillance policies: for Internet and e-mail, 36; recognition, 691–92
Survey of Workplace Violence Prevention, 257
Suspected Terrorist Watch List Search, definition, 710
Suspicious activity, 253
Swanson, Marianne, 425–528
System, definition, 527
System Operational Status, definition, 527

Tabletop exercises, 365–66; definition, 710
Tags, definition, 710

Teams: collaboration, 352; Community Emergency Response, 98–99; conflict and, 345–46; crisis communications, 182–84; crisis management, 200–203, 700; developing, 344–46; emergency response, 91, 92, 702; for exercise and training basics, 374–75; improving team effectiveness, 343–46; problem assumptions, 344; skills, 345; threat assessment, 263–64; training, 202; United States Computer Emergency Readiness, 54–57; virtual crisis management, 202; vulnerability assessment, 141–49
Technical Controls, definition, 527
Technical Cyber Security Alerts, 54
Technical Error Industrial Accidents, definition, 710
Technical experts, 183
Template, definition, 710
Terms/definitions, 25
Terrorism, 167; agroterrorism and, 65–72, 70; chemical facilities and, 77–81; customs-trade partnership against, 81–84; definition, 59–60, 710; dynamics, 60; ecoterrorism, 62–65; food security, 75–77; growing concerns about, 60–61; terrorist watch list, 283; types, 60; watch list, 282; web sites, 421
Testing, business continuity plan validation, 131
Theft, 43, 242–44. *See also* Social engineering
Third-party allies, 183
Threats, 2. *See also* Crises; Crisis management; assessment team, 263–64; business impact assessment matrix and, 143; categories, 105–7; definition, 527, 710; insider, 40–42; web sites, 419–20; outside, 706; vulnerability assessment teams and, 142, 144–46
Ticketing System, definition, 710
Title, 23–24
Topics, 25
Tornadoes, 566–68
Trading partner security, 325–26
Training: awareness, 324; for the crisis management team, 202; for emergency response, 100; for employees, 544; exercise and training basics, 361–75;

INDEX

schedule, 541; testing as validation, 101–2; vendors, 423
Trains, 337, 684–87
Transparency, definition, 710
Transparency International, 214
TrapWire, 291
Travel, 335–38; at the airport, 336; by car, 337; employees living abroad, 614–15; at the hotel, 336–37; overseas, resources, 420; personal conduct overseas, 688–89; Personal Security Guidelines for the American Business Traveler Overseas, 673–96; preparation, 335–36; security, 654–58; Security Guidelines for American Enterprises Abroad, 633–72; targeting recognition, 337–38; by train, 337
Triton Energy, 215–16
Trojan horses, 2; definition, 710
Trusted Subject, definition, 710
TWA, 171
Tyco, 178–79, 231, 242
Type A personalities, workplace aggression and, 279–80

UN. *See* United Nations
Unauthorized Access, definition, 711
Unauthorized Disclosure, definition, 711
Uniform Federal Sentencing Guidelines for Organizations, 232
United Nations (UN), 214
United States Computer Emergency Readiness Team (US-CERT), 40, 42, 54–57; sample cyber security alert, 54–56
United Way of America, 231
U.S. Chemical Safety Board, 157, 179–80; definition, 711
U.S. Customs and Border Protection (CBP), definition, 711
U.S. Department of Agriculture (USDA), 68, 73
U.S. Department of Commerce, 209, 226
U.S. Department of Defense (DOD), 9
U.S. Department of Health and Human Services, 286, 314
U.S. Department of Homeland Security (DHS), 61, 73, 88, 165; definition, 711
U.S. Department of Labor, 257

U.S. Drug Enforcement Administration (DEA), 286; guidance, 286
U.S. Federal Sentencing Guidelines for Organizations (USFSGO), 233
U.S. Foreign Corrupt Practices Act, 211–12
U.S. Secret Service, 40
U.S. State Department, 335, 336
US-CERT. *See* United States Computer Emergency Readiness Team
USDA. *See* U.S. Department of Agriculture
User, definition, 711
User Registration, definition, 711
USFSGO. *See* U.S. Federal Sentencing Guidelines for Organizations

Valdez accident, 162
Van Riel, Cees, 248
Vehicular security, 654–58
Vendors, 422–23
Vendor Vulnerability Identification and Control, definition, 711
Verification, definition, 711
Version, 24
Video surveillance, 288–90. *See also* Countersurveillance; benefits of digital IP surveillance, 288–89; evolution, 288; policy, definition, 711; problems with, 289–90
Violations, reporting, 25
Violence. *See* Workplace violence
Virtual Teams, definition, 711
Viruses, 2; definition, 711
Visas, 674
Vishing, 44; definition, 711
Visitors, 658–64
Vulnerability: analysis, 535–38; definition, 527, 711
Vulnerability assessment team, 141–49; assembly, 146–48; definition, 712; description, 142–46; focus, 142–44
Vulnerability Discovery, definition, 712

Waddock, Sandra, 249
Wal-Mart, 246
Warm Site, definition, 712
WATQ. *See* Workplace Aggression Tolerance Questionnaire
Web sites, as resources, 411–21

Wendy's, 172–73
What's in it for me (WIIFM), 350–51
WHO. *See* World Health Organization
WiFi. *See* Wireless connection
WIIFM. *See* What's in it for me
Wireless connection (WiFi), 39
Workers' compehnsation, 282
Workers Compensation Record Search, definition, 712
Workplace aggression, 272–78, 383–86; assessment, 276–77; consequences, 274–75; contributors, 278–80; definition, 272–73, 712; effect/danger ratio, 274–75; evaluation, 383; facilitators and inhibitors of, 275; individual traits, 279–80; justice and, 278–79; types, 273–74; WATQ assessment, 276–77
Workplace Aggression Tolerance Questionnaire (WATQ), 276–77; application, 276–77; definition, 712; evaluation, 383–85
Workplace violence, 167; definition, 712; prevention and policies, 257–72; employee security survey, 270–71; incident report form, 268–70; plan design, 261–62; policy statement, 262–63; self-inspection security checklist, 264–68; threat assessment team, 263–64; programs and policies, 260–61
World Bank, 214, 215
WorldCom, 231
World Health Organization (WHO), 304, 312
World Trade Center: attacks: 1993, 60–61; 2001, 61, 77, 86, 104, 163
Worms, 2; definition, 712

Y2K, 348

ABOUT THE EDITOR AND CONTRIBUTORS

W. TIMOTHY COOMBS, Ph.D., is an associate professor in the Department of Communication Studies at Eastern Illinois University. He is the 2002 recipient of the Jackson, Jackson and Wagner Behavioral Science Prize from the Public Relations Society of American for his crisis research. His research has led to the development and testing of the Situational Crisis Communication Theory (SCCT). He has published nationally and internationally in the areas of crisis management and preparedness. His works have been translated into Chinese, Dutch, and German. Dr. Coombs, who received his Ph.D. from Purdue University, has published twenty-five professional articles, over thirty academic journal articles, and thirty book chapters. He has also published two coauthored books and two of which he is the sole author, including the award-winning *Ongoing Crisis Communication* (now in its second edition) and *Code Red in the Boardroom: Crisis Management as Organizational DNA* (Praeger, 2006). He was part of a forum at the Batten Institute at the Darden School of Management, "Defining Leadership: A Forum to Discuss Crisis Leadership Competency" and has a chapter in the related *Executive Briefing on Crisis Leadership*. Dr. Coombs has lectured at various venues in the United States, Europe, and Australia on the subject of crisis management and related topics. He has also consulted with companies in the construction, airlines, petrochemical, and health care industries on crisis-related topics. Dr. Coombs does consulting work through Communications Northwest, LLC.

ROBERT C. CHANDLER, Ph.D., is professor and chair of communication in the Communication Division at the Center for Communication and Business at Pepperdine University. Dr. Chandler is a member of numerous academic associations including the National Communication Association, International Communication Association, Western States Communication Association, American Forensics Association, and the American Academy of Experts in Traumatic Stress. He is an expert in organizational and business communication; crisis communication; communication during emergencies, crises, and disasters; communication priorities for pandemics and other public health crises; risk communication; behavioral and psychometric assessment and appraisal; leadership; multicultural diversity; organizational integrity; employee ethical conduct;

and business ethics. He is an acclaimed speaker, presenter, and trainer in a wide range of organizational and corporate settings including national Webinars, DRJ World Conference, Contingency Planning and Management, Continuity Insights Management Conference, RSA Conference, and the International Security Conference. Dr. Chandler is an accomplished researcher and scholar with more than one hundred academic and professional papers, including widely circulated white papers on communication during disasters. The author of more than fifty publications, he is the author or coauthor of several books including *Crisis Communication Planning; Terrorism: How Can Business Continuity Cope?; Disaster Recovery and the News Media;* and *Managing the Risks for Corporate Integrity: How to Survive an Ethical Misconduct Disaster.*

DOUGLAS G. CONORICH is the global solutions manager for IBM Global Services' Managed Security Services (MSS). In this capacity, he has responsibility for developing new security offerings, ensuring that the current offerings are standardized globally. He oversees all training of new members of the MSS team worldwide in how to do "ethical hacking" and service delivery. Mr. Conorich has over thirty years' experience with computer security, holding a variety of technical and management positions. He has expertise in the areas of systems security management, including security policy generation and review, security implementation, audit verification procedures, encryption management, and security product design. Mr. Conorich is a networking and UNIX expert, with more than fifteen years' experience in these areas. He has undergraduate degrees in physics, computer science, and meteorology and a master's degree in physics from the University of New Mexico.

ÁGNES HUFF, Ph.D., is president and CEO of Ágnes Huff Communications Group, LLC. She has over twenty years' experience in public relations, marketing communications, crisis management, and strategic counseling. In 1995, she founded Ágnes Huff Communications Group, LLC, a full-service marketing communications firm located in Los Angeles, specializing in integrated communications. The firm is a certified MBE/SBE with the state of California and the county and city of Los Angeles. Through her relationships with the media, Ms. Huff's clients have appeared on the covers of *Fortune* and *CFO Magazine* and have been prominently featured in *Forbes*, the *L.A. Times*, *USA TODAY*, and the *Wall Street Journal*, as well as on ABC, CBS, CNN, and E! Entertainment. With her background in psychology, Ms. Huff brings a depth of understanding and dimension to her work that is generally not found in a traditional agency setting. With crisis management as an agency Center of Excellence, her firm works with prominent international clients including Bahamasair, Philips Semiconductors, Xerox Corporation, and World Airways on a variety of reputation management and communications assignments.

BETTY A. KILDOW, CBCP, FBCI, has been an emergency management and business continuity consultant for more than fifteen years. As a Certified Business

Continuity Professional (CBCP) and a Fellow of the Business Continuity Institute (FBCI), she has developed emergency management courses for the American Management Association and the University of California, Berkeley Extension. Ms. Kildow is author of the book *Front Desk Security and Safety: An On-the-Job Guide to Handling Emergencies, Threats, and Unexpected Situations* (AMACOM Publishing, 2004). She has written numerous articles for professional publications in the United States and the United Kingdom, and is frequently called upon as a speaker.

MICHAEL SEESE, M.S., M.A., CISSP, is an assistant vice president in the Corporate Security Services division of National City Corporation of Cleveland, Ohio, specializing in information security policy development and privacy. Prior to his current assignment, Mr. Seese served NCC as a business contingency planning consultant. He holds a master's of science in information security, a master's of arts in psychology, and recently earned his CISSP. With over twenty years' experience as an IT professional and journalist, he has had numerous articles published in professional journals. Recent speaking engagements include NetSec 2004, CPM 2005 West, and infosecurity New York 2006. He can be reached at Michael.Seese@NationalCity.com or mail@MichaelSeese.com.

GEARY SIKICH is the author of *It Can't Happen Here: All Hazards Crisis Management Planning, Emergency Management Planning Handbook* (available in both English and Spanish), and *Integrated Business Continuity: Maintaining Resilience in Uncertain Times*. Mr. Sikich is the founder and a principal with Logical Management Systems, Corporation (www.logicalmanagement.com), based in Munster, Indiana. He has extensive experience in management consulting in a variety of fields. Mr. Sikich consults on a regular basis with companies worldwide on business continuity and crisis management issues. He has a bachelor's of science degree in criminology from Indiana State University and a master's of education in counseling and guidance from the University of Texas, El Paso.

KRISTA VARNEY is a security program coordinator in the Corporate Security Services (CSS) division of National City Corporation. The CSS division incorporates both physical and logical security. She started her career in the Information Services division of National City and has spent the past several years in the Security division, specializing in ISO 27001 and 17799. She holds an undergraduate degree from Miami University (Ohio) and a master's of business administration from Baldwin Wallace College. In addition, Ms. Varney is a Certified Information Systems Security Professional (CISSP) and Project Management Professional (PMP). Ms. Varney was also a speaker at the RSA Conference in 2006 and 2007.